AF559783

TEXTBOOK OF BIOMATHEMATICS

TEXTBOOK OF BIOMATHEMATICS

By
Dr. Sharad Srivastava
Deptt. of Zoology
Vikram University
Ujjain

DISCOVERY PUBLISHING HOUSE PVT. LTD.
NEW DELHI-110 002

First Published - 2011

Reprinted - 2017

ISBN: 978-81-8356-831-9

Textbook of Biomathematics

Published by:

DISCOVERY PUBLISHING HOUSE PVT. LTD.

4383/4B, Ansari Road, Darya Ganj

New Delhi-110 002 (India)

Phone: +91-11-23279245, 43596064-65

Fax: +91-11-23253475

E-mail: discoverypublishinghouse@gmail.com

sales@discoverypublishinggroup.com

web: www.discoverypublishinggroup.com

Printed at:

Infinity Imaging Systems

Delhi

Preface

This book "Biomathematics" has been written for the students of microbiology of various universities. The aim of this book is to present the subject in such a way that even an average student may understand it without difficulty. The language is very simple and easily understandable. A exercise is given at the end of main topics so that student may try the problem and assess their confidence of answering the problem after studying the topics of the book. The book contains complete syllabus prescribed by various universities so it can be safely recommended as a text and reference book.

The author also thankful to many other authors whose book has been consulted with during the preparation of this book.

Suggestion for improvement of the book will gratefully acknowledge.

Author

Contents

Pages

Preface

1. **Fluid Mechanics** 1

Introduction, What is Fluid, Fluid Measurement Units, Fluid Mechanics Development, Classification of Fluid, Surface Tension and Capillarity, Pressure Intensity Inside a Droplet, Fluid Properties, Momentum Theory of Propellers, Specific Gravity, Equation of State : The Perfect Gas, Compressibility and Elasticity, Applications of the Impulse-momentum Equation, Momentum Correction Factor, Force on a Pipe Bend, Angular Momentum Principle–moment of Momentum Equation, Jet Propulsion-reaction of Jet, Solved Examples, Exercises.

2. **Properties of Multiplication of Matrices** 65

Introduction, Post-Multiplication and Pre-Multiplication of Matrices, Multiplication of Matrices is Associative, Multiplication of Matrices is Disruptive with Respect to Matrix Addition, Positive Integral Power of a Square Matrix, Hermitian Matrix, Theorems on Hermitian and Skew-Hermitian Matrices, Analysis of Matrix, Commutative Matrices, Theorems on Unitary Matrices, Identically Partitioned Matrices, Symmetric Idempotent Matrix, Solved Examples, Exercises.

3. **Basic Concept of Integration** 166

Introduction, Some Integrals Forms, Methods of Integration, Integrals of e^{ax} cos bx and e^{ax} sin bx, Integral of the Product of two Functions, Successive Integration by Parts, Some More Standard Integrals, Hyperbolic Functions, Constant of Integration, Two Simple Theorems, Extended Forms of

Fundamental Formulae, Fundamental Formulae, Solved Examples.

4. Analysis of Determinants **251**

Introduction, Determinant of a Square Matrix, Cofactor of an Element, Properties of Determinants, Minor of An Element, An Important Property of the Determinant, Solved Examples, Exercises.

5. Forms of the Differential Equations **306**

Introduction, Integrating Factors, Rules for Finding Integrating Factors, Linear Equations of the Form dx/dy+ Rx = S, R and S Being Functions of y Alone, Non-Homogeneous Equations, Linear Equations, Homogeneous Equations, Differential Equations of First Order and First Degree, Equations Reducible to the Linear Form, Rules for Finding Integrating Factors, Equations With Separated Variables, Solved Examples, Exercises.

1

Fluid Mechanics

INTRODUCTION

In solids the molecules are very closely spaced, but in liquids the spacing between the molecules is relatively large and in gases the space between the molecules is still larger. As such in a given volume a solid contains a large number of molecules, a liquid contains relativity less number of molecules and a gas contains much less number of molecules. It thus follows that in solids the force of attraction between the molecules is large on account of which there is very little movement of molecules within the solid mass and hence solids possess compact and rigid form. In liquids the force of can move freely within the liquid mass, but the force of attraction between the molecules is sufficient to keep the liquid together in a definite volume. In gases the force of attraction between the molecules is much less due to which the molecules of gases have greater freedom of movement so that the gases fill completely the container in which they are placed.

It may, however, be stated mechanical analysis a fluid is considered to be *continuum*, i.e., a continuous distribution of matter with no voids or empty spaces. This assumption is justifiable because ordinarily the fluids involved in most of the engineering problems have large number of the engineering problems have large number of molecules and the distances between them are small. The two classes of fluids, viz., gases and liquids also exhibit quite different characteristics. Gases can be compressed much readily under the action of external pressure and when the external pressure is removed the gases tend to expand indefinitely. On the other hand under ordinary conditions liquids are quite difficult to compress and therefore they may for most purposes be regarded as incompressible. Moreover, even if pressure acting on a liquid mass is removed, still the cohesion between particles holds them together so that the liquid does not expand indefinitely and it may have a free surface, that is a surface from which all pressure except atmospheric pressure is removed.

A solid can resist tensile, compressive and shear forces upto a certain limit. A fluid has no tensile strength or very little of it, and it can resist the compressive forces only when it is kept in a container. When subjected to a shearing force, a fluid defer continuously as long as this force is applied. The inability of the fluids to resist shearing stress gives them their characteristic property to change shape or to flow. This, however, does not mean that the fluids do not offer any resistance to shearing forces. In fact as the fluids flow there exists shearing (or tangential) stresses between the adjacent fluid layers which result in opposing the movement of one layer over the other. The amount of shear stress in a fluid depends on the magnitude of the rate of deformation of the fluid element. However, if a fluid is at rest no shear force can exist in it.

WHAT IS FLUID

A *fluid* may be defined as a substance which is capable of flowing. It has no definite shape of its own, but conforms to the shape of the containing vessel. Further even a small amount of shear force exerted on a fluid will cause it to undergo a deformation which continues as long as the force continues to be applied.

A *gas* is a fluid, which possesses a definite volume, which varies only slightly with temperature and pressure. Since under ordinary conditions liquids are difficult to compress, they may be for all practical purposes regarded as incompressible. It forms a free surface or an interface separating it from the atmosphere or any other gas present.

A *gas* is a fluid, which is compressible and possesses no definite volume but it always expands until its volume is equal to that of the container. Even a slight change in the temperature of a gas has a significant effect on its volume and pressure. However, if the conditions are such that a gas undergoes a negligible change in its volume, it may be regarded as incompressible. But if the change in volume is not negligible the compressibility of the gas will have to be taken into account in the analysis.

A *vapour* is a gas whose temperature and pressure are such that it is very near the liquid state. Thus steam may be considered a vapour because its state is normally not far from that of water.

The fluids are also classified as ideal fluids and practical or real fluids.

Ideal fluids are those fluids which have no viscosity and surface tension and they are incompressible. As such for ideal fluids no resistance is encountered as the fluid moves. However, in nature the ideal fluids do not exist and therefore. These are only imaginary fluids. The existence of these imaginary fluids was conceived by the mathematicians in order to simplify the mathematical analysis of the fluids in motion. The fluids which have low viscosity such as air,' water etc., may however be treated as ideal fluids without much error.

Practical or real fluids are those fluids which are actually available in nature. These fluids possess the properties such as viscosity, surface tension and compressibility and therefore a certain amount of resistance is always offered by these fluids when they are set in motion.

FLUID MEASUREMENT UNITS

There are in general four systems of units, two in metric (C.G.S. or M. K. S.) system and two in the English (F. P. S.) system. Of the two, one is known as the absolute or physicist's system and the other as the gravitational or engineer's system.

The difference between the absolute and gravitational systems is that in the former the standard is the unit of mass. The unit of force is then derived by Newton's law. In the gravitational system the standard is the unit of force and the unit of mass is derived by Newton's law. Table 4 below lists the various units of measurement for some of the basic or fundamental quantities.

The units of measurement of the various other quantities may be readily obtained with the help of the below table. Further Table 1 below illustrates all the four systems of units in which the units are defined so that one unit of force equals one unit of mass times one unit of acceleration.

Table 1

Systems of Units	*Relationship*
Metric Absolute	1 dyne = 1 gram $\times \frac{1\text{cm}}{\text{sec}^2}$
Metric Gravitational	1 kilogram (f) = 1 metric slug $\times \frac{1\text{m}}{\text{sec}^2}$
English Absolute	1 poundal = 1 pound $\times \frac{1\text{ft}}{\text{sec}^2}$
English Gravitational	1 pound (f) = 1 slug $\times \frac{1\text{ft}}{\text{sec}^2}$

Further the following relationships may be utilized to affect the conversion from one system to another.

1 gram-wt. = 981 dynes : 1 metric slug = 9810 gm (mass)

1 lb-wt. = 32.2 poundals : 1 slug = 32.2 lb (mass)

The use of the different systems of units by the scientists and the engineers and also by the different countries of the world often leads to a lot of confusion. Therefore, it was decided at the Eleventh General Conference of Weights and Measures held in Paris in 1960 to adopt a unified, systematically constituted, coherent system of units for international use.

This system of units is called the *International System of Units* and is designated by the abbreviation *SI Units.* More and more countries of the world are now adopting this system of units. There are six base units in SI system of units which are given in Table 2.

Table 2

Quantity	*Unit*	*Symbol*
Length	meter	m
Mass	kilogram	kg
Time	second	s
Electric Current	ampere	A
Thermodynamic temperature	kelvin	K
Luminous intensity	candela	cd

In addition to the above noted base units there are two supplementary units which are given in Table 3.

Table 3

Quantity	*Unit*	*Symbol*
Plane angle	radian	rad
Solid angle	steradian	sr

The unit of a derived quantity is obtained by taking the physical law connecting it with the basic (or primary or fundamental) quantities and then introducing the corresponding units for the basic quantities. Thus in SI units the unit of force is newton (N) which according to Newton's second law of motion is expressed as 1 N=1 kg × 1 m/s^2, *i.e.*, a force of 1 N is required to accelerate a mass of 1 kg by 1m/s^2. The units for some of the derived quantities have been assigned special names and symbols.

Units may be defined as those standards in terms of which the various physical quantities like length, mass, force, area, volume, velocity, acceleration etc., are measured. The system of used in mechanics are based upon Newton's second law of motion, which states that force equals mass times acceleration or F = M =a, where F is the force, m is the mass and a is the deceleration.

Table 4

Quantity	*Metric Units*		*English Units*	
	Gravitational	*Absolute*	*Gravitational*	*Absolute*
Length	Meter (m)	Meter (m)	Foot (ft)	Foot (ft)
Time	Second (sec)	Second (sec)	Second (sec)	Second (sec)
Mass (msl)	Metric slug [gm(mass)]	Gram (sl)	Slug [lb(mass)]	Pound
Force	kilogram [kg(f)]	Dyne	Pound [lb (f)]	Poundal (pdl}
Temperature	°C	°C	°F	°F

FLUID MECHANICS DEVELOPMENT

The problems, man encountered in the fields of water supply, irrigation, navigation and water power, resulted in the development of the fluid mechanics. However, with the exception of Archimedes (250B.C.) principle which is considered to be as true today as some 2200 years ago, little of the scant knowledge of the ancients appears in modem fluid mechanics.

After the fall of Roman Empire (476 A.D) there is no record of progress made in fluid mechanics until the time of Leonardo da Vinci (1500 A.D.), who designed the first chambered canal lock. However, upto da Vinci's time, concepts of fluid notion must be considered to be more art than science. Fluid mechanics is that branch of science which deals with the behaviour of the fluids at rest as well as in motion. In general the scope of fluid mechanics is very wide which includes the study of all liquids and gases. But usually it is confined to the study of liquids and those gases for which the effects due to comprehensibility may be neglected. The gases with appreciable however dealt with under the subject of Gas dynamics. Some two hundred years ago mankind's centuries of experience with the flow of water began to crystallize in scientific form. Two distinct Schools of thought gradually evolved in the treatment of fluid mechanics.

One, commonly known as Classical Hydrodynamics, deals with theoretical aspects of the fluid flow, which assumes that shearing stresses are non-existent in the fluids, that aspects of fluid flow which has been developed from experimental findings and is, therefore, more of empirical nature. Notable contributions to theoretical hydrodynamics have been made by Euler, D' Alembert, Navier, Lagrange, Stokes, Kirchoff, Rayleigh, Rankine, Kelvin,

Lamb and many others. Many investigators have contributed to the development of experimental hydraulics, notable amongst them being Chezy, Venturi, Bazin, Hagen, Poiseuille, Darcy, Weisbach, Kutter, Manning, Francis and several others.

Although the empirical formulae developed in hydraulics have found useful application in several problems, it is not possible to extend them to the flow of fluids other than water and in the advanced field of aerodynamics. As such there was a definite need for a ne approach to the problems of fluid flow-an approach which relied on classical hydrodynamics for its analytical development and at the same time on experimental means for checking the validity of the theoretical analysis. The modern Fluid Mechanics provides this new approach taking a balanced view of both the theorists and the experimentalists. The generally recognized founder of the modern fluid mechanics is the German Professor, Ludwig prandtl. His most notable contribution being the boundary layer theory which has had a tremendous influence upon the understanding of the problems involving fluid motion. Other notable contributors to the modern fluid mechanics are Blasius, Bakhmeteff, Nikuadse, Von-karman, Reynolds, Rouse and many others.

The fundamental principles of fluid mechanics applicable to the problems involving the motion of a particular class of fluids called Newtonian fluids (such as water, air, kerosene, glycerine etc.) have been discussed along with the relevant portions of the experimental hydraulics.

CLASSIFICATION OF FLUID

Perfect Fluid

The *ideal fluid* is one which is incapable of sustaining any tangential stress or action in the form of a shear but the normal force (pressure) acts between the adjoining layers of the fluid. The pressure at every point of an ideal fluid is equal in all directions, whether the fluid be at rest or in motion. This theory defines some concepts of the flow such as wave motion, the lift and the induced drag of an airfoil etc., but it fails to define the phenomena such as skin friction, drag of a body etc.

Actual Fluid

The *real fluid* is one in which both the tangential and normal forces exist. Viscosity, which is also known as an internal friction, of a fluid is that characteristic of real fluid which is capable to offer resistance to shearing stress. The resistance is, comparatively, small (not negligible) for fluids such as water and gases but it is quite large for other fluids such as oil, glycerine, paints, varnish, molasses, coal-tar etc.

SURFACE TENSION AND CAPILLARITY

Due to molecular, liquids process certain properties such as cohesion and adhesion. *Cohesion* means inter-molecular attraction between molecules of the same liquid. That means it is a tendency of the liquid to remain as one assemblage of particles. *Adhesion* means attraction between the molecules of a liquid and the molecules of a solid boundary surface in contact with the liquid. The property of cohesion enables a liquid to resist tensile stress, while adhesion enables it to stick to another body. *Surface tension* is due to cohesion between liquid particles at the surface, whereas *capillarity* is due to both cohesion adhesion.

Surface Tension

A liquid molecule on the interior of the liquid body has other molecules on all sides of it, so that the forces of attraction are in equilibrium and the molecule is equally attracted on all the sides, as a molecule at point *A* shown in Fig. 1. On the other hand a liquid molecule at the surface of the liquid, (*i.e.*, at the interface between a liquid and a gas) as at point *B*, does not have any liquid molecule above it, and consequently there is a net downward force on the molecule due to the attraction of the molecules below it. This force on the molecules at the liquid surface, is normal to the liquid surface. Apparently owing to the attraction of liquid molecules below the surface, which is in tension and special layer seems to form on the liquid at the surface, which is in tension and small loads can be supported over it. For example, a small needle placed gently upon the water surface will not sink but will be supported by the tension at the water surface.

The property of the liquid surface film to exert a tension is called the surface tension. It is denoted by σ (Greek 'sigma') and it is the force required to maintain unit length of the film in equilibrium. In SI units surface tension is expressed in N/m. In the metric gravitational system of units it is expressed in kg(f)/cm or kg(f)/m. In the English gravitational system of units it is expressed in lb(f)/in or lb(f)/ft.

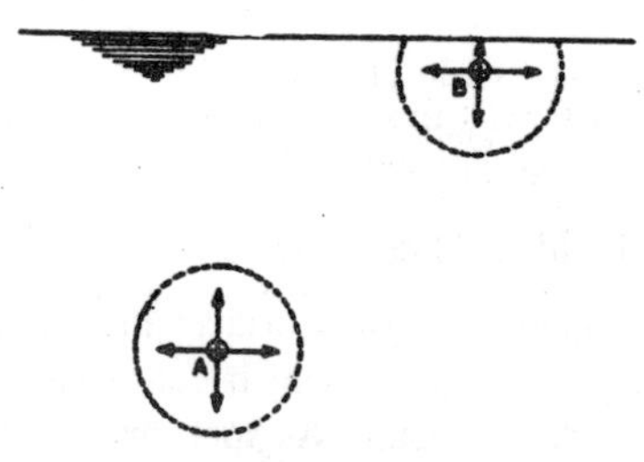

Fig 1

As surface tension is directly dependent upon inter-molecular cohesive forces, its magnitude for all liquid decreases as the temperature rises. It is also dependent on the fluid in contact with the liquid surface; thus surface tensions are usually quoted in contact with air. The surface tension of water in contact with air varies from 0.0736 N/m [or 0.0075 kg (f)/m] at 19°C to 0.0589 N/m [or 0.006 kg (f)/m] at 100°C. More organic liquids have values of surface tension between 0.0206 N/m [or 0.0021 kg (f)/m] and 0.0304 N/m [or 0.0031 kg (f)/m] and mercury has a value of about 0.4944 N/m [or 0.0504 kg (f)/m], at normal temperature and the liquid in each case being in contact with air.

The effect of surface tension is illustrated in the case of a droplet as well as a liquid jet. When a droplet is separated initially from the surface of the main body of liquid, then due to surface tension there is a net inward force exerted over the entire surface of the droplet which causes the surface of the droplet to contract from all the sides and results in increasing the internal pressure within the droplet. The contraction of the droplet continues till the inward force due to surface tension is in balance with the internal pressure and the droplet forms into sphere which is the shape for minimum surface area. The internal pressure intensity within a droplet and a jet of liquid in excess of the outside pressure intensity may be determined by the expressions derived below.

PRESSURE INTENSITY INSIDE A DROPLET

Consider a spherical droplet of radius r having internal pressure intensity p in excess of the outside pressure intensity. If the droplet is cut into two halves, then the forces acting on one half will be those due to pressure intensity p on the projected area (μr^2) and the tensile force due to surface tension σ acting around the circumference ($2\mu r$). These two forces will be equal and opposite for equilibrium and hence we have

$$p(\mu r^2) = \sigma(2\mu r)$$

or $$p = \frac{2\sigma}{r} \qquad ...(1)$$

Equation 4 indicates that the internal pressure intensity increases with the decrease in the size of droplet.

Pressure Intensity Inside a Soap Bubble

A spherical soap bubble had two surfaces in contact with air, one inside and the other outside, each one of which contributes the same amount of tensile force due to surface tension. As such on a hemispherical section of a soap bubble of radius r the tensile force due to surface tension is equal to

2σ (2μr). However, the pressure force acting on the hemispherical section of the soap bubble is same as in the case of a droplet and it is equal to *p* (μr²). Thus equating these two forces for equilibrium, we have

$$p(\pi r^2) = 2\sigma(2\pi r)$$

or $$p = \frac{4\sigma}{r} \qquad ...(2)$$

Pressure Intensity Inside a Liquid Jet

Consider a jet of liquid of radius r, length l and having internal pressure intensity p in excess of the outside pressure intensity. If the jet is cut into two halves, then the forces acting on one half will be those due to pressure intensity p on the projected area ($2rl$) and the tensile force due to surface tension σ acting along the two sides ($2l$). These two forces will be equal and opposite for equilibrium and hence we have

$$p(2rl) = \sigma(2l)$$

or $$p = \frac{\sigma}{r} \qquad ...(3)$$

Capillarity

If molecules of certain liquid possess, relatively, greater affinity for solid molecules, or in other words the liquid has grater adhesion than cohesion then it will wet a solid surface with which it is in contact and will tend to rise at the point of contact, with the result that the liquid surface is concave upward and the angle of contact θ is less than 90°. If a glass tube of small diameter is partially immersed in water, the water will wet the surface of the tube and it will rise in the tube to some height, above the normal water surface, with the angle of contact θ, being zero. The wetting of solid boundary by liquid results in creating decrease of pressure within the liquid, and hence the rise in the liquid surface takes place, so that the pressure within the column at the elevation of the surrounding liquid surface is the same as the pressure at this elevation outside the column.

On the other hand, if for any liquid there is less attraction for solid molecule or in other words the cohesion predominates, then the liquid will not wet the solid surface and the liquid surface will be depressed at the point of contact, with the result that the liquid surface is concave downward and the angle of contact θ is greater than 90°. For instance if the same glass tube is now inserted in mercury, since mercury does not wet the solid boundary in contact with it, the level of mercury inside the tube will be lower than the adjacent mercury level, with the angle of contact θ equal to about 130°. The tendency of the liquids which do not adhere to the solid surface, results in creating an increase of pressure across the liquid surface, (as in the case of a drop of liquid). It is because of the increased internal pressure, the elevation of the meniscus (curved liquid surface) in the tube is lowered to the level where the pressure is the same as that in the surrounding liquid.

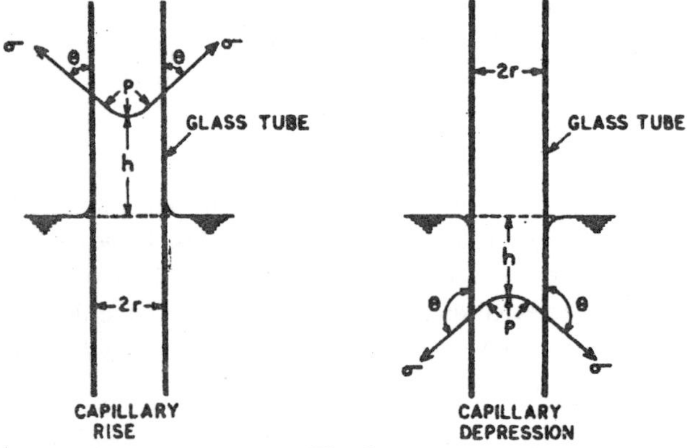

Fig. 2

Such a phenomenon of rise of fall of liquid surface relative to the adjacent general level of liquid is known as *capillarity*. Accordingly the rise of liquid surface is designated as capillary rise and the lowering of liquid surface as capillary depression, and it is expressed in terms of m or mm of liquid in SI units, in terms of cm or mm of liquid in the metric system of units and in terms of inch or ft of liquid in the English system of units.

The capillary rise (or depression) can be determined by considering the conditions of equilibrium in a circular tube of small diameter inserted in a liquid. It is supposed that the level of liquid has risen (or fallen) by h above (or below) the general liquid surface when a tube of radius r is inserted in the liquid, see Fig. 2.

For the equilibrium of vertical forces acting on the mass of liquid lying above (or below) the general liquid level, the weight of liquid column h (or the total internal pressure in the case of capillary depression) must be balanced by the force, at surface of the liquid, due to surface tension σ. Thus equating these two forces we have

$$sw\mu r^2 h = 2\mu r\sigma \cos\theta$$

where w is the specific weight of water, s is specific gravity of liquid, and θ is the contact angle between the liquid and the tube. The expression for h the capillary rise (or depression) then becomes

$$h = \frac{2\sigma \cos\theta}{swr} \qquad ...(4)$$

As stated earlier, the contact angle θ for water and glass is equal to zero. Thus the value of cos θ is equal to unity and hence *h* is given by-the expression

$$h = \frac{2\sigma}{wr}$$

For capillary rise (or depression) indicates that the smaller the radius *r* the greater is the capillary rise (or depression).

The above obtained expression for the capillary rise (or depression) is based on the assumption that the meniscus or the curved liquid surface is a section of a sphere. This is, however, true only in case of the tubes of small diameters ($r < 2.5$ mm) and as the size of the tube becomes larger, the meniscus becomes less spherical and also gravitational forces become more appreciable. Hence such simplified solution for computing the capillary rise (or depression) is possible only for the tubes of small diameters.

However, with increasing diameter of tube, the capillary rise (or depression) becomes much less. It has been observed that for tubes of diameters 6 mm more the capillary rise (or depression) is negligible. Hence in order to avoid a correction for the effects of capillarity in manometer, used for measuring pressures, a tube of diameter 6 mm or more should be used.

Another assumption made in deriving equation 4 is that the liquids and tube surfaces are extremely clean. In practice, however, such cleanliness is virtually never encountered and *h* will be found to be considerably smaller than that given by equation 4. In respect of this, equation 4 will provided a conservative estimate of capillary rise (or depression).

If a tube of radius *r* is inserted in mercury (sp. gr. s_1) above which a liquid of sp. gr.s_2 lies then by considering the conditions of equilibrium it can be shown that the capillary depression *h* is given by

$$h = \frac{2\sigma\cos\theta}{rw(s_1 - s_2)} \qquad ...(5)$$

in which σ is the surface tension of mercury in contact with the liquid and rest of the notation are same as defined earlier.

Further if two vertical parallel plates *t* distance apart and each of width *l* are held partially immersed in a liquid of surface tension σ and sp. gr. s, then the capillary rise (or depression) *h* may be determined by equating the weight of the liquid column *h* (or the total internal pressure in the case of capillary depression) (*swhlt*) to the force due to surface tension (2σ/cosθ). Thus we have

$$swhlt = 2\sigma l \cos\theta$$

or $$h = \frac{2\sigma \cos\theta}{swt} \qquad ...(6)$$

Properties of water and air respectively at different temperatures are listed. Properties of some of the common liquids such as glycerine, kerosene, alcohol, mercury etc., at 20°C.

FLUID PROPERTIES

Certain characteristics of a continuous fluid are independent of the motion of the fluid. These characteristics are known as the basic properties of the fluid. We shall discuss some of the properties of a fluid.

Density

The *density* ρ represents a quantitative expression of the idea of mass. It is defined as the mass of the fluid contained within a unit volume. Consider δm is the mass of the fluid in a small volume δv surrounding that point, then, mathematically the density at a point is defined as

$$\rho = \lim_{\delta v \to 0} \frac{\delta \text{m}}{\delta v}.$$

In physical sense $\delta v \to \varepsilon^3$ in which ε is large in comparison with the mean distance between molecules. In other words ε^3 is the infinitesimal volume over which the substance can be taken as a continuum. In fact, the limit $\delta v \to 0$ implies that after a certain stage the continuum hypothesis will breakdown and so the limit does not exists and the ratio will starts fluctuating rapidly. The density is an index of the inertial characteristics. The density of the fluid depends on the space coordinates and the temperature *i.e.*, $\rho = f(x, y, z, t)$ The density of water at 4°C, is 1 g/cm^3 or 1000 kg/m^3.

For gases, the density is a function of pressure and temperature. Under ideal conditions, the equation of state for an ideal gas $\rho = \rho RT$ provides a solution for density.

Specific Weight

The *specific weight* γ of a fluid is the weight per unit volume. It occurs either dealing with fluid static or with liquids with a free surface. The specific weight depends on the mass density and the gravitational acceleration. It varies from place to place due to changes in density and acceleration due to gravity g. It has units of force per unit volume. The specific weight can be determined from the density by the relation.

$$\rho = \gamma/g$$

$$\Rightarrow \gamma = \rho g, \text{ in absolute CGS and SI.}$$

The specific weight of water under standard acceleration is

$$\gamma = 1 \times 981 = 981 \text{ dyn/em}^3 \text{ (cgs)}$$
$$= 1000 \times 9.81 = 9810 \text{ N/m}^3 \text{ (SI)}.$$

Specific Volume

The *specific volume* v_s is the volume occupied by a unit mass of fluid. The reciprocal of the density is called the *specific volume.* It is commonly used in problems involving gas flows

$$v_s = 1/\rho.$$

Specific Gravity

The *specific gravity S* of a substance is the ratio of its specific weight of a fluid to the specific weight of an equal volume of water at standard conditions (4°C or 68°F).

Pressure

The *pressure p* at a point in the fluid is the limit of the ratio of normal force δA by the surrounding fluid particles as the area approaches zero. It is defined as

$$p = \lim_{\delta A \to 0} \frac{\delta F}{\delta A}.$$

We know that the pressure at every point of an ideal fluid is equal in all directions whether the fluid be at rest or in motion It does not depend on the orientation of the plane. If it varies with the orientation (as in real fluids in motion) then the average of all such values at that point is taken. It follows that an element δA of a very small aream free to rotate about its centre will have a force of constant magnitude acting on either sides of it.

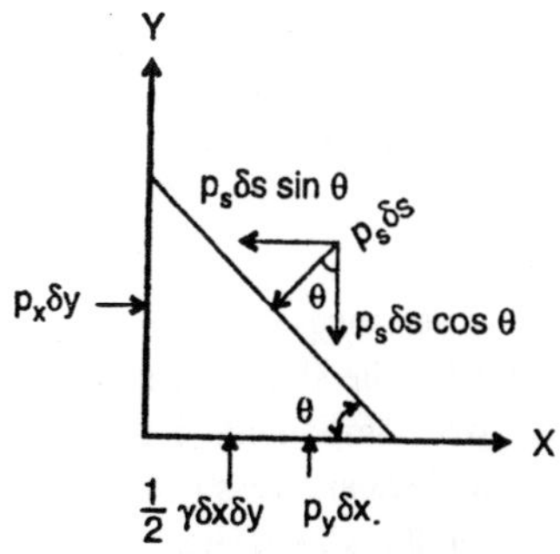

Fig 3

Consider a small wedge-shaped free body of unit length at the point (x, y) in a fluid at rest. Let P_x, P_y and P_s are the average pressures on the three faces; ρ, its density and γ, the specific weight of the fluid. The only forces that act on the body are the normal surface and gravity. The equation of equilibrium in the X and Y directions are, respectively, given by

$$\sum F_x = p_x \delta y - p_s \delta s \sin\theta$$

$$\sum F_y = p_y \delta x - p_y \delta s \cos\theta - \frac{1}{2}\gamma\ \delta x\ \delta y$$

Since $\sin\theta = \delta y/\delta s$ and $\cos\theta = \delta x/\delta s$.

From (2) and (3), we have

and $P_y \delta x$ - $P_s\ \delta x - \frac{1}{2}\gamma\ \delta x\ \delta y = 0$.

The last term $\left(\frac{1}{2}\gamma\ \delta x\ \delta y\right)$ in the equation (5) is neglected, being an infinitesimal. Thus, we obtain

$$P_x = P_y = p_s.$$

It follows that *the pressure is same at a point in all directions in a fluid at rest.*

If the fluid is in motion then the shear stress occur and, in general, the normal stresses are no longer the same in all directions at a point. Thus the pressure is defined as the average of any three mutually perpendicular normal stresses at a point.

$$p = \frac{1}{3}(p_x + p_y + p_s).$$

Pressure has the units of force per unit area. In general, the pressure varies with space and time *i.e.*,

$$P = f(x, y, z, t).$$

The dimensions of the pressure are

$$(MLt^{-2})\ L^{-2} = ML^{-1}T^{-2}\ \text{N/m}^2.$$

Viscosity

Each element of fluid experiences stress exerted on it by other elements of the fluid which surround it. The stress at each part of the surface of the element is resolved into two components : normal and tangential to the surface, known as pressure and shear stress respectively. Pressure are exerted whether the fluid is moving or at rest, but shear stresses occur only in moving fluids. This feature is the one which enables fluids to be distinguished from solids.

The property which gives rise to shear stresses is called *viscosity.* Viscosity arises when there is a relative motion between different fluid layers. It is possessed by all real fluids. Its magnitude is expressed by a coefficient which relates the size of the shear stress at a point in a fluid to the rate of shear strain which causes it.

Consider two parallel plates placed at a distance y_0 apart. The space between in is filled with the fluid. The lower plate is fixed and a force *F* is applied to the upper plate which is moving relative to the lower on with a velocity *U.* Let *A* is the area of the upper plate. The force per unit area is proportional to the velocity decrease in the distance y_0.

$$i.e., \quad \frac{F}{A} \propto \frac{U}{y_0} \Rightarrow F = \mu \frac{AU}{y_0},$$

where μ is the constant of proportionality.

The particles of the fluid in contact with each plate will adhere to the surfaces. The fluid in immediate contact with a solid boundary has the same velocity as the boundary *i.e.*, there is *no slip* at the boundary. The shear stress ι, between any two thin sheets of fluid is

$$\tau = \frac{F}{A}(\text{Force / unit area}) = \frac{\mu U}{y_0},$$

which represents the rate of angular deformation of the fluid. The equation may be generalized for the case of two adjacent layers of fluid separated by a distance *dy* both moving parallel to the *X*-direction with the velocities *u* and *u* + *du* respectively. The shear stress exerted by one layer of fluid on the other is proportional to its velocity relative to that of the other layer and inversely proportional to the distance between them. $\frac{U}{y_0}$ may be replaced by the velocity gradient $\frac{du}{dy}$ which holds for situations where angular velocity and shear stress change with y. Thus the shear stress is directly proportional to the transverse velocity gradient

$$\tau = \mu \frac{du}{dy}. \qquad ...(1)$$

The proportionality factor μ is called the *dynamic viscosity* or *viscosity* of the fluid. The equation (1) is known Newton's law of viscosity and the fluid obeying this is called *Newtonian fluids.*

(i) If ι = 0 then μ = 0 the equation (1) represents and *ideal fluid or perfect fluid.*

(ii) If $\frac{du}{dy} = 0$ then $\mu = \infty$ the equation (1) represents the *elastic bodies.*

(iii) A fluid for which the constant of proportionality does not change with the rate of deformation (shear strain) is said to be *Newtonian* and is represented by a straight line.

(iv) If the viscosity varies with the rate of deformation, then it represent *Non-Newtonian* fluids. Non-Newtonian fluids are those in which there is no shear stress and there exists a non-linear relation between ι and $\frac{du}{dy}$. The main classes of non-Newtonian fluids are Binghan plastics, Pseudoplastic and Dilatants.

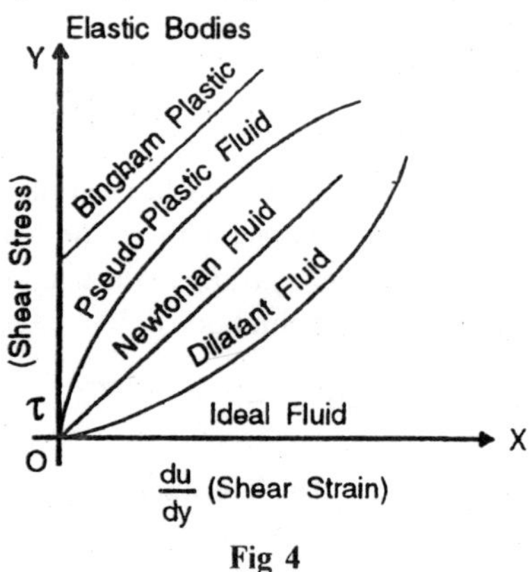

Fig 4

The viscosity of a gas increases while the velocity of a liquid decrease with increase in temperature. The dimensions of the coefficient of viscosity μ are determined as

$$\Rightarrow \mu = \frac{\text{Force / unit area}}{\text{(Velocity / Length)}}$$

$$= \frac{100 \times 7.7 \times 10^6}{2.2 \times 10^{10}} = \frac{7}{200}$$

$$\Rightarrow \mu = \frac{\text{Force / unit area}}{\text{(Velocity / Length)}}$$

$$\Rightarrow \mu = \frac{ML/T^2 - T^2}{(L/T)(1/L)}$$

$$\Rightarrow \mu = ML^{-1}T^{-1} \quad \text{N-m/s}^2,$$

where M is the mass, T is the time and L is the length. The ratio of the coefficient of absolute viscosity μ to the density ρ of the fluid is known as the coefficient of *kinematic viscosity* ν and is represented as

$$\nu = \frac{\mu}{\rho} = \frac{ML^{-1}T^{-1}}{ML^{-3}} = L^2T^1$$

Viscosity of a fluid is practically independent of pressure and depends upon the temperature only.

Temperature

When two bodies of different heat content are brought into contact there is a flow of heat until the bodies are in thermal equilibrium. The two bodies are said to have a common property, known as *temperature*. The magnitude of temperature may be related to the molecular activity arising from heat transfer. The temperature difference in a body is reduced by heat flowing from regions of higher temperature to those of lower temperature. This process takes place in all substances which occur in nature, in liquid, gases, as well as in solid bodies.

Thermal Conductivity

When a fluid in static equilibrium is heated non-uniformly, heat may be transferred from regions of higher temperature to those of lower temperature. Consider a surface element situated at some point in the fluid. The heat flux (rage of flow per unit area) in the direction of the normal to the element is proportional to the rate of change of temperature at that points. The heat flow occurs in the direction of decreasing temperature. Let q_n denotes the heat flux and $\partial T/\partial n$ denotes the rate of increase of temperature with distance in the direction of the normal (*Fourier heat conduction law*), then we have

$$q_n \propto -\frac{\partial T}{\partial n} \Rightarrow q_n = -K\frac{\partial T}{\partial n},$$

where K is a positive proportionality factor known as the coefficient of *thermal conductivity*. This is a material property of the fluid which varies from fluid to fluid. The thermal conductivity, in general, is a function of temperature and pressure. It is independent of the pressure gradient if this is small.

Specific Heat

The *specific heat*, C is defined as the amount of heat required to raise the temperature of a unit mass of medium by one degree, *i.e.*

$$C = \partial Q / \partial t,$$

where Q is the quantity of heat added per unit mass of the gas. The specific heat depends on the process in which the heat is added. Usually, the two common processes considered are either constant volume process or constant pressure process.

$$C_v = \left(\frac{\partial Q}{\partial t}\right)_v, \text{ and } C_p = \left(\frac{\partial Q}{\partial t}\right)_p$$

The ratio of the two specific heats is usually denoted by the relation

$$\gamma = \frac{C_p}{C_v},$$

which is a measure of the relative-internal complexity of the molecules.

Vapour Pressure

All liquids have a tendency to evaporate when exposed to the atmosphere. The rate at which the evaporation occurs is dependent on the molecular activity of the liquid, which is a function of temperature and the condition of the atmosphere adjoining the liquid. Consider a closed bottle partly filled with a liquid and maintained at a constant temperature. The number of vapour molecules in the air above the liquid increases when the liquid evaporates, simultaneously a small number of vapour molecules re-enter the liquid. Thus the concentration of vapour molecules above the liquid surface increases, with the passage of time, to such an extent that the rate at which molecules enter the liquid is equal to the rate at which molecules leave the liquid is equal to the rate at which molecules leave the liquid. Hence the air above the liquid surface is saturated with vapour molecules. The pressure on the liquid surface exerted by the vapour molecules is called *Vapour pressure.* The vapour pressure is dependent on temperature. The phenomenon of boiling a liquid is closely related to the vapour pressure. When the pressure above a liquid equals the vapour pressure of the liquid, boiling occurs.

In the flow of liquids, it is possible that very low pressures are produced at certain locations in the system. Under such circumstances the pressures may be equal to or less than the vapour pressures. When the pressure of a liquid is reduced to a value slightly less than the saturated vapour pressure at the temperature of the liquid, the liquid is in unstable state, and normally tends to form vapour pockets distributed throughout the liquid. The appearance of such pockets are named as *cavitation.* For examples when the water is heated slowly, bubbles formed near the bottom are known cavities formed in the water. Vortices in rivers are called cavities.

MOMENTUM THEORY OF PROPELLERS

A propeller is a revolving mechanism which uses the torque of a shaft to produce axial thrust. The conversion of torque into axial thrust is done by propeller by changing the momentum of the fluid in which it is submerged. When a propeller submerged in an undisturbed fluid rotates, it exerts a force on the fluid and pushes the fluid backwards. The reaction to this force on the fluid provides a forward force on the propeller itself and this force is the so-called 'propeller thrust' which is used for propulsion. Although the complete design of a propeller cannot be done according to the momentum theory, yet the application of this theory leads to some useful results as indicated by simple analyses of the problem below.

Propeller moving to the left with velocity V through still fluid. By applying a velocity equal in magnitude to that of the propeller but in opposite direction (*i.e.*, –V) on the entire system, the propeller may be considered to be a stationary 'actuating rotor' occupying a fixed position while the fluid on which it directly acts). The fluid enters the slip stream with (that is, the fluid on which it directly acts).

The fluid enters the slip stream with velocity V at section 1 upstream from the propeller where the flow is undisturbed, and the fluid velocity increases as it approaches and leaves the propeller. At section 2 some distance behind the propeller, the fluid leaves the slip stream with velocity V_j. For a propeller to operate in a body of still fluid, the pressure at some distance a head of and behind the propeller (*i.e.*, at sections 1 and 2) and the pressure over the slip stream boundary are the same being equal to the pressure of the undisturbed fluid. As shown in the lower portion the pressure decreases from the value p_0, rises at the propeller, and then drops to p_0 again. It is assumed that all fluid elements passing through the propeller have their pressure increased by exactly the same amount Δp. The rotational effect of the propeller is neglected. Therefore, the thrust on the propeller is equal to

$$T_p = \frac{\pi D^2}{\Delta}(\Delta p) \qquad \text{....(1)}$$

By applying the momentum equation to the free body of fluid between sections 1 and 2 the slip stream boundary, the only force acting on it in the flow direction is propeller thrust T_p, since the outer boundary of the body is everywhere at the same pressure.

$$\text{Therefore} \quad T_p = \rho Q(V_j - V) \qquad \text{...(2)}$$

where Q is the rate of flow through the slip stream.

$$p_0 + \rho\frac{V^2}{2} + \Delta p = p_0 + \rho\frac{V_j^2}{2} \qquad \text{...(3)}$$

in which Δp is the work which the propeller performs on the fluid in the slip stream. Simultaneous solution of the above equations yields

$$Q = \frac{\pi D^2}{4}\frac{V + V_j}{2} \quad ...(4)$$

Equation 9 shows that the velocity of flow at the propeller is the average of the approaching velocity V and the exit velocity V_j. This result known as Froude's theorem after William Froude (1810–79) is one of the principle assumptions in propeller design.

If the undisturbed fluid be considered stationary, the propeller advances through it at velocity V. The rate at which the useful work is done by the propeller is then equal to the product of the propeller thrust T_p and the velocity V. That is

$$\text{Power output} = T_p\, V = \rho Q\,(V_j - V)V \quad ...(5)$$

In addition to the useful work, some power is lost in increasing the kinetic energy of flow in the slip stream, which is given as

$$\text{Power lost} = \rho Q\,\frac{(V_j - V)^2}{2} \quad ...(6)$$

since $(V_j - V)$ is the velocity of the downstream fluid relative to earth. The power supplied to the propeller by the engine is the sum of power output and power lost. Thus

$$\text{Power input} = \left[\rho Q(V_j - V)V + \rho Q\frac{(V_j - V)^2}{2}\right] \quad ...(7)$$

The theoretical propulsive efficiency η_{th}, sometimes known as the Froude efficiency is given by the ratio of the equations 7 and 8:

$$\eta_{th} = \frac{\text{Power output}}{\text{Power input}}$$

$$= \frac{\rho Q(V_j - V)V}{\left[\rho Q(V_j - V)V + \rho Q\frac{(V_j - V)^2}{2}\right]}$$

$$= \frac{2}{1 + (V_j / V)} \quad ....(8)$$

It may however be noted that this efficiency does not account for friction or the effects of the rotational motion imparted by the propeller to the fluid, and hence it is considered to be theoretical. Further the propulsive efficiency, which is a function of the velocity ratio (V_j/V), increases as the ratio (V_j/V) decreases. As may be seen the efficiency will have a limiting value of 100 per cent if (V_j/V) is equal to one. This condition is however impossible, because such a propeller will produce no thrust due to zero velocity change. In practice, an aircraft propeller may have a maximum efficiency of about 0.85 to 0.9 times the value given by equation 13. However, owing to compressibility effects, the efficiency of an aircraft propeller drops rapidly with speeds above 640 km per hour ship propeller efficiencies are usually less, owing to restrictions in diameter.

SPECIFIC GRAVITY

Specific gravity (sp. gr.) is the ratio of specific weight (or mass density) of a fluid to the specific weight (or mass density) of a standard fluid. For liquids, the standard fluid chosen for comparison is pure water at 4°C (39.2°F). For gases, the standard fluid chosen is either hydrogen or air at some specified temperature and pressure. As specific weight and mass density of a fluid vary with temperature, temperatures must be quoted when specific gravity is used in precise calculations of specific weight or mass density. Being a ratio of two quantities with same units, specific gravity is a pure number independent of the system of units used.

The specific gravity of water at the statdard temperature (*i.e.*,4°C), is therefore, equal to 1.0. The specific gravity of mercury varies from 13.5 to 13.6. Knowing the specific gravity of any liquid, its specific weight may be readily calculated ny the following relation, *viz.*,

$$w = \text{Sp. gr of liquid} \times \text{Specific weight of water}$$

$$= (\text{Sp. gr of liquid}) \times 9\ 810 \text{N/m}^3$$

EQUATION OF STATE : THE PERFECT GAS

The density ρ of a particular gas is related to its absolute pressure p and absolute temperature T by the *equation of state,* which for a perfect gas takes the form.

$$p = \rho RT; \text{ or } pV = mRT \qquad ...(1)$$

in which R is a constant called the *gas constant,* the value of which is constant for the gas concerned, and V is the volume occupied by the mass m of the gas. The absolute pressure is the pressure measured above absolute zero (or complete vacuum) and is given by.

$$p = p_{gage} + p_{atm} \text{ (see also section 2.5)}$$

The absolute temperature is expressed in 'kelvin' *i.e.,* K, when the temperature is measured in °C and it is given by

$$T_{(abs)} = T\ K = 273.15 + t°\text{C}$$

No actual gas is perfect. However, most gases (if at temperatures and pressures well away both from the liquid phase and from dissociation) obey this relation closely and hence their pressure, density and (absolute) temperature may, to a good approximation, be related by equation 1.2. Similarly air at normal temperature and pressure behaves closely in accordance with the equation of state.

It may be noted that the gas constant R is defined by equation 2.2 as $p/\rho T$ and, therefore, its dimensional expression is $(FL/M\theta)$. Thus in SI units the gas constant R is expressed in newton-metre per kilogram per kelvin, *i.e.,* (N.m/kg. K). Further, since 1joule = 1 newton × 1meter, the unit for R also becomes joule per kilogram per kelvin, *i.e.,* (J/kg. K). Again, since 1 N= 1kg × 1 m/s^2, the unit for R becomes (m^2/s^{2}K).

In metric gravitational and absolute systems of units, the gas constant R is expressed in kilogram (f)-metre per metric slug per degree C absolute, *i.e.,* [kg(f)-m/msl deg.C abs.], and dyne-centimetre per gram (m) per degree C absolute, *i.e.,* [dyne-cm/gm(m) deg. C abs.] respectively. For air the value of R is 287 N-m/kg K, or 287 J/kg K, or 287 m^2/s^2 K.

In metric gravitational system of units the value of R for air is 287 kg(f)-m/msl deg. C abs. Further, since 1 msl = 9.81 kg (m), the value of R for air becomes (287/9.81) or 29.27 kg(f)-m/kg deg. C abs. Since specific volume may be defined as reciprocal of mass density, the equation of state may also be expressed in terms of specific volume of the gas as

$$p\vartheta = RT \qquad \text{...(2)}$$

in which ϑ is specific volume.

The equation of state may also be expressed as

$$p = wRT \qquad \text{...(2b)}$$

in which w is the specific weight of the gas. The unit for the gas constant R then becomes (m/K) or (m/deg. C abs). It may, however, be shown that for air the value of R is 29.27 m/27 m/K.

For a given temperature and pressure, equation 1.2 indicates that ρR = constant. By Avogadro's hypothesis, all pure gases at the same temperature and pressure have the same number of molecules per unit volume. The density is proportional to the mass of an individual molecule and so the product of R and the 'molecular weight' M is constant for all perfect gases. This product MR is known as the *universal gas constant.* For real gases it is not strictly constant but for monatomic and diatomic gases its variation is slight. If M is the ratio of the mass of the molecule to the mass of a hydrogen atom, MR = 8310 J/kg K.

COMPRESSIBILITY AND ELASTICITY

All fluids may be compressed by the application of external force, and when the external force is removed the compressed volumes of fluids expand to their original volumes. Thus fluids also possess elastic characteristics like elastic solids. Compressibility of a fluid is quantitatively expressed as inverse of the bulk modulus of elasticity K of the fluid, which is defined as :

$$K = \frac{\text{Stress}}{\text{Strain}} = -\frac{dp}{\left(\frac{dV}{V}\right)} = \frac{\textit{Change} \text{ in pressure}}{\left(\frac{\text{Change in volume}}{\text{Original volume}}\right)} \quad ...(1)$$

Thus bulk modulus of elasticity K is a measure of the incremental change in pressure dp which takes place when a volume V of the fluid is changed by an incremental amount dV. Since a rise in pressure always causes a decrease in volume, dV is always negative, and the minus sign is included in the equation to give a positive value of K.

For example, consider a cylinder containing a fluid of volume V, which being compressed by a piston. Now if the piston is moved so that the volume V decreases by a small amount dV, then the pressure will increase by amount dp, the magnitude of which depends upon the bulk modulus of elasticity of the fluid, as expressed in equation 1.

In SI units the bulk modulus of elasticity is expressed in N/m^2. In the metric gravitational system of units it is expressed in either kg(f)/cm^2 or kg(f)m^2. In the English system of units it is expressed either in lb(f)/in^2 or lb(f)/ft^2. The bulk modulus of elasticity for water and air at normal temperature and pressure is approximately 2.06×10^9 N/m^2 [or 2.1×10^8 kg (f)/m^2] and 1.03×10^5 N/m^2 [or 1.05×10^4 kg (f)/m^2] respectively. Thus air is about 20,000 times more compressible than water. The bulk (volume) modulus of elasticity of mild steel is bout 2.06×10^{11} N/m^2 [or 2.1×10^{10} kg (f)/m^2] which shows that water is about 100 times more compressible than steel.

However, the bulk modulus of elasticity of a fluid is not constant, but it increases with increase in pressure. This is so because when a fluid mass is compressed, its molecules becomes close together and its resistance to further compression increases, *i.e.,* K increases. Thus for example, the bulk modulus of water roughly doubles as the pressure is raised from 1 atmosphere to 3500 atmospheres.

The temperature of the fluid also affects the bulk modulus of elasticity of the fluid. In the case of liquids there is a decrease of K with increases of temperature. However, for gases since pressure and temperature are interrelated and as temperature increases, pressure also increases, an increase in temperature results in an increase in the value of K.

For liquids since the bulk modulus of elasticity is very high, the change of density with increase of pressure is very small even with the largest pressure change encountered.

Accordingly in the case of liquids the effects of compressibility can be neglected in most of the problems involving the flow of liquids. However, in some special problems such as rapid closure of valve or water hammer, where the changes of pressure are either very large or very sudden, it is necessary to consider the effect of compressibility of liquids.

On the other hand gases are easily compressible and with the change in pressure the mass density of gases changes considerably and hence the effects of compressibility cannot ordinarily be neglected in the problems involving the flow of gases. However, in a few cases where there is not much change in pressure and so gases undergo only very small changes of density, the effects of compressibility may be disregards, *e.g.*, the flow of air in a ventilating system is a case where air may be treated as incompressible.

APPLICATIONS OF THE IMPULSE-MOMENTUM EQUATION

The impulse-momentum equation, together with the energy equation and the continuity equation provides the basic mathematical relationships for solving various engineering problems in fluid mechanics. Since the impulse momentum equation relates the resultant external forces on a chosen free body of fluid or control volume in a flow passage, to the change of momentum flux at the two end sections, it is especially valuable in solving those problems in fluid mechanics in which detailed information about the flow process within the control volume may be either not available or rather difficult to evaluate. Thus in order to apply the impulse-momentum equation, a control volume is first chosen which includes the portion of the flow passage which is to be studied. The boundaries of the control volume are usually extended upto such an extent that its end sections lie in the region of uniform flow. All the external forces acting on this control volume are then considered and the momentum equations in the corresponding directions of reference are applied to evaluate the unknown quantities.

In general the impulse-momentum equation is used to determine the resultant forces exerted on the boundaries of a flow passage by a stream of flowing fluid as the flow changes its direction or the magnitude of velocity or both. The problems of this type include the pipe bend, jet propulsion, propellers and stationary and moving plates or vanes. The application of impulse-momentum equation to the problems of pipe bends, jet propulsion and propellers is illustrated in the following paragraphs. However the problems of stationary and moving vanes are discussed in the chapter of *Impact of Free jets*. Furthermore, the impulse-momentum equation may also be applied to solve the problems of non-uniform flow in which an abrupt change of flow section occurs. The problems of this type include sudden enlargement in pipes, hydraulic jump in open channels etc.

MOMENTUM CORRECTION FACTOR

The above derived impulse-momentum equations in terms of the mean velocities of flow are based on the assumption that the velocity of flow at each section is uniform, that is the velocity is same at different points on the same section. However, in actual practice the velocity is not uniform over a cross section of the flow passage, on account of which the momentum flux computed on the basis of the mean velocity of flow at any section is not equal to the actual momentum flux flowing through the section. The actual momentum flux flowing through any section maybe obtained by integrating the momentum flux flowing through different elementary areas of the cross-section. Thus if v is the velocity of flowing fluid at any point through an elementary area 'dA' of the cross-section, then the mass of fluid flowing per unit time is (ρvdA) and the corresponding momentum flux flowing through this elementary area is ($\pi v^2 dA$). The total momentum flux flowing through the entire cross section Ais equal to

$$\int_A \rho v^2 dA = \frac{w}{g} \int_A v^2 dA$$

which may be evaluated from the known velocity distribution at the cross section.

It is however more convenient to express the momentum flux flowing through any cross-section in terms of the mean velocity of flow. But the actual momentum flux is always greater than that computed by using the mean velocity of flow. Hence in order to account for this difference in the values of the momentum flux due to the non-uniform velocity distribution at any cross-section a factor called *momentum correction* factor represented by β (Greek beta) is introduced, so that the momentum flux computed by using the mean velocity V may be expressed as ($\beta \rho A V^2$) and it is then equal to the actual total momentum flux flowing through the entire cross-section. Thus equating the two, the value of the momentum correction factor may be obtained as

$$\beta \pi A V^2 = \rho \int_A v^2 dA$$

Therefore $$\beta = \frac{1}{AV^2} \int_A v^2 dA \qquad ...(1)$$

Mathematically the square of the average is less than the average of the squares, that is $V < \frac{1}{A}\int_A v^2 dA$, the numerical value of β will always be greater than 1. The actual value of β depends on the velocity distribution at the flow section. If the velocity is uniform over the entire cross-section, β will be equal to 1. For turbulent flow in pipes the value of β lies between 1.02 and 1.05. However, for laminar flow in pipes the value of b is 1.33.

In view of the above discussion, if the velocity distribution is non-uniform, then the momentum correction factor will be required to be introduced in the impulse-momentum equations expressed in terms of mean velocity at each section. Thus equations 5, 7 and 8 are modified as follows:

$$\Sigma F_x = \rho Q\,[\beta_2 (V_2)_x - \beta_1 (V_1)_x]$$

$$\Sigma F_y = \rho Q\,[\beta_2 (V_2)_y - \beta_1 (V_1)_y]$$

$$\Sigma F_z = \rho Q\,[\beta_2 (V_2)_z - \beta_1 (V_1)_z]$$

in which β_1 and β_2 are the momentum correction factors at sections 1–1 and 2–2 respectively. However, in most of the problems of turbulent flow, the value of β is nearly equal to 1 and therefore it may be assumed as one, without any appreciable error being introduced.

FORCE ON A PIPE BEND

As state earlier the impulse-momentum equation is applied to determine the resultant force (or thrust) exerted by a flowing fluid on a pipe bend. Reducing pipe bend through which a fluid of density ρ flows steadily from section 1 to 2. It is desired to find the force exerted by the flowing fluid on the pipe bend. For this the portion of the bend lying between sections 1 and 2 may be chosen as a control volume and all the external forces acting on this may be considered as indicated below.

1. The fluid in the control volume will be subjected to pressure forces p_1A_1 and p_2A_2 by the fluid adjacent to these sections, where p_1, p_2 and A_1, A_2 are the mean pressure and cross-sectional areas at sections 1 and 2 respectively.
2. The boundary surface of the bend will exert forces on the fluid in the control volume. These forces will be distributed non-uniformly over the curved surface of the bend. But for the ease of computation it is assumed that these distributed forces are equivalent to a single concentrated force R, which has R_x and R_y as its components along x and y directions respectively. It may however be stated that

according to Newton's third law of motion, the force exerted by the flowing fluid on the bend will be equal and opposite to R, which is required to be determined.

3. The self-weight W of the fluid in the control volume will be acting in the vertical downward direction.

Thus by applying the impulse-momentum equation in both x and y directions the following expressions are obtained.

$$p_1A_1 \cos \theta_1 - p_2A_2 \cos \theta_2 - R_x = \rho Q (V_2 \cos \theta_2 - V_1 \cos \theta_1) \quad ...(1)$$

For y direction:

$$p_1A_1 \sin \theta_1 + p_2A_2 \sin \theta_2 + R_y - W = \rho Q (-V_2 \sin \theta_2 - V_1 \sin \theta_1) \quad ...(2)$$

From the above equations R_x and R_y can be determined from which the magnitude and direction of the force R exerted by the bend on the fluid can be computed. The force (or thrust) exerted by the fluid on the bend is equal and opposite to R.

Often the continuity equation, the energy equation and the equation of state are required to be used to determine the unknown flow characteristics to be used in the above equations.

For a horizontal bend in equation 2 the term W, representing the weight of the fluid in the bend will be eliminated.

The quantity on the right hand side of equations 1 and 2 is often termed as *dynamic thrust* exerted by the flowing fluid on the bend and vice-versa, in order to distinguish it from the static pressure forces appearing on the left hand side of these equations.

ANGULAR MOMENTUM PRINCIPLE–MOMENT OF MOMENTUM EQUATION

The angular momentum principle states that the torque exerted on any body is equal to the rate of change of angular momentum. The torque is defined as the moment of the force and the angular momentum is defined as the moment of momentum; the moments being taken about the axis of rotation.

Consider a fluid mass δm which is rotating about the z-axis. Let V_x and V_y be its velocity components in x any y directions respectively.

If $a_x = \dfrac{dV_x}{dt}$ and $a_y = \dfrac{dV_y}{dt}$ represent the acceleration components of the fluid mass, one obtains

$$\delta F_x = \frac{dV_x}{dt}\delta m, \delta F_y = \frac{dV_y}{dt}\delta m$$

where δF_x and δF_y are the components of external forces causing the acceleration.

The moment of the external forces about z-axis (counter-clockwise being considered positive) or the torque δT_z, is then obtained as

$$\delta T_z = (x\delta F_y - y\ \delta F_x) = \left(x\frac{dV_y}{dt} - y\frac{dV_x}{dt}\right)\delta m$$

By the rules of differentiation

$$\frac{d}{dt}(xV_y - yV_x) = \frac{dx}{dt}V_y - \frac{dy}{dt}V_x + x\frac{dV_y}{dt} - y\frac{dV_x}{dt}$$

$$= V_xV_y - V_yV_x + x\frac{dV_y}{dt} - y\frac{dV_x}{dt}$$

Therefore, since $(V_xV_y - V_yV_x) = 0$ and δm is constant,

$$\delta T_z = \frac{d}{dt}(xV_y - yV_x)\delta m$$

$$= \frac{d}{dt}[(xV_y - yV_x)\delta m] \qquad ...(3)$$

The quantities $(\delta mV_y)_x$ and $(\delta mV_x)y$ represent the "moments of momentum" or "angular momentum" about the z-axis. Therefore the right hand side of the above expression represents the rate of change of angular momentum about z-axis, and this equal to the torque.

In the above derivation, since z-axis is arbitrarily chosen, a torque equation for the x or y axis may also be similarly obtained.

Hence, it may be stated that the resultant external torque about any axis is equal to the rate of change of angular momentum about that axis. This is the law of moment of moment (or law of angular momentum).

It is usually convenient to express $(xV_y - yV_x)$ in terms of V_t and r, where V_t is the tangential velocity and r is the radial distance as defined

$x = r \cos \theta;\ y = r \sin \theta$

$V_x = V \cos \phi;\ V_y = V \sin \phi$

Hence

$(xV_y - y\ V_x) = r \cos \theta\ (V \sin \phi) - r \sin \theta\ (V \cos\phi)$

$= r\ V \sin (\phi - \theta)$

$= rV_t$

Thus by substituting in equation 4

$$\delta T_z = \frac{d(rV_t\delta m)}{dt} \quad ...(4)$$

Applying equation 4 or 5 to each of the several small fluid masses of a system and summing all the resulting equations, the resultant external torque T_z for a steady flow system is obtained as

$$\Sigma(\delta T_z) = \frac{\Sigma d(rV_t\delta m)}{dt}$$

or $$T_z = \rho Q(r2Vt2 - r1Vt1) \quad ...(5)$$

in which r_2 and V_{t2} and radial distance and tangential velocity at section 2 and r_1 and V_{t1} are same quantities at section 1 of the control volume.

By rewriting equation 5 in the form

$$T_z - \rho Q r_2 V_{t2} + \rho Q r_1 V_{t1} = 0 \quad ...(6)$$

it can be shown that the moment of the momentum flux across an area about any axis equals the moment of all the external forces applied at the centre of the area about the same axis. Further, it may be seen from equation 6 that it the external forces that act on the fluid mass exert no net moment about a fixed axis (*i.e.*, Tz = 0), the moment of momentum of the fluid mass with respect to that axis remains constant. This principle is known as the *law of conservation of moment of momentum or the law of conservation of angular momentum.*

The concept of angular momentum is applied in analyzing the flow problems, such as flow through turbo machinery, where torques are more significant in the analysis than forces. The work done by the flowing fluid on a wheel of a radial flow hydraulic turbine which has radially fixed curved vanes has been evaluated by applying the principle of angular momentum.

JET PROPULSION- REACTION OF JET

When a jet of fluid issues form an opening and strikes an obstruction placed in its path, it exerts a force on the obstruction. This force exerted by the jet is known as the action of the jet. Recalling Newton's third law of motion, since every action is accompanied by an equal and opposite reaction, the jet while coming out of opening exerts a force on the opening in the form of back kick. This force exerted by the jet on the source from which it is issued is known as the reaction of the jet and is therefore equal in magnitude but opposite in direction to the action of the jet. Further if the source issuing the jet is free to move, it will start moving in the direction opposite to that of jet. Thus the reaction of jet can be utilized for the propulsion of various bodies. The principle of jet propulsion, which is applied in the propulsion of surface ships, aircrafts, rockets etc., may be explained by some of the examples described below.

Jet Propulsion of Orifice Tank

Consider a jet of fluid of area a being issued under a constant head H from an orifice provided in the side of a tank, which is large enough so that the velocity within the tank may be neglected. Let the velocity of the jet be V assumed to be given by $V = C_v \sqrt{2gH}$; and Q be the discharge of fluid coming out of orifice. Applying the impulse-momentum equation, the force on the fluid to change its velocity from O to V is

$$F = \frac{wQ}{g}(V - O) = \frac{waV^2}{g}$$

$$= 2waC_v^2H \qquad ...(1)$$

This issuing jet will exert a force on the tank which will be the reaction of the jet and its value will be equal to that of F given by equation 1 but in a direction opposite to V. A physical explanation for the existence of the reaction is that at the vena-contract a the pressure of the fluid is reduced to zero gage pressure and there is also a reduction of pressure of the fluid is reduced to zero gage pressure and there is also a reduction of pressure on the tank walls immediately adjacent to the orifice where the velocity of fluid becomes appreciable.

However, on the opposite side of the tank at the same depth the pressure over a corresponding area is wH and the difference of pressure between the two sides of the tank gives rise to the reaction force. Further, it is seen from equation 1 that in the ideal case with no friction since $C_v = 1$, the reaction of the jet is twice the hydrostatic force exerted upon an area of the same size as the jet and at the same depth below the surface.

Now if the tank considered above is mounted on a frictionless trolley the orifice is initially kept plugged, then as soon as the orifice is opened the jet will be issued and due to the reaction of the jet the tank will start moving with some velocity say u in the direction opposite to the direction of the issuing jet. Thus, the jet issuing from the orifice exerts a propelling force on the tank.

As the tank starts moving with a velocity u, the actual velocity of the issue of the jet will be V_r which is the velocity of the issuing jet relative to the moving tank. Thus V_r will be equal to the vectorial difference of the absolute velocity V of the jet and the velocity of propulsion u of the tank, *i.e.*, $V_r = [V - (-u) = (V + u)$. The negative sign for u has been considered because its direction is opposite to that of V. Thus applying impulse-momentum equation :

Propelling force, F = Reaction of jet

$$= \frac{W}{g}[(V+u)-u]$$

where W is the weight of fluid actually coming out per second. Since W = wa (V + u), as the effective velocity of the efflux of jet is V_r = (V + u), the expression for F becomes

$$F = \frac{wa(V+u)}{g}V \qquad ...(2)$$

Work done by the jet on the moving tank = (F × u)

Actual kinetic energy of the issuing jet

$$= \frac{WV_r^2}{2g} = \frac{W(V+u)^2}{2g}$$

Hence the this case the efficiency of propulsion

$$\eta = \frac{F \times u}{\frac{WV_r^2}{2g}} = \frac{2Vu}{(V+u)^2} \qquad ...(3)$$

Again for a given value of V the efficiency will be maximum if (dη/du) = 0. Thus

$$\frac{d\eta}{du} = 2\left[\frac{(V+u)^2 V - 2(V+u)Vu}{(V+u)}\right] = 0$$

Substituting the value of u in equation 16 the value of maximum efficiency is obtained as

$$\eta_{max} = 0.5 \text{ or } 50\%$$

Jet Propulsion of Ships

The principle of jet propulsion was earlier used for the propulsion of small ships. The ship carries centrifugal pumps which lift water from the surrounding sea and discharge it in the form of a jet by forcing through the orifice provided at the back of the ship.

The reaction produced by the jet entering the sea propels the ship in the direction opposite to that of the jet. The pump intakes may have two alternative arrangements. In one case the intakes may face in the same direction as that of the issuing jet or the intakes may be on the sides of the ship. In the second arrangement the pump intakes may face in the direction of the motion of the ship.

The main difference in the two arrangements is that if the pump intakes face in the direction of the jet then the water has to be sucked by the pumps against the motion of the ship, according more work will be required to be done by the pumps. On the other hand if the pump intakes face in the direction of motion of the ship then water will enter the pipe intakes due to the movement of the ship itself and hence less work will be required to be done by the pumps.

Let V be the absolute velocity of the issuing jet and u be the velocity of the moving ship. Thus, the velocity of the jet relative to the motion of the ship will be $V_r = (V + u)$. Since the effective velocity of the issue of the jet is V_r the kinetic energy available with the water

$$= \frac{WV_r^2}{2g}$$

where W represents the weight of water issuing from the jet per second. If a represents the area of the issuing jet, then $W = wQ = (waV_r)$. Applying the impulse-momentum equation in the direction of the jet,

$$\text{Propelling force } F = \frac{W}{g}[V + u) - u] = \frac{W}{g}V \qquad ...(4)$$

The above expression for the propelling force may be readily derived by bringing the ship to a stationary state before the impulse-momentum equation is applied. For this a velocity equal in magnitude to that of the ship but in opposite direction, *i.e.*, –u, is applied to the whole system. Thereby bringing the ship to rest, but making the effective velocity of the jet as (V + u) and also developing a velocity equal to u in the same direction as that of the jet for the water in the surrounding sea. The application of the impulse-momentum equation will then provide the expression for the propelling force as given by equation 5.

The work done per second on the ship by the reaction of the jet is equal to

$$(F \times u) = \frac{WV_u}{g} = \frac{waV_r(V_r - u)u}{g} \qquad ...(5)$$

which is the output of the system.

Now if it is assumed that the pump intakes face in the same direction as that of the issuing jet, then the energy required to be supplied will be equal to the kinetic energy of the jet. Thus energy supplied per second

$$\frac{WV_r^2}{2g} = \frac{waV_r^3}{2g}$$

∴ Efficiency of the propulsion

$$\eta = \frac{\dfrac{waV_r(V_r - u)u}{g}}{\dfrac{waV_r^3}{2g}}$$

$$= \frac{2(V_r - u)u}{V_r^2}$$

$$= \frac{2Vu}{(V + u)^2}$$

For a given jet velocity V, the condition for maximum efficiency of propulsion is given by $\left(\dfrac{d\eta}{du}\right) = 0$

$$\text{Thus } \frac{d\eta}{du} = \frac{2V[(V+u)^2 - 2(V+u)u]}{(V+u)^4} = 0$$

$u = V$, (since $V \neq 0$ and also $V \neq -u$)

Hence for maximum efficiency of propulsion $u = V$ and by substitution

$$\eta_{max} = \frac{2u^2}{(2u)^2} = 0.5 \text{ or } 50\%$$

In the above derivation the loss of head due to friction etc. in the intake and ejecting pipes has been neglected. But if this loss of head is to be considered and is equal to H_L, then the corresponding loss of energy per sec $= (WH_L)$. In which case the work done by the pump or the total energy supplied per second.

$$= \left(\frac{WV_r^2}{2g} + WH_L\right)$$

and then the efficiency of jet propulsion

$$\eta = \frac{\dfrac{WVu}{g}}{\left(\dfrac{WV_r^2}{2g} + WH_L\right)} = \frac{2Vu}{[V_r^2 + 2gH_L]}$$

In the above case it was assumed that the pump intake faces in the direction of the jet or is on one side of the ship. But if the pump intake faces in the direction of the motion of the ship, then since the water is possessing an initial kinetic energy equal to $\left(\dfrac{Wu^2}{2g}\right)$, corresponding to the velocity of the moving ship, the energy required to be supplied is reduced by this amount. Hence in this case the energy supplied per second

$$= \left[\left(\frac{W}{2g}V_r^2\right)-\left(\frac{W}{2g}u^2\right)\right] = \frac{W}{2g}(V_r^2 - u^2)$$

However in this case also the work done by this jet on the ship will be same as represented by equation 18. Accordingly the efficiency of propulsion will be

$$\frac{\frac{W}{g}(V_r - u)u}{\frac{W}{g}(V_r^2 - u^2)} = \frac{2u}{(V_r + u)} = \frac{2u}{(V + 2u)}$$

In this case, however it is not possible to derive a practical condition for maximum efficiency. But for u = V, which is the condition for the maximum efficiency in the previous case, corresponding value of the efficiency for this case will be

$$\eta = \frac{2u}{u + 2u} = \frac{2}{3} = 0.667 \text{ or } 66.7\%$$

Since in actual practice the velocity of the ship u will normally be less than the velocity of the jet V, and therefore the limiting value of u is equal to V. Accordingly the above obtained value of the of the efficiency may be considered as the maximum possible efficiency for this case.

Again in this case also if the head loss due to friction etc. in the intake and ejecting pipes is equal to H_L, then total energy supplied per second.

$$= \left[\frac{W}{2g}(V_r^2 - u^2) + WH_L\right]$$

According the efficiency of jet propulsion becomes

$$\eta = \frac{\frac{WVu}{g}}{\left[\frac{W}{2g}(V_r^2 - u^2) + WH_L\right]}$$

$$= \frac{2Vu}{[(V_r^2 - u^2) + 2gH_L]}$$

It may however be stated that the jet propulsion in ships is now not commonly adopted because the overall efficiency for such units is much lower than that of screw propeller units.

SOLVED EXAMPLES

Example 1:

A boat travelling at 12 m/s in fresh water has a 0.6 m diameter propeller which takes 4.25 m³ of water per second between its blades. Assuming that the effects of the propeller hub and the boat hull on flow conditions are negligible, calculate the propeller hub and the boat hull on flow conditions are negligible, calculate the thrust on the boat, the theoretical efficiency of the propulsion, and the power input to the propeller.

Solution:

We have

$$Q = \frac{\pi D^2}{4}\frac{V + V_j}{2}$$

$$\text{or } 4.25 = \frac{\pi}{4}(0.6)^2\frac{12 + V_j}{2}$$

$\therefore V_j = 18.06$ m/s

From equation 8.20 we have

$$T_p = \rho Q\,(V_j - V)$$

$$\text{or} \quad T_p = 1000 \times 4.25\,(18.06 - 12)$$

$$T_p = 25755 \text{ N} = 25.755 \text{ kN}$$

Again from equation 26, we have

$$\eta_{th} = \frac{2}{1 + (V_j / V)}$$

$$= \frac{2}{1 + (18.06/12)} = 0.798 \text{ or } 79.8\%$$

From equation 25

$$\text{Power input} = \left[\rho Q(V_j - V)V + \rho Q\frac{V_j - V)^2}{2}\right]$$

$$= \rho Q\,(V_j - V)\left[V + \frac{V_j - V}{2}\right]$$

$$= 1000 \times 4.25\,(18.06 - 12)\left[12 + \frac{18.06 - 12}{2}\right]$$

$$= 387098 \text{ W} = 387.098 \text{ kW}.$$

Example 2:

A water sprinkler has 10 mm diameter nozzles at either end of a rotating arm, each of which is discharging water in opposite direction at right angle to the rotating arm, at a velocity of 8 m/s. If the axis of rotation is at a distance of 0.15 m from one end and 0.2 m from the other, determine the torque required to hold the arm stationary. If friction is neglected, determine the constant angular speed of the arm.

Solution:

The rate of change of moment of momentum is the required to hold the arm stationary.

Initial moment of momentum is zero. Final moment of momentum

$$= \rho Q(V_2 r_2 + V_1 r_1)$$

$$\therefore \quad \text{Torque } T = \rho Q(V_2 r_2 + V_1 r_1)$$

$$= 1000 \times \left[\frac{\pi}{4} \times (0.01)^2 \times 8\right][8 \times 0.2 + 8 \times 0.15]$$

$$= 1.759 \text{ N-m}$$

If the angular velocity of the sprinkler is ω, then the absolute velocities of flow through the nozzle are

$$V_1 = 8 - 0.15\,\omega$$

and $V_2 = 8 - 0.2\,\omega$

Since the moment of momentum of flow entering is zero and there is no friction, the moment of momentum leaving the sprinkler must also zero.

Thus $\rho Q[8 - 0.15\omega) \times 0.15 + (8 - 0.2\omega) \times 0.2] = 0$

or $\omega = \dfrac{2.8}{0.0625} = 44.8\text{rad/s.}$

Example 3:

Through a very narrow gap of height h, a thin plate of large extent is pulled at a velocity V. On one side of the plate is oil of viscosity m_1. and on the other side oil of viscosity m_2. Calculate the position of the plate so that (i) the shear force on the two sides of the plate is equal; (ii) the pull required to drag the plate is minimum.

Solution:

Let y be the distance of the thin plate from one of the surface as shown in the accompanying figure.

(i) Force per unit area on the upper surface of the plate

$$= \mu_1 \left(\frac{dv}{dy} \right) = \mu_1 \left(\frac{V}{h-y} \right)$$

Force per unit area on the bottom surface of the plate

$$= \mu_2 \left(\frac{dv}{dy} \right) = \mu_2 \frac{V}{y}$$

Equating the two, we get

$$\mu_1 \frac{V}{h-y} = \mu_2 \frac{V}{y}$$

$$\therefore \quad y = \frac{\mu_2 h}{(\mu_1 + \mu_2)}$$

(ii) Let F be the pull per unit area required to drag the plate, then

$$F = \mu_1 \left(\frac{V}{h-y} \right) + \mu_2 \left(\frac{V}{y} \right)$$

F = Sum of the shear forces per unit area on both the surfaces of the plate.

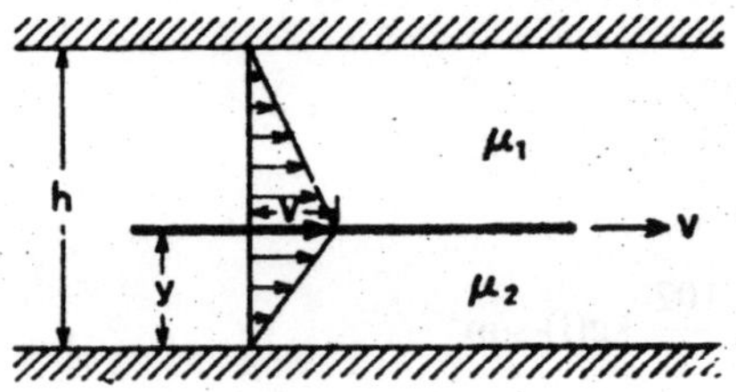

Fig. 5

For the force F to be minimum

$$\frac{dF}{dy} = 0$$

or $$\frac{dF}{dy} = \frac{\mu_1 V}{(h-y)^2} - \frac{\mu_2 V}{y^2} = 0$$

or $$y = \frac{h}{1 + \sqrt{(\mu_1/\mu_2)}}$$

Example 4:

If the equation of a velocity profile over a plate is $v = 2y^{2/3}$; in which v is the velocity in m/s at a distance of y metres above *the plate, determine the shear stress at y = 0 and y = 0.075 m (or 7.5 cm). Given m = 0.835 N.s/m² (or 8.35 poise).*

Solution:

The velocity profile over the plate is given us follows

$$v = 2y^{2/3}$$

$$\therefore \quad \frac{dv}{dy} = 2 \times \frac{2}{3} \times y^{-1/3} = \left(\frac{4}{3}\right) y^{-1/3}$$

Shear stress $\tau = \mu\left(\frac{dv}{dy}\right)$

(*a*) SI units $\tau = 0.835 \times \frac{4}{3} y^{-1/3}$

At $y = 0$, $\tau = \infty$(infinite)

At $y = 0.075$ m

$$\tau = 0.835 \times \frac{4}{3} \times \frac{1}{(0.075)^{1/3}}$$

$$= 2.64 \text{ N/m}^2$$

(*b*) Metric gravitational units

$$\mu = 8.35 \text{ poise}$$

$$= \frac{8.35 \times 0.102}{10} \text{ kg(f)-s/m}^2$$

$$= 8.517 \times 10^{-2} \text{ kg(f)-s/m}^2$$

$$= 8.517 \times 10^{-2} \times \frac{4}{3} \text{y}^{-1/3}$$

At $y = 0$, $\tau = \infty$ (infinite)

At y = 7.5 cm = 0.075 m, $\tau = 8.517 \times 10^{-2} \frac{4}{3} \times \frac{1}{(0.075)^{1/3}}$

$= 0.269$ kg(f)/m² **Ans.**

Example 5:

If the pressure of a liquid is increased from 75 kg(f)/cm² to 140 kg(f)/cm², the volume of the liquid decreases by 0.147 percent. Determine the bulk modulus of elasticity of the liquid.

Solution:

From equation 1.5. bulk modulus of elasticity

$$K = -\frac{dp}{(dV/V)}$$

$$dp = (140 - 75) = 65 \text{ kg(f)/cm}^2$$

and $$\frac{dV}{V} = -\frac{0.147}{100} = -0.00147$$

$$\therefore \quad K = \frac{65}{0.00147} = 4.42 \times 10^4 \text{ kg(f)/cm}^2$$ **Ans.**

Example 6:

A liquid compressed in a cylinder has a volume of 0.0113 m³ at 6.87´10⁶ N/m² (6.87 MN/m²) pressure and a volume of 0.0112 m³ at 13.73´10⁶ N/m² (13.73 MN/m²) pressure. What is its bulk modulus of elasticity?

Solution:

Here we have bulk modulus of elasticity

$$K = -\frac{dp}{(dV/V)}$$

$$dp = (13.73 \times 10^6 - 6.87 \times 10^6) = 6.86 \times 10^6 \text{ N/m}^2$$

$$dV = (0.0112 - 0.0113) = -0.0001/\text{m}^2$$

and $V = 0.0113 \text{ m}^3$

$$\therefore \quad K = \frac{6.86 \times 10^6 \times 0.0113}{0.0001}$$

$$= 7.75 \times 10^8 \text{ N/m}^2 (0.775 \text{ GN/m}^2)$$ **Ans.**

Example 7:

At a depth of 2 kilometres in the ocean the pressure is 840 kg(f)/cm². Assume the specific weight at surface as 1025 kg(f)/m³ and that the average bulk modulus of elasticity is 24 ´ 10³ kg(f)/cm² for that pressure range. (a) What will be the change in specific volume between that at the surface and at that depth ? (b) What will be the specific volume at that depth? (c) What will be the specific weight at that depth ?

Solution:

Bulk modulus of elasticity

$$K = -\frac{dp}{(dV/V)}$$

$$K = 24\times10^3 \text{ kg(f)/cm}^2$$

and $dp = 840 \text{ kg/cm}^2$

$$\therefore \quad \frac{(dV)}{V} = -\frac{840}{24\times10^3} = -0.035$$

The negative sign corresponds to a decrease in the volume with increase in pressure.

The specific volume of the water at the surface of the ocean

$$= \frac{1}{1025} \text{m}^3/\text{kg(f)}$$

$\therefore$ The change in specific volume between that at the surface and at that depth is

$$dV = \frac{0.035}{1025} = 3.41\times10^{-5} \text{m}^3/\text{kg(f)}$$

The specific volume at that depth will be thus equal to

$$V_1 = \left(\frac{1}{1025} - \frac{0.035}{1025}\right)$$

$$= 9.41\times10^{-4} \text{m}^3/\text{kg(f)}.$$

The specific weight of water at that depth is

$$\frac{1}{V_1} = \frac{1}{9.41\times10^{-4}} = 1063 \text{ kg(f)/m}^3 \quad \textbf{Ans.}$$

Example 8:

What should be the diameter of a droplet of water, If the pressure inside is to be 0.0018 kg(f)/cm² grater than the outside ? Given the value of surface tension of water in contact with air at 20°C as 0.0075 kg (f)/m.

Solution:

The internal pressure intensity p in excess of the outside pressure is given by equation 1.6 as

$$p = \frac{2\sigma}{r}$$

or $$r = \frac{2\sigma}{p}$$

or $$2r = \frac{4\sigma}{p}$$

and $\sigma = 0.0075$ kg(f)/m

By substitution, we get

$$2r = d = 4 \times \frac{0.0075}{100} \times \frac{1}{0.0018} \text{cm}$$

$$= \frac{4 \times 0.0075 \times 10}{100 \times 0.0018} \text{mm}$$

$$= 1.67 \text{ mm} \quad \textbf{Ans.}$$

Example 9:

What is the pressure within a droplet of water 0.05 mm in diameter at 20℃, if the pressure outside the droplet is standard atmospheric pressure of 1.03 kg(f)/cm² ? Given σ = 0.0075 kg(f)/m for water at 20℃.

Solution:

Here we have internal pressure intensity p in excess of the outside pressure is given as

$$p = \frac{2\sigma}{r}$$

$$\sigma = 0.0075 \text{ kg(f)/m}$$

$$= \frac{0.0075}{100} \text{kg(f)/cm}$$

$$r = \frac{0.05}{2} = 0.025 \text{ mm} = \frac{0.025}{10} \text{cm}$$

By substitution, we get

$$p = 2 \times \frac{0.0075}{100} \times \frac{10}{0.025}$$

$$= 0.06 \text{ kg(f)/cm}^2$$

The pressure intensity outside the droplet of water

$$= 1.03 \text{ kg(f)/cm}^2$$

$\therefore$ The pressure intensity within the droplet of water

$$= (1.03 + 0.06) = 1.09 \text{ kg(f)/cm}^2. \quad \textbf{Ans.}$$

Example 10:

Calculate the capillary rise in a glass tube of 2 mm diameter when immersed in (a) water, (b) mercury. Both the liquids being at 20℃ and the values of the surface tensions for water and mercury at 20℃ in contact with air are respectively 0.0075 kg(f)/m and 0.052 kg(f)/m

Solution:

Here we have the capillary rise (or depression) is given as

$$h = \frac{2\sigma \cos\theta}{swr}$$

(a) For water $\theta = 0, \cos\theta = 1$

$\sigma = 0.052$ kg(f)/m

$= \frac{0.052}{100}$ kg(f)/cm

$sw = (13.6 \times 1000)$ kg(f)/m^3

$= (13.6 \times 0.001)$ kg(f)/cm^3

$r = 1$ mm $= 0.1$ cm

By substitution, we get

$$h = 2 \times \frac{0.0075}{100} \times \frac{1}{0.001 \times 0.1}$$

$= 1.5$ cm.

(b) For mercury $\theta = 130°, \cos\theta = -0.6428$

$\sigma = 0.052$ kg(f)/m $= \frac{0.052}{100}$ kg(f)/cm

$sw = (13.6 \times 1000)$ kg(f)/m^3

$= (13.6 \times 0.001)$ kg(f)/cm^3

$r = 1$ mm$=0.1$ cm

By substitution, we get

$$h = 2 \times \frac{0.052}{100} \times \frac{(-0.6428)}{13.6 \times 0.001 \times 0.1}$$

$= 0429$ cm.

The negative sign in the case of mercury indicates that there is capillary depression.

Note: often the value of contact angle for mercury is taken as 180°, in which case cosθ = -1 and the capillary depression becomes.

$$h = -\frac{2 \times 0.052 \times 1}{100 \times 13.6 \times 0.001 \times 0.1}$$

$= -0765$ cm. **Ans.**

Example 11:

Determine the minimum size of glass tubing that can be used to measure water level, if the capillary rise in the tube is not to exceed 0.25 cm. Take surface tension of water in contact with air as 0.0075 kg (f)/m.

Solution:

Capillary rise

$$h = \frac{2\sigma}{swr}$$

$$\sigma = 0.0075 \text{ kg(f)/m}$$

$$= \frac{0.0072}{100} \text{kg(f)/cm}$$

$$sw = 1000 \text{ kg(f)/m}^3 = 0.001 \text{ kg(f)/cm}^3$$

$$h = 0.25 \text{ cm}$$

By substitution, we get

$$0.25 = \frac{2 \times 0.0075}{0.001 \times 100 \times r}$$

$$\therefore \quad r = 0.6 \text{ cm}$$

The minimum diameter of the tube is 1.2 cm. **Ans.**

Example 12:

In measuring the unit surface energy of a mineral oil (sp.gr. 0.85) by the bubble method, air is forced to form a bubble at the lower end of a tube of internal diameter 1.5 mm immersed at a depth of 1.25 cm in the oil. Calculate the unit surface energy if the maximum bubble pressure is 15 kg(f)/ m².

Solution:

The effective pressure attributable to surface tension is

$$p = \left[15 - \frac{(0.85 \times 1000 \times 1.25}{100}\right] = 4.375 \text{ kg(f)/m}^2$$

$$p = \frac{2\sigma}{r}$$

The radius of the bubble is taken equal to that of the tube, thus by substitution, we get

$$4.375 - \frac{2 \times \sigma}{0.75 \times 10^{-3}}$$

$\therefore \quad \sigma = 0.001\,64$ kg (f)/m. **Ans.**

Example 13:

Calculate the capillary effect in mm in a glass tube 3 mm in diameter when immersed in (a) water (b) mercury. Both the liquids are at 20℃ and the values of the surface tensions for water and mercury at 20℃ in contact with air are respectively 0.0736 N/m and 0.51 N/m.

Contact angle for water = 0° and for mercury = 130°

Solution:

From capillary rise (or depression) is given as

$$h = \frac{2\sigma\cos\theta}{swr}$$

(*a*) For water $\theta = 0, \cos\theta = 1$

$$\sigma = 0.0736 \text{ N/m}$$

$$sw = 9810 \text{ N/m}^3$$

$$r = \frac{3}{2} = 1.5 \text{ mm} = 1.5 \times 10^{-3}\text{m}$$

By substitution, we get

$$h = \frac{2 \times 0.0736 \times 1}{9810 \times 1.5 \times 10^{-3}}$$

$$= 1.00 \times 10^{-2} \text{ m} = 10 \text{ mm}$$

(*b*) For mercury $\theta = 130°, \cos\theta = -0.6428$

$$\sigma = 0.51 \text{ N/m}$$

$$sw = (13.6 \times 9810) \text{ N/m}^3$$

$$r = \frac{3}{2} = 1.5 \text{ mm} = 1.5 \times 10^{-3}\text{m}$$

By substitution, we get

$$h = \frac{2 \times 0.51 \times (-0.6428)}{13.6 \times 9810 \times 1.5 \times 10^{-3}}$$

$$= -3.276 \times 10^{-3} \text{m}$$

$$= -3.276 \text{ mm}$$

The negative sign in the case of mercury indicates that there is capillary depression.

Example 14:

If 5 m^2 of a certain oil weighs 4000 kg (f). Calculate the specific weight. mass density and specific gravity of this oil.

Solution:

Here we have

$$\text{Specific weight of oil} = \frac{\text{Weight}}{\text{Volume}}$$

$$= \frac{4000 \text{ kg(f)}}{5 \text{ m}^3} = 800 \text{ kg (f)/m}^3$$

$$\text{Mass density of oil} = \frac{\text{Specific Weight of oil}}{\text{Acceleration due to gravity}}$$

$$= \frac{800 \text{ kg(f)/m}^3}{9.81 \text{ m/sec}^2}$$

$$= 81.55 \text{ msl/m}^3$$

$$\text{Specific gravity of oil} = \frac{\text{Specific weight of oil}}{\text{Specific weight of water}}$$

$$= \frac{800 \text{ kg(f)/m}^3}{1000 \text{ kg(f)/m}^3} = 0.8 \textbf{ Ans.}$$

Example 15:

If 5 m^3 of a certain oil weighs 40 kN, Calculate the specific weight, mass density and specific gravity of this oil.

Solution:

Here we have

$$\text{Specific weight of oil} = \frac{\text{Weight}}{\text{Volume}} = \frac{400 \times 1000 \text{ N}}{5 \text{ m}^3} = 800 \text{ N/m}^3$$

$$\text{Mass density of oil} = \frac{\text{Specific weight of oil}}{\text{Acceleration due to gravity}}$$

$$= \frac{800 \text{ N/m}^3}{9.81 \text{ m/s}^2} = 815.49 \text{ kg/m}^3$$

$$\text{Specific gravity of oil} = \frac{\text{Specific weight of oil}}{\text{Specific weight of water}}$$

$$= \frac{800 \text{ N/m}^3}{9810 \text{ N/m}^3} = 0.815 \textbf{ Ans.}$$

Example 16:

Carbon-tetra chloride has a mass density of 1594 kg/m³. Calculate its mass density, specific weight and specific volume in the metric, and the English gravitational systems of units. Also calculate its specific gravity.

Solution:

Here we have

Mass density of carbon-tetra chloride

$$= 1594 \text{ kg/m}^3$$

Since $1 \text{ kg} = \frac{1}{9.81} \text{msl}$

∴ Mass density of carbon tetra chloride in the metric gravitational system of units

$$= \frac{1594}{9.81} = 162.49 \text{ msl/m}^3$$

Acceleration due to gravity = 9.81 m/sec²

∴ Specific weight of carbon-tetra chloride in the metric gravitational system of units.

$$= 162.49 \times 9.81 = 1594 \text{ kg(f)/m}^3$$

Specific volume of carbon-tetra chloride in the metric gravitational system of units

$$= \frac{1}{\text{Specific weight}}$$

$$= \frac{1}{1594} = 6.274 \times 10^{-4} \text{m}^3 / \text{kg(f)}$$

Since 1 kg(f) = 2.205 lb(f)

and 1 m = 3.281 ft

∴ Specific weight of carbon-tetra chloride in the English gravitational system of units

$$= \frac{1594 \times 2.205}{(3.281)^3} = 99.51 \text{ lb(f)/ft}^3$$

Acceleration due to gravity = 32.2 ft/sec²

∴ Mass density of carbon-tetra chloride in the English gravitational system of units

$$= \frac{99.51}{32.2} = 3.09 \text{ slugs/ft}^3$$

Specific volume of carbon-tetra chloride in the English gravitational system of units

$$= \frac{1}{\text{Specific weight}}$$

$$= \frac{1}{99.51} = 1.005 \times 10^{-2} \text{ft}^3 / \text{lb(f)}$$

$$\text{Specific gravity} = \frac{\text{Mass density of carbon tetra chloride}}{\text{Mass density of water}}$$

Mass density of carbon tetra chloride in SI units

$= 1594 \text{ kg/m}^3$

Mass density of water in SI units

$= 1000 \text{ kg/m}^3$

$$\therefore \text{Specific gravity} = \frac{1594 \text{ kg/m}^3}{1000 \text{ kg/m}^3} = 1.594$$ **Ans.**

Example 17:

A plate 0.0254 mm distant from a fixed plate, moves at 61 cm/sec and requires a force fo 0.2 kg(f)/m² to maintain this speed. Determine the dynamic viscosity of the fluid between the plates.

Solution:

Here we have

Shear stress

$$\tau = \frac{F}{A} = \mu \frac{dv}{dy} = \mu \frac{V}{Y}$$

$$\tau = \frac{F}{A} = 0.2 \text{ kg(f)/m}^2$$

$V = 61$ cm/sec $= 0.61$ m/sec

and $Y = 0.0254$ mm $= 2.54 \times 10^{-5}$ m

By substituting in the above equation, we get

$$0.2 = \mu \times \frac{0.61}{2.54 \times 10^{-5}}$$

$$\mu = \frac{0.2 \times 2.54 \times 10^{-5}}{0.61} \text{kg(f)-sec/m}^2$$

$$= 8.328 \times 10^{-6} \text{kg(f)-sec/m}^2$$

$$= 8.328 \times 10^{-10} \text{kg(f)-sec/cm}^2 \textbf{ Ans.}$$

Example 18:

At a certain point in castor oil the shear stress is 0.216 N/m² the velocity gradient 0.216s⁻¹. If the mass density of castor oil is 959.42 kg/m³, find kinematic viscosity.

Solution:

From $\tau = \mu\left(\frac{dv}{dy}\right)$

$$\tau = 0.216 \text{ N/m}^2; \left(\frac{dv}{dy}\right) = 0.216\text{s}^{-1}$$

By substitution, we get

$0.216 = \mu\ (0.216)$

$\therefore\ \mu = 1 \text{ N.s/m}^2$

∴ Kinematic viscosity

$$\upsilon = \frac{\mu}{\rho} = \frac{1}{959.42} = 1.042 \times 10^{-3} \text{m}^2/s \textbf{ Ans.}$$

Example 19:

If a certain liquid has viscosity 4.9×10^{-4} kg(f)-sec/m² and kinematic viscosity 4.9×10^{-2} stokes, what is its specific gravity ?

Solution:

Here we have

Kinematic viscosity $\upsilon = \frac{\mu}{\rho}$

$\Rightarrow$ Mass density $\rho = \frac{\mu}{\upsilon}$

$$\mu = 4.9 \times 10^{-4} \text{kg(f)-sec/m}^2$$

$$\upsilon = 3.49 \times 10^{-2} \text{stokes}$$

$$= 3.49 \times 10^{-6} \text{m}^2/\text{sec}$$

$$\therefore\ \rho = \frac{4.9 \times 10^{-4}}{3.49 \times 10^{-6}} = 140.0 \text{msl/m}^3$$

$$\therefore \text{ sp.gr of the liquid} = \frac{\text{Mass density of liquid}}{\text{Mass density of water}}$$

$$= \frac{140.4}{102} = 1.38 \textbf{ Ans.}$$

Example 20:

Water flows through a 0.9 m diameter pipe at the end of which there is a reducer connecting to a 0.6 m diameter pipe. If the gage pressure at the entrance to the reducer is 412.02 kN/m²/4.2 kg (f) cm²] and the velocity is 2 m/s, determine the resultant thrust on the reducer, assuming that the frictional loss of head in the reducer is 1.5 m.

Solution:

For continuity of flow

$$\frac{\pi}{4}(0.9)^2 \times 2 = \frac{\pi}{4}(0.6)^2 \times V_2$$

$$V_2 = 4.5 \text{ m/s}$$

SI units

Applying Bernoulli's equation, we have

$$\frac{p_1}{w} + \frac{V_1^2}{2g} = \frac{p_2}{w} + \frac{V_2^2}{2g} + \text{jf}$$

$$\frac{412.02 \times 10^3}{9810} + \frac{(2)^2}{2 \times 9.81} = \frac{P_2}{w} + \frac{(4.5)^2}{2 \times 9.81} + 1.5$$

$$\text{or } \frac{p_2}{w} = (42.0 + 0.24 - 1.032 - 1.5) = 39.672 \text{ m}$$

$$\therefore\ p_2 = (39.672 \times 9810)$$

$$= 389182 \text{ N/m}^2 = 389.182 \text{ kN/m}^2$$

Let F_x be the force exerted by the reducer on the fluid, acting opposite to the direction of flow, then applying equation 12, we get

$$412.02 \times 10^3 \times \frac{\pi}{4}(0.9)^2 - 389.182 \times 10^3 \times \frac{\pi}{4}(0.6)^2 - F_x$$

$$= 100 \times \frac{\pi}{4}(0.9)^2 \times 2(4.5 - 2)$$

or F_x = 148896 N =148.896 kN

∴ The resultant thrust exerted by the fluid on the reducer = 148.896 kN, which is acting in the direction of flow.

Metric units

Applying Bernoulli's equation, we have

$$\frac{p_1}{w}+\frac{V_1^2}{2g}=\frac{p_2}{w}+\frac{V_1^2}{2g}+h_f$$

or $$\frac{4.2\times10^4}{1000}+\frac{(2)^2}{2\times9.81}=\frac{P_2}{w}+\frac{(4.5)^2}{2\times9.81}+1.5$$

or $\frac{p_2}{w}$ = (42.0 + 0.24 –1.032 – 1.5) = 39.672 m

∴ p_2 = (39.672× 1000)

= 39.672 × 10^3 kg(f) m^2

Let F_x be the force exerted by the reducer on the fluid, acting opposite to the direction of flow, then applying equation 12, we get

$$4.2\times104\times\frac{\pi}{4}(0.9)^2-39.672\times10^3\times\frac{\pi}{4}(0.6)^2-F_x$$

$$=\frac{1000}{9.81}\times\frac{\pi}{4}(0.9)^2\times2(4.5-2)$$

or F_x = 15178kg (f)

∴ The resultant thrust exerted by the fluid on the reducer = 15178 kg(f). Which is acting in the direction of flow.

Example 21:

A tank 1.5 m high stands on a trolley and is full of water. It has an orifice of diameter 0.1 m at 0.3 m from the bottom of thank tank. If the orifice is suddenly opened, what will be the propelling force on the trolley? Coefficient of discharge of the orifice is 0.60.

Solution:

Discharge from the orifice

$= \text{Cda}\sqrt{2gH}$

$= 0.60\times\frac{\pi}{4}(0.1)^2(2\times9.81\times1.2)^{1/2}$

= 0.023m^2/s

Velocity of the jet issuing from the orifice

$$=\frac{Q}{a}=\frac{0.023\times4}{\pi\times(0.1)^2}=2.93\text{m/s}$$

Propelling force $F=\frac{wQV}{g}=\frac{9810\times0.023\times2.39}{9.81}=67.39\text{N}$.

Example 22:

The kinematic viscosity and specific gravity of a certain liquid are 5.58 stokes (5.58×10^{-4} m^2/s) and 2.00 respectively. Calculate the viscosity of this liquid in both metric gravitational and SI units.

Solution:

Here we have

(a) Metric gravitational units.

Sp.gr. of the liquid = 2.00

Mass density of water = 102 msl/m^3

∴ Mass density of the liquid = (2 × 102) = 204 msl/m^3

Kinematic viscosity of the liquid

= 5.58 stokes

= 5.58×10^{-4} m^2/sec

∴ Viscosity of the liquid

$$\mu = \upsilon \times \rho$$

$$= (5.58 \times 10^{-4} \times 204) \text{ kg(f)-sec/m}^2$$

$$= 0.114 \text{ kg(f)-sec/m}^2$$

(b) SI units

Specific gravity of the liquid = 2.00

Mass density of water = = 1000 kg/m^3

∴ Mass density of the liquid = (2 × 1000) = 2000 kg/m^3

Kinematic viscosity of the liquid

= 5.58×10^{-4} m^2/s

∴ Viscosity of the liquid

$$\mu = \upsilon \times \rho$$

$$= (5.58 \times 10^{-4} \times 2000) \text{ N-s/m}^2$$

$$= 1.116 \text{ N.s/m}^2$$ **Ans.**

Example 23:

A rectangular plate of size 25 cm by 50 cm and weighing 25 kg(f) slides down a 30° inclined surface at a uniform velocity of 2m/sec. If the uniform 2 mm gap between the plate and the inclined surface is filled with oil determine the viscosity of the oil.

Solution:

When the plate is moving with a uniform velocity of 2 m/sec, the viscous resistance to the motion is equal to the component of the weight of the plate along the sloping surface. Component of the weight of the plate along the slope = 25 sin 30° = 12.5 kg(f)

Viscous resistance = $(\tau \times A)$

$$= \mu \frac{dv}{dy} \times A$$

$$= \mu \frac{V}{Y} \times A$$

$V = 2$ m/sec; $y = 2 \times 10^{-3}$ m; and $A = (0.25 \times 0.5)\text{m}^2$

By substituting these values, we get

$$\text{Vesxous resistance} = \mu \times \frac{2}{2 \times 10^{-3}} \times (0.25 \times 0.5)$$

$$= 125\ \mu\ \text{kg(f)}$$

Equating the two, we get

$$125\ \mu = 12.5$$

$$\mu = 0.1\ \text{kg(f)-sec/m}^2$$ **Ans.**

Example 24:

A cubical block of 20 cm edge and weight 20 kg(f) is allowed to slide down a plane inclined at 20° to the horizontal on which there is thin film of oil of viscosity 0.22× 10⁻³ kd(f)-s/m². What terminal velocity will be attained by the block if the fill thickness is estimated to be 0.025 mm ?

Solution:

The force causing the downward motion of the block is

$$F = W \sin 20° = (20 \times 0.3420) = 6.84\ \text{kg(f)}$$

which will be equal and opposite to shear resistance.

$$\therefore \quad \tau = \frac{F}{A} = \frac{6.84}{(0.20 \times 0.20)} 171\ \text{kg(f)/m}^2$$

Further from equation 1.3 we have

$$\tau = \mu \frac{dv}{dy} = \mu \frac{V}{Y}$$

$\mu = 0.22 \times 10^{-3}$ kg(f)-s/m²; y = 0.025 mm = 0.025×10^{-3}m

Thus by substitution we get

$$171 = \frac{0.22 \times 10^{-3} V}{0.025 \times 10^{-3}}$$

$\therefore \quad V =$ 19.43 m/sec. **Ans.**

Example 25:

A cylinder of 0.30 m diameter rotates concentrically inside a fixed cylinder of 0.31 m diameter. Both the cylinders are 0.3 m long. Determine the viscosity of the liquid which fills the space between the cylinders if a torque of 0.98 N.m is required to maintain an angular velocity of 2π rad/s (or 60 r.p.m., since angular velocity $\omega = \frac{2\pi N}{60}$, where N is speed of rotation in r.p.m.).

Solution:

Tangential velocity of the inner cylinder

$$V = r\omega$$

$$= 0.15 \times 2\pi$$

$$= 0.942 \text{ m/s}$$

For the small space between the cylinders the velocity profile may be assumed to be linear, then

$$\frac{dv}{dy} = \frac{V}{Y}$$

$$= \frac{0.942}{(0.155 - 0.15)} = 188.4 \text{ s}^{-1}$$

The torque applied to maintain the constant angular velocity is equal to the torque resisted due to shear stress.

Torque resisted $= \tau \times (2\pi \times 0.15 \times 0.30) \times 0.15$

Thus $\quad 0.98 = \tau \times (2\pi \times 0.15 \times 0.30) \times 0.15$

$\therefore \quad \tau = 23.11 \text{ N/m}^2$

From equation 1.3

$$\tau = \mu \frac{dv}{dy}$$

$$\therefore \quad \mu = \frac{\tau}{(dv / dy)}$$

$$= \frac{23.11}{188.4} = 0.123 \text{ N.s/m}^2$$ **Ans.**

Example 26:

Velocity distribution for laminar flow of real fluid in a pipe is given as $v = V_{max} [1-r^2/R^2)]$, *where* V_{max} *is velocity at the centre of the pipe, R is pipe radius, and v is velocity at radius r from the centre of the pipe. Determine the momentum correction factor.*

Solution:

From equation 10, Momentum correction factor is given as

$$\beta = \frac{1}{AV^2}\int_A v^2 dA$$

$$\text{Mean velocity } V = \frac{Q}{A} = \frac{\int vdaA}{A}$$

$$= \frac{\int^R 0V_{max}\left(\frac{R_2 - r^2}{R^2}\right)(2\pi rdr)}{\pi R^2} = \frac{V_{max}}{2}$$

$$\text{Thus} \quad \beta = \frac{1}{(\pi R^2)\left(\frac{V_{max}}{2}\right)^2}\int_0^R V_{max}^2\left(\frac{R^2 - r^2}{R^2}\right)^2 (2\pi rdr)$$

$$= \frac{8}{R^6}\left(\frac{1}{2}R^6 - \frac{1}{2}R^6 + \frac{1}{6}R^6\right) = \frac{4}{3} = 1.33.$$

Example 27:

The resistance to motion of a vessel is 24.525 kN at a velocity of 4.5 m/s. The jet efficiency is to be 80% and the mechanical efficiency of the pumps is 75% Hydraulic losses in the ducts are 5% of the relative kinetic energy at exit. Determine (a) the velocity of the jet; (b) the orifice area at exit; (c) the power required to drive the pumps for the given speed of the vessel, assuming that the water is drawn in through the intakes facing in the direction of the motion of the ship.

Solution:

(a) u = 4.5 m/s and the efficiency = 80%

But efficiency

$$\eta = \frac{2u}{V_r + u}$$

or $$0.8 = \frac{2 \times 4.5}{V_r + 4.5}$$

$$V_r = 6.75 \text{ m/s}$$

and $V = 2.25$ m/s

(b) Work done per second

$$= \text{resistance} \times \text{velocity of vessel}$$

$$= (24.525 \times 4.5) = 110.363 \text{ kN m/s}$$

$$= \frac{W}{g}(V_r - u)u$$

$$= \frac{W}{g}(6.75 - 4.5)\, 4.5$$

$$W = \frac{110.363 \times 9.81}{4.5 \times 2.25}$$

$$= 106.929 \text{ kN/s} = 106929 \text{ N/s}$$

$$W = waV_r$$

$$\therefore a = \frac{W}{wV_r} = \frac{106929}{9810 \times 6.75} = 1.615 \text{ m}^2$$

Area of orifice at exit

$$= 1.615 \text{ m}^2$$

(c) Power supplied to jet

$$= \frac{W}{2g}(V_r^2 - u^2)$$

Loss of energy in ducts

$$= 0.05\frac{V_r^2}{2g} \times W$$

$\therefore$ Total power required from pumps

$$= \frac{W}{2g}[(V_r^2 - u^2) + 0.05V_r^2]$$

$$= \frac{106929}{2 \times 9.81}[(6.75^2 - 4.5^2) + 0.05 \times 6.75^2]\text{W}$$

$$= 150369 \text{ W} = 150.396 \text{ kW}$$

The mechanical efficiency of the pumps is 75%

∴ Power required to drive pumps

$$= \frac{150.369}{0.75} = 200.5 \text{ kW}$$

Example 28:

A ship whose resistance is 24.525 kN is to be driven at 5 m/s by means of a jet of water directed under water. The velocity of the jet is to be 7.5 m/s relative to the ship. The efficiency of the pump operating the jet is estimated to be 80%, the frictional resistance of the pipes being equal to 3 m of water. Calculate :

(a) The power required to drive the pump;

(b) The overall efficiency of the system in the following cases:

(i) the water enters the ship through an inlet facing a head:

(ii) the water enters through an inlet in side of ship.

Solution:

$$u = 5 \text{ m/s};$$

$$V_r = 7.5 \text{ m/s};$$

$$R = 24.525 \text{ kN}$$

$$\eta_p = 0.8;$$

and $H_L = 3$ m

The reaction of the jet should be just equal to the resistance to the motion of the ship

$$F = R = 24.525 \text{ kN}$$

But $F = \frac{W}{g}V = \frac{W}{g}(V_r - u)$

or $24.525 = \frac{W}{9.81}(7.5 - 5)$

∴ $W = 96.236 \text{ kN} = 96236 \text{ N}$

(a) (i) when the water enter the ship through an inlet facing ahead, the output of the pump

$$= \left[\left(\frac{WV_r^2}{2g} - \frac{Wu^2}{2g}\right) + WH_L\right]$$

$$= W\left[\frac{(V_r^2 - u^2)}{2g} + H_L\right]$$

∴ Output of the pump

$$= 96236\left[\frac{(7.5^2 - 5^2)}{2 \times 9.81} + 3.0\right]$$

$$= 441989 \text{ W} = 441.989 \text{ kW}$$

∴ Input of the pump

$$= \frac{\text{Output}}{\eta_p}$$

$$= \frac{441.989}{0.8} = 552.486 \text{ kW}$$

∴ Power required to drive the pump

$$= 552.486 \text{ kW}$$

(ii) When the water enters through an inlet in the side of the ship, the output of the pump

$$= \left[\frac{WV_r^2}{2g} + WH_L\right]$$

$$= W\left[\frac{V_r^2}{2g} + H_L\right]$$

∴ Output of the pump

$$= 96236\left[\frac{7.5^2}{2 \times 9.81} + 3.0\right]$$

$$= 564614 \text{ W} = 564.614 \text{ kW}$$

∴ Input of the pump

$$= \frac{\text{Output}}{\eta} = \frac{564.614}{0.8}$$

$$= 705.768 \text{ kW}$$

(b) (i) The overall efficiency of the system for this case

$$\eta = \frac{F \times u}{\text{Imput of the jump}}$$

$$= \frac{24.525 \times 10^3 \times 5}{552.486 \times 10^3}$$

$$= 0.222 \text{ or } 22.2\%$$

(ii) The overcall efficiency of the system for this case

$$\eta = \frac{F \times u}{\text{Imput of the pump}}$$

$$= \frac{24.525 \times 10^3 \times 5}{705.768 \times 10^3}$$

$$= 0.174 \text{ or } 17.4\%.$$

Example 29:

The diameter of a pipe bend is 0.3 m at inlet and 0.15 m at outlet and the flow is turned through 120° in a vertical plane. The axis at inlet is horizontal and the centre of the outlet section is 1.5 m below the centre of the inlet section. The total volume of fluid contained in the bend is 0.085m³. Neglecting friction, calculate the magnitude and direction of the force exerted on the bend by the water flowing through it at 225 l/s when the inlet pressure is 137.34 kN/m².

Solution:

For continuity of flow

$$Q = A_1V_1 = A_2V_2$$

$$\text{or } 225 \times 10^{-3} = \frac{\pi}{4}(0.3)^2 V_1 = \frac{\pi}{4}(0.15)^2 V_2$$

$$\therefore \quad V_1 = 3.18 \text{ m/s};$$

and $V_2 = 12.73$ m/s

Neglecting friction losses, by applying Bernoulli's equation, we get

$$\frac{p_1}{w} + \frac{V_1^2}{2g} + Z_1 = \frac{p_2}{w} + \frac{V_2^2}{2g} + Z_2$$

$$\text{or } \frac{137.34 \times 10^3}{9810} + \frac{(3.18)^2}{2 \times 9.81} + 1.5 = \frac{p_2}{w} + \frac{(12.73)^2}{2 \times 9.81} + 0$$

$$\text{or } \frac{p_2}{w} = (14.0 + 0.515 + 1.5 - 8.26) = 7.755 \text{ m}$$

$$\therefore \quad p_2 = (7.755 \times 9810)$$

$$= 76077 \text{ N/m}^2 = 76.077 \text{ kN/m}^2$$

By applying equation 12, we get

$$137.34 \times 10^3 \times \frac{\pi}{4}(0.3)^2 - 76.077 \times 10^3 \times \frac{\pi}{4}(0.15)^2 \cos 120° - F_x$$

$$= 1000 \times 225 \times 10^{-3} (12.73 \cos 120° - 3.18)$$

$$\text{or } F_x = \frac{\pi}{4}(0.5)^2[549.36 \times 10^3 + 38.039 \times 10^3] + 225 \ (6.37 + 3.18)$$

or $F_x = 12529 \text{ N} = 12.529 \text{ kN}$

Similarly by applying equation 13, we get

$$F_y + (0.085 \times 9810) - 76.077 \times 10^3 \times \frac{\pi}{4}(0.15)^2 \sin 60°$$

$$= 100 \times 225 \times 10^{-3} (12.73 \sin 60°)$$

$$\text{or} \quad F_y = (2480.51 - 833.85 + 1164.28)$$

$$= 2810.94 \text{ N} = 2.811 \text{kN}$$

$$\text{Thus} \quad F = \sqrt{F_x^2 + F_y^2}$$

$$= \sqrt{(12.529)^2 + (2.811)^2} = 12.84 \text{ kN}$$

$$\tan \alpha = \frac{2.811}{12.529} = 0.2244$$

$\therefore \alpha = 12°39'$

$\therefore$ Force of 12.84 kN acts on the bend at an angle of 12°39' upwards from inlet axis.

Example 30:

A plate at a distance of 0.2 cm. from the fixed plate moves at 2m/sec. and requires a force of 40 dyne/cm² to maintain this speed. Determine the coefficient of viscosity of the fluid between the plates.

Solution:

The velocity gradient becomes

$du/dy = 2\times 100/0.2 = 10^3$, $F = 40$ dynes/cm²

$$\mu = \frac{F}{du/dy} = \frac{40}{10^3} = 4 \times 10^{-2} \text{poise.} \qquad \textbf{Ans.}$$

Example 31:

Determine the coefficient of viscosity μ of a fluid, the rate at which the fluid is moving out of a circular pipe of radius a and length l is measured when the pressures on the two sides of the pipe is p_1 and p_2. The formula giving flux is

$$Q = \pi a4 (P1 - P2)/8\mu l,$$

where $Q = 800$ cc/sec.l $= 60$cm, $a = 0.5$ cm,

and $P_1 - P_2 = 4.9 \times 10^3$ dynes/sec^2.

Solution:

$$\text{Since } Q = \frac{\pi a^4 (P_1 - P_2)}{8\pi l} \Rightarrow \mu = \frac{\pi a^4 (P_1 - P_2)}{8Ql}$$

$$\text{or} \quad \mu = \frac{22 \times (0.5)^4 \times 4.9 \times 10^3}{7 \times 800 \times 60} = \frac{77}{3840}$$

$$\text{or} \quad \mu = 77/3840 = 0.02 \text{ poise.}$$ **Ans.**

Example 32:

Determine the coefficient of viscosity of a fluid, the fluid is made to rotate between two long co-axial cylinders of radius r_1 and $r_2 (r_2 > r_1)$. If the inner cylinder rotates with angular velocity ω while the outer is at rest, then the torque T on a unit length of each cylinder is

$$T = 4\pi \ \omega \ \mu \ r_1^2 r_2^2 / (r_2^2 - r_1^2),$$

where radii of the cylinders are 3 cm and 3.5 cm, the inner cylinder rotates at a speed of 120 rpm. and the torque is 5.35×10^2 dynes-cm.

Solution:

$$\text{Since } T = 4\pi \ \omega \ \mu \ \frac{r_1^2 r_2^2}{(r_1^2 - r_2^2)}$$

$$\text{or} \quad \mu = \frac{T(r_2^2 - r_1^2)}{(4\mu \ \omega \ r_1^2 r_2^2)}$$

or $\omega = 120/60 = 2$ rps.

or $\mu = [5.35 \times 10^2 \times (3.5^2 - 3^2) \times 7]/4 \times 22 \times 2 \times (3.5)^2 \times 3^2$

or $\mu = 535 \times 13 / 176 \times 63 = 0.60$ poise. **Ans.**

Example 33:

The mass flux through a rectangular channel of length l, breadth b and height h ($b \ll l$, $h \ll l$) is given by

$$Q = bh^2 (P_1 - P_2) / 12\mu l,$$

where p_1 and p_2 are the pressures at the ends of the channel, and μ is the coefficient of viscosity. Determine the coefficient of viscosity if

$Q = 20$ *cc./sec.,* $l = 100$*cm.,* $b = 6$ *cm.,* $h = 0.5$*cm.,*

$P_1 - P_2 = 1.2 \times 10^3$ *dynes/cm.*

Solution:

Since $Q = bh^2 (P_1 - P_2) / 12\mu l$,

or $\mu = bh^2 (P_1 - P_2) / 12Ql$,

or $\mu = 6 \times (0.5)2 \times 1.2 \times 10^3/12 \times 20 \times 10$

or $\mu = 3/40 = 0.075$ poise. **Ans.**

Example 34:

A plate weighing 150 N measures 80 × 80 cm. *It slides down an inclined plane over an oil film of 1.2 mm thick. for an inclination of* µ/6 *and a velocity of 20 cm/s, calculate the velocity of the fluid.*

Solution:

Shear stress

$$\tau \frac{\text{Force}}{\text{Area}} = \frac{150 \sin \mu / 6}{0.80 \times 0.80} = 117.19 \text{ N/m}^2.$$

Rate of deformation $du / dy = 20/0.12 = 175$ rad/s.

From Newton's law of viscosity, we have $\tau = \mu \frac{du}{dy}$

$$\Rightarrow \quad \mu = \frac{\tau}{du / dy} = 117.19 / 175 = 0.67 \text{ N-s/m}^2. \quad \textbf{Ans.}$$

Example 35:

A liquid compressed in a cylinder has a volume of 0.4 cc. at 6.8×10^7 dynes/cm^2 *and a volume of 0.396 cc. at* 1.36×10^8 *dynes/cm^2. What is its bulk modulus ?*

Solution:

Bulk modulus of liquid is given by

$$k = -\frac{dp}{(dv / v)},$$

where dp is the change in pressure and (*dv/v*) is the relative change in the volume.

Thus $k = 6.8 \times 10^7 \times 0.4 / 0.004$

$k = 6.8 \times 10^9$ dynes/cm^2.

Example 36:

Bulk modulus of water is 2.2 × 10^{10} dynes/cm^2. Find the change in the volume when 100 cc. of water is subjected to an increase of pressure by 7.7 × 10^6 dynes/cm^2.

Solution:

Bulk modulus of a liquid is given by

$$k = -\frac{dp}{dv/v}$$

$$\Rightarrow \quad \delta v = -(v/k)\delta p.$$

Thus the decrease in the volume $= \dfrac{100 \times 7.7 \times 10^6}{2.2 \times 10^{10}} = \dfrac{7}{200}$

= 0.035 cc.

Example 37:

Find the shape of the surface of a fluid under a gravitational field and bounded on one side by a vertical plane wall.

Solution:

Let the plane z = 0 be the surface of the fluid far from the wall. The equation of the shape of the interface is given by

$$\sigma/R = -\rho g z + \text{constant}, \qquad \text{...(1)}$$

where R is radius of curvature and σ is the surface tension.

The equation (1) becomes

$$\frac{2z}{a^2} - \frac{z''}{(1+z'^2)^{3/2}} = \text{const.};$$

$$R = -\frac{(1+z'^2)^{3/2}}{z''}, a^2 = \frac{2\sigma}{\rho g}$$

The constant vanishes as $R = \infty$ at z = 0.

By integrating, we have

$$\frac{1}{\sqrt{(1+z'^2)}} = B - \frac{z^2}{a^2},$$

where B is an integration constant.

Subce z' = 0 at z = 0; B =1

$$\Rightarrow \quad z' = \left[-1 + (1 - z^2/a^2)^{-2}\right]^{1/2} \qquad ...(2)$$

By integrating (2), we have

$$x = -\frac{a}{\sqrt{2}}\cosh^{-1}\left(\frac{a\sqrt{2}}{z}\right) + a\sqrt{(2 - z^2/a^2)} + x_0 \qquad ...(3)$$

where $x = 0,\ z = h$;

$$x_0 = \frac{a}{\sqrt{2}}\cos^{-1}\left(\frac{a\sqrt{2}}{z}\right) - a\sqrt{(2 - h^2/a^2)}$$

The relation (3) gives equation of the interface. **Ans.**

EXERCISES

1. A hydraulic lift consists of a 25 cm diameter ram which slides in a 25.015 cm diameter cylinder, the annular space being filled with oil having a kinematic viscosity of 0.025 cm^{2}sec and specific gravity of 0.85. If the rate of travel of he ram is 9.15 m/min, find the frictional resistance when 3.05 m of the ram is engaged in the cylinder.

   [1.055 kg(f)]

2. The water level in a steel tank is measured with a piezometer of diameter 5 mm. If the reading of water surface in the tube is 90 cm. Find the actual depth of water in the tank.

3. A cylinder 0.1 m diameter rotates in an annular sleeve 0.102 m internal diameter at 100 r.p.m. The cylinder is 0.2 m long. If the dynamic viscosity of the lubricant between the two cylinders is 1.0 poise, find the torque needed to drive the cylinder against viscous resistance. Assume that Newton's Law of viscosity is applicable and the velocity profile is linear.

   [0.1645 N.m]

4. A fluid compressed in a cylinder has a volume of 0.011 32 m^3 at a pressure of 70.30 kg(f)/cm^2. What should be the new pressure in order to make its volume 0.011 21 m^3? Assume bulk modulus of elasticity K of the liquid as 703 0 kg(f)/cm^2.

   [138.61 kg(f)/cm^2]

5. If the volume of a liquid decreases by 0.2 per cent for an increase of pressure from 6.867 MN/m^2 to 15.696 MN/m^2, what is the value of the bulk modulus o the liquid ?

   [44.145 × 108 N/m^2]

6. If a certain liquid has a mass density of 129 msl/m^3, what are the values of its specific weight, specific gravity and specific volume in metric gravitational and metric absolute systems of units.

7. If 5.27 m^3 of a certain oil weighs 44 kN, calculate the specific weight, mass density and specific gravity of the oil.

   [8349 N/m^3; 851.09 kg/m^3; 0.851]

9. The specific gravity of a liquid is 3'0, what are its specific Weight, specific mass and specific volume.

   [3000 kg(f)/m^3; 305.8 mslAn^3; 0.33×10^{-3} m^3/kg(f); 29.43 kN/m^3; 3000 kg/m^3; 3.398×10^{-5} m^3/N]

10. A certain liquid has a dynamic viscosity of 0.073 poise and specific gravity of 0.87. Compute the kinematic viscosity of the liquid in stokes and also in m^2/s.

    (0.0839 stokes; 0.0839×10^{-4} m^2/s]

11. If a certain liquid has a viscosity of 0.048 poise and kinematic viscosity 3.50×10^{-2} stokes, what is its specific gravity? [1.371]

12. In a stream of glycerine in motion at a certain point the velocity gradient is 0.25 s^{-1}. If for fluid $\rho = 129.3$ msl/m^3 (1268.4 kg/m^3) and $\upsilon = 6.30 \times 10^{-4}$ m^2/s, calculate the shear stress at the point.

    [0.02036 kg(f)/m^2; 0.19977 N/m^2]

13. A body weighing 441.45 N with a flat surface area of 0.093 m^2 slides down lubricated inclined plane making a 30° angle with the horizontal. For viscosity of 0.1 N.s/m^2 and body speed of 3m/s, determine the lubricant film thickness. [0.126 mm]

2

Properties of Multiplication of Matrices

INTRODUCTION

Definition: *Let $A = [a_{ij}]$ $m \times n$ and $B = B = [b_{ik}]$ $n \times p$ be two matrices such that the number of column in A is equal to the number of row in B.*

Then the matrix $[c_{ik}]$, $m \times p$ such that $c_{ik} = \sum_{j=1}^{n} a_{ij} b_{ik}$ is called the product of the matrices A ande B in that order and we write $= AB$.

As an example, consider the matrices

$$A = \begin{bmatrix} 1 & 2 & 3 \\ 4 & 5 & 6 \end{bmatrix} \text{ and } B \begin{bmatrix} 7 & 8 \\ 9 & 10 \\ 11 & 12 \end{bmatrix}$$

Here the number of columns in A = 3 = then number of rows in B and thus we can evaluate AB.

Let Ab = $[c_{ij}]$, where $[c_{ij}]$ is 2×2 matrix.

Now to write c_{11}, we take the elements of the first row of A *viz.*, 1, 2, 3 in this order and the elements of the first column of B *viz.*, 7, 9, 11 in this order and form the products 1.7, 2.9, 3.11 and finally add them.

i.e., $c_{11} = 1.7 + 2.9 + 3.11 = 58$

Similarly $c_{12} = 1.8 + 2.10 + 3.12 = 64;$

$c_{21} = 4.7 + 5.9 + 6.11 = 139$

and $c_{22} = 4.8 + 5.10 + 6.12 = 154$

Hence $AB = [c_{ij}] = \begin{bmatrix} c_{11} & c_{12} \\ c_{21} & c_{12} \end{bmatrix} = \begin{bmatrix} 58 & 64 \\ 139 & 154 \end{bmatrix}.$

Note: The product AB can be calculated only if the number of columns in A be equal to the number of rows in b. The two matrices A and B satisfying this condition are called conformable to multiplication.

POST-MULTIPLICATION AND PRE-MULTIPLICATION OF MATRICES

The matrix AB is the matrix A post-multiplied by B whereas the matrix BA is the matrix A pre-multiplied by B.

In the product AB, the matrix A is know as the pre-factor and the matrix B is know as the post-factor.

The product in both the above the above cases viz. AB and BA may or may not exist and may be equal or different.

i.e., we say AB $\neq$ BA in general.

The same is discussed on the next page:

Case 1: If the matrix A is $m \times n$ and the matrix B is $n \times k$, then the product AB exists whereas BA does not exist, since we know that AB can be calculated only if the numbers of columns in A is equal to the number of rows in B.

Case 2: If the matrix A is $m \times n$ and the matrix B is $n \times m$, then both AB and BA exist, but the matrix AB is $m \times m$ while the matrix BA is $n \times n$.

Hence AB $\neq$ BA thought AB and BA exist.

Case 3: If both A and B are square matrices of the same order, then AB as well as BA exist but are not necessarily equal *i.e.,* if

$$A = \begin{bmatrix} 1 & 2 \\ 3 & 4 \end{bmatrix} \text{ and } B = \begin{bmatrix} 3 & 1 \\ 4 & 7 \end{bmatrix}$$

then

$$AB = \begin{bmatrix} 1 & 2 \\ 3 & 4 \end{bmatrix} \times \begin{bmatrix} 3 & 1 \\ 4 & 7 \end{bmatrix} = \begin{bmatrix} 1.3 + 2.4 & 1.1 + 2.7 \\ 3.3 + 4.4 & 3.1 + 4.7 \end{bmatrix}$$

$$= \begin{bmatrix} 11 & 25 \\ 25 & 31 \end{bmatrix}$$

and

$$BA = \begin{bmatrix} 3 & 1 \\ 4 & 7 \end{bmatrix} \times \begin{bmatrix} 1 & 2 \\ 3 & 4 \end{bmatrix} = \begin{bmatrix} 3.1 + 1.3 & 3.2 + 1.4 \\ 4.1 + 7.3 & 4.2 + 7.4 \end{bmatrix}$$

$$= \begin{bmatrix} 6 & 10 \\ 25 & 36 \end{bmatrix}$$

$\therefore \quad AB \neq BA.$

But if $A = \begin{bmatrix} 1 & 0 \\ 0 & -2 \end{bmatrix}$ and $B = \begin{bmatrix} 1 & 0 \\ 0 & 4 \end{bmatrix}$

then $\quad AB = \begin{bmatrix} 1 & 0 \\ 0 & -2 \end{bmatrix} \times \begin{bmatrix} 1 & 0 \\ 0 & 4 \end{bmatrix} = \begin{bmatrix} 1.1 + 0.0 & 1.0 + 0.4 \\ 0.1 - 2.0 & 0.0 - 2.4 \end{bmatrix}$

$$= \begin{bmatrix} 1 & 0 \\ 0 & -8 \end{bmatrix}$$

and $\quad BA = \begin{bmatrix} 1 & 0 \\ 0 & 4 \end{bmatrix} \times \begin{bmatrix} 1 & 0 \\ 0 & -2 \end{bmatrix} = \begin{bmatrix} 1.1 + 0.0 & 1.0 + 0\,(-2) \\ 0.1 + 4.0 & 0.0 + 4\,(-2) \end{bmatrix}$

$$= \begin{bmatrix} 1 & 0 \\ 0 & -8 \end{bmatrix}$$

$\therefore \quad AB = BA.$

Hence in general $AB \neq BA$.

Note 1: If AB = BA, then matrices A and B are said to commit. If AB = – BA, the matrices A and B are said to anticommute.

Note 2: If $A = \begin{bmatrix} 1 & 1 \\ 1 & 1 \end{bmatrix}$ and $B = \begin{bmatrix} 1 & 0 \\ -1 & 0 \end{bmatrix}$,

then $\quad AB = \begin{bmatrix} 1 & 1 \\ 1 & 1 \end{bmatrix} \times \begin{bmatrix} 1 & 0 \\ -1 & 0 \end{bmatrix}$

$$= \begin{bmatrix} 1.1 + 1.(-1) & 1.0 + 1.0 \\ 1.1 + 1.(-1) & 1.0 + 1.0 \end{bmatrix} = \begin{bmatrix} 0 & 0 \\ 0 & 0 \end{bmatrix}$$

i.e., AB is zero matrix (or null matrix) whereas neither A nor B is a zero matrix.

$\therefore \quad AB = O$ does not imply that either $A = 0$ or $B = O$.

Here $\quad BA = \begin{bmatrix} 1 & 0 \\ -1 & 0 \end{bmatrix} \times \begin{bmatrix} 1 & 1 \\ 1 & 1 \end{bmatrix}$

$$= \begin{bmatrix} 1.1 + 0.1 & 1.1 + 0.1 \\ -1.1 + 0.1 & -1.1 + 0.1 \end{bmatrix} = \begin{bmatrix} 1 & 1 \\ -1 & -1 \end{bmatrix}$$

i.e., $\quad BA \neq O$

If $A = \begin{bmatrix} 4 & 4 \\ 3 & 3 \end{bmatrix}$ and $B = \begin{bmatrix} -1 & 1 \\ 1 & -1 \end{bmatrix}$, then

$$AB = \begin{bmatrix} 4 & 4 \\ 3 & 3 \end{bmatrix} \times \begin{bmatrix} -1 & 1 \\ 1 & -1 \end{bmatrix}$$

$$= \begin{bmatrix} 4(-1) + 4(1) & 4(1) + 4(-1) \\ 3(-1) + 3(1) & 3(1) + 3(-1) \end{bmatrix} = \begin{bmatrix} 0 & 0 \\ 0 & 0 \end{bmatrix} = 0$$

i.e., the product of two non-zero square matrices can be a zero matrix.

and $$BA = \begin{bmatrix} -1 & 1 \\ 1 & -1 \end{bmatrix} \times \begin{bmatrix} 4 & 4 \\ 3 & 3 \end{bmatrix}$$

$$= \begin{bmatrix} (-1).4 + 1.3 & (-1).4 + 1.3 \\ 1.4 + (-1).3 & 1.4 + (-1).3 \end{bmatrix} = \begin{bmatrix} -1 & -1 \\ 1 & 1 \end{bmatrix} \neq 0$$

MULTIPLICATION OF MATRICES IS ASSOCIATIVE

Let $A = [a_{ij}]$, $B = [b_{jk}]$ and $C = [c_{kr}]$ be three $m \times n$, $n \times p$ and $p \times l$ matrices respectively, then (AB). C = A. (BC).

Proof:

Let $AB = [d_{ik}]$, where $d_{ik} = \sum_{j=1}^{n} a_{ij} b_{jk}$ ...(1)

Then (AB). $C = [d_{ik}] \times [c_{kr}] = [e_{ir}]$,

Where $$e_{ir} = \sum_{k=1}^{p} d_{ik} b_{kr}$$

$$= \sum_{k=1}^{p} \left(\sum_{j=1}^{p} a_{ij} b_{jk} \right) . C_{jr}, \text{ from (1)}$$

i.e., (i, r)th element of (AB). $C = \sum_{k=1}^{p} \sum_{j=1}^{n} a_{ij} b_{jk} c_{kr}$ ...(2)

And let $BC = [g_{jr}]$, $g_{jr} = \sum_{k=1}^{p} b_{jk} c_{kr}$ ...(3)

Then A. $BC = [a_{ij}] \times [g_{jr}] = [h_{ir}]$,

Where $h_{ir} = \sum_{j=1}^{n} a_{ij} g_{jr}$

$$= \sum_{j=1}^{n} a_{ij} \left(\sum_{k=1}^{p} b_{jk}\, c_{kr} \right), \text{ form (3)}$$

i.e., (i, r)th element of A. (BC) $= \sum_{k=1}^{p}\sum_{j=1}^{n} a_{ij}\, b_{jk}\, c_{kr}$, ...(4)

since the summation can be interchanged.

A From (3) and (4) we can conclude the (i, r)th elements of (AB)· C and A· (BC) are the same and their orders are also $m \times l$.

Hence (AB)· C = A· (BC).

MULTIPLICATION OF MATRICES IS DISRIBUTIVE WITH RESPECT TO MATRIX ADDITION

(a) Let $A = [a_{ij}]$, $B = [b_{jk}]$ and $C = [c_{jk}]$ be three $m \times n$, $n \times p$ and $n \times p$ matrices respectively, then

$$A (B + C) = AB + AC$$

Proof:

$$A (B + C) = [a_{ij}] \times \{[b_{jk}] + [c_{jk}]\}$$
$$= [a_{ij}]\, [b_{jk} + c_{jk}] = [b_{ik}], \text{ say,}$$

Where $d_{ik} = \sum_{j=1}^{n} a_{ij}\, (b_{jk} + c_{jk})$

$\Rightarrow$ (i, k)th element of A (B + C) $= \sum_{j=1}^{n} a_{ij}\, b_{jk} + \sum_{j=1}^{n} a_{ij}\, c_{jk}$ ...(1)

Again $AB = [a_{ij}]\, [b_{jk}] = [e_{ik}]$ say,

Where $e_{ik} = \sum_{j=1}^{n} a_{ij}\, b_{jk}$ *i.e.,*)i, k)th element of AB $= \sum_{i=1}^{n} a_{ij}\, b_{jk}$ ...(2)

Similarly we can prove that

(i, k)th element of AC $= \sum_{j=1}^{n} a_{ij}\, c_{jk}$ ...(3)

$\therefore$ From (2) and (3) we have

(i, k)th element o AB + AC $= \sum_{j=1}^{n} a_{ij}\, b_{jk} + \sum_{j=1}^{n} a_{ij}\, c_{jk}$ ...(4)

Hence from (1) and (4) we conclude that A (B + C) = AB + AC.

(b) Let $A = [a_{ij}]$, $B = [b_{jk}]$ and $C = [c_{jk}]$ be three $n \times p$, $m \times n$ and $m \times n$ matrices respectively.

The $(B + C) = A = BA + CA$.

(Note: If A and be $m \times n$ and $n \times p$ matrices then BA can not exist whereas AB exists).

Proof:

Its proof is simlar to that of part (a)

POSITIVE INTEGRAL POWER OF A SQUARE MATRIX

We find that if A is square matrix, then only the product AA is defined and we write A^2 for AA. Also by associative law

$$A^2A = (AA)\,A = A\,(AA) = AA^2$$

So A^2A or AA^2 is written as A^3.

In general AAA...A is denoted by A^n if there are n factors.

Definition: *In general AAA...A upto m factors is denoted by A^m.*

Theorem 1:

If A be a square matrix ($n \times n$ say), then

$$A^p \cdot A^q = A^{p+q}\text{, for any pair of positive integers } p \text{ and } q.$$

Proof:

We shall prove that by the method of induction.

From definition we know that $A^p \cdot A = A^{p+1}$, where p is any positive integer.

$\therefore$ $A^p \cdot A^q = A^{p+q}$ holds when $q = 1$, whatever p may be.

We shall now prove that if it holds for a particular value m say of q, for all values of p.

Now $A^p \cdot A^{m+1} = A^p \cdot (A^m \cdot A)$, by definition given above

$= (A^p \cdot A^m) \cdot A$, by associative law

$= (A^{p+m}) \cdot A$, by hypothesis

$= A^{p+m+1}$, by definition given above

$= A^{p+(m+1)}$, by associative law of addition of numbers.

i.e., $A^p \cdot A^q = A^{p+q}$ holds for the value $m + 1$ of q, whatever p may be if it holds for $q = m$.

Hence the result is established by mathematical induction.

Theorem 2:

If A be a square matrix, then

$(A^p)^q = A^{pq}$, for every pair of positive integers p and q.

Proof:

The proof is similar to the above theorem.

KINDS OF HERMITIAN AND SKEW-HERMITIAN MATRICES

Hermitian Matrix

Definition: *A square matrix A such that $\overline{\mathbf{A}}' = A$ is called Hermitian i.e., the matrix $[a_{ij}]$ is Hermitian provided $a_{ij} = a_{ji}$, for all values of i and j.*

For example: $A = \begin{bmatrix} 1 & \alpha + i\beta & \gamma + i\delta \\ \alpha - i\beta & m & x + iy \\ \gamma - i\delta & x - iy & n \end{bmatrix}$

It A is Hermitian Matrix then $a_{ij} = \bar{a}_{ij}$ (by definition)

$\therefore$ a_{ij} is real for a_{ij}. Thus, every diogonal element of a Hermitian Matrix must be real. A Hermitian Matrix over the field of real numbers is nothing but a real symmetric matrix.

Skew-Hermitian Matrix

Definition: *A square matrix A such that $\overline{\mathbf{A}}' = -A$ is called Skew-Hermitian i.e., the matrix $[a_{ij}]$ is skew Hermitian provided $a_{ij} = -a_{ji}$ for all values of i and j.*

For example: $A = \begin{bmatrix} 2i & \alpha + i\beta & 3 \\ \alpha - i\beta & -i & \gamma + i\delta \\ 3 & \gamma - i\delta & 0 \end{bmatrix}$.

It A is Skew Hermitian Matrix then

$$a_{ij} = -\bar{a}_{ij}$$

$$\therefore \quad a_{ij} = -\bar{a}_{ij} = 0$$

i.e., a_{ij} must be either a pure imaginary must be number or zero. Thus, the diagonal elements of a Skew Hermitian Matrix must be pure imaginary number or zero

THEOREMS ON HERMITIAN AND SKEW-HERMITIAN MATRICES

Theorem 1:

The diagonal elements of a skew-hermition matrix are either purely imaginary or zero.

Proof:

Let $[a_{ij}]$ be an n × n skew-Hermitian matrix, then according to definition we have

$$a_{ij} = \bar{a}_{ij}, \text{ for all } 1 \le i \le n,\ 1 \le j \le n \qquad ...(1)$$

Now the diagonal elements are a_{ij}, where $1 \le i \le n$.

$\therefore$ From (1), we have $a_{ij} = \bar{a}_{ij}$, for all $1 \le i \le n$ ...(2)

If $a_{ij} = \alpha + i\beta$, where α and β are real,

then $\bar{a}_{ij} = \alpha - i\beta$.

$\Rightarrow$ From (2), we get $\alpha + i\beta = (\alpha - i\beta)$

$$\Rightarrow \qquad \alpha + i\beta = -\alpha + i\beta \ \Rightarrow\ 2\alpha = 0 \ \Rightarrow\ \alpha = 0$$

$\therefore$ $a_{ij} = 0 + i\beta = i\beta$, which is purely imaginary and can be zero if $\beta = 0$

Hence, the diagonal elements of a Skew Hermitan Matrix are either purely imaginarly or zero.

Theorem 2:

Every square matrix (with complex elements) can be uniquely expressed as the sum of a Hermitian and a skew-hermitian matrices.

Proof:

Let **A** be a square matrix. Then we can write

$$A = \frac{1}{2}(A + A^{\Theta}) + \frac{1}{2}(A - A^{\Theta}) \qquad ...(1)$$

Now $\left(\overline{A + A^{\Theta}}\right) = \bar{A} + \overline{A^{\Theta}}$

$$\therefore \quad \left\{\left(\overline{A + A^{\Theta}}\right)\right\} = \left\{\bar{A} + \overline{A^{\Theta}}\right\} = (\bar{A})' - (\overline{A^{\Theta}})',$$

$$= A^{\Theta} + (\overline{A^{\Theta}})', \text{ by def } (\bar{A}) = A^{\Theta},$$

$= A^{\Theta} + (\overline{A^{\Theta}})'$ ...(2)

Now $(\overline{\mathbf{A}^{\Theta}})'$ = transposed conjugate of $\mathbf{A}^{\Theta}$

= transposed conjugate of $(\overline{\mathbf{A}})'$

= transposed matrix of **(A)'**

since conjugate of $\overline{\mathbf{A}}$ is **A**

$= \mathbf{A}, \quad \because \quad (\mathbf{A}')' = \mathbf{A}$

$\therefore$ From (2) we get, $\left\{\left(\overline{\mathbf{A} + \mathbf{A}^{\Theta}}\right)\right\} = \mathbf{A}^{\Theta} + \mathbf{A} = \mathbf{A} + \mathbf{A}^{\Theta}$

as addition lf matrices obey commutative law.

$\therefore$ By definition we find that $\mathbf{A} + \mathbf{A}^{\Theta}$ is a Hermitian matrix.

Again $\left\{\left(\overline{A + A^{\Theta}}\right)\right\}' = \left\{\overline{A} - \overline{A}^{\Theta}\right\}' = (A)' - \left(A^{\Theta}\right)$

$= \mathbf{A}^{\Theta} - A$, as above

$= -(A - \mathbf{A}^{\Theta})$.

$\therefore$ By definition we find that $\mathbf{A} - \mathbf{A}^{\Theta}$ is a Skew-Hermitian matrix

$\therefore$ From we conclude that the square matrix A is the sum of a Hermitian and a Skew-Hermitian Matrices.

Theorem 3:

The diagonal elements of a Hermitian matrix are necessary real.

Proof:

Let $[a_{ij}]$ be an n × n Hermitian matrix, then according to definition we have

$a_{ij} = \overline{a}_{ij}$, for all $1 \le i \le n$, $1 \le j \le n$ ...(1)

Now the diagonal elements are a_{ij}, where $1 \le i \le n$.

$\therefore$ From (1), we have $a_{ij} = \overline{a}_{ij}$, for all $1 \le i \le n$ ...(2)

If $a_{ij} = \alpha + i\beta$, where α and β are real,

then $\overline{a}_{ij} = \alpha - i\beta$

$\Rightarrow$ From (2), we get $\alpha + i\beta = \alpha - i\beta$

$\Rightarrow$ $2i\beta = 0 \Rightarrow \beta = 0$

$\therefore$ $a_{ij} = \alpha + i\,(0) = \alpha$, which is purely real.

Hence the diagonal elements of a Hermitian matrix, are necessarily real

ANALYSIS OF MATRIX

Let us consider the system of equations:

$$x + y + z = 1$$
$$x + 2y\ 3 + z = 4$$
$$x + 3y + 5z = 7.$$

Here x, y, z are all unknown and their co-efficient are numbers. Arranging the coefficients in the order in which they occur in the equations and enclosing them in square brackets. We get a rectangular array of the form,

$$\begin{vmatrix} 1 & 1 & 1 \\ 1 & 2 & 3 \\ 1 & 3 & 5 \end{vmatrix}$$

The rectangular array is an example of a matrix. The horizontal lines ($\rightarrow$) are called rows or row vectors, and the vertice lines are called column. There are three rows and three column in this matrix. Hence, it is a matrix of order 3×3.

We shall use the capital letters to denote matrix,

Thus $$A = \begin{bmatrix} 5 & 1 & 2 \\ 2 & 1 & 3 \end{bmatrix}_{2\times 3}$$

and $$B = \begin{bmatrix} 1 & 2 & 3 \\ 3 & 2 & 1 \\ 2 & 1 & 3 \end{bmatrix}_{3\times 3}$$

are both matrix. They are of the type 2×3 and 3×3 respectively.

MATRIX DEFINITION

A system of any mn numbers arranged in a rectangular array of m rows and n columns is called a matrix of order $m \times n$ or an $m \times n$ matrix (which is read as m by n matrix).

Or

A set of mn elements of a set S arranged in a rectangular array of m rows and n columns is called an $m \times n$ matrix over S.

A m × n matrix is usually written as

$$\begin{bmatrix} a_{11} & a_{12} & a_{13} & \cdots & a_{1n} \\ a_{21} & a_{22} & a_{22} & \cdots & a_{2n} \\ \cdots & \cdots & \cdots & \cdots & \cdots \\ a_{m1} & a_{m2} & a_{m3} & \cdots & a_{mn} \end{bmatrix} \text{ is an } m \times n \text{ matrix.}$$

where the symbols a_{ij} represent any numbers (a_{ij} lies in the ith row and jth column).

Note 1: A compact form of the above matrix may be represented by the symbols $[a_{ij}]$, (a_{ij}) $\| a_{ij} \|$ or by a single capital letter A, say.

The element a_{ij} belong to ith row and th column and sometime called the (;)th element of the matrix.

Note 2: Each of the mn numbers constituting an m × n matrix is known as an *element of the matrix.*

The elements of matrix may be scalar or vector quantities.

Note 3: The plural of 'matrix' is 'matrices'.

Example 1:

The results of a music competition are given in the following matrix:

$$\begin{bmatrix} 3 & 2 & 1 & 0 \\ 0 & 3 & 2 & 4 \\ 5 & 0 & 3 & 0 \\ 2 & 1 & 4 & 3 \end{bmatrix}$$

Here the rows represent the teams A, B, C, D in heat order and the columns represent the number of wins, first place, second plane third place and fourth place scored by the teams.

From the above matrix find (a) How many events did the team A win? (b) How may first places did the team B win? (c) How many third places did the team C win? (d) what does 0 represent in second row?

Solution:

(a) ∵ The first row represents the team A. So the required number = sum of the elements of first row = 3 + 2 + 1 = 6. **Ans.**

(b) As first element of second row (which represents the team B) is zero, os the team B did not win any first place. **Ans.**

(c) The third row represents the team C and third column represents the third place scored by the teams, so the number of third places won by the team C is 3.

(d) The second row represents the team B and the first column represents the first place scored by teams. So 0 in the second row represents that the team B did not score any first place.

Example 2:

Write down the orders of the matrices:

(a) $\begin{bmatrix} 2 & 3 & 5 \\ 1 & 0 & 3 \end{bmatrix}$; *(b)* $\begin{bmatrix} 2 \\ 3 \end{bmatrix}$,

(c) [3, 4, 5]; *(d)* [1].

Solution:

The order of the given matrix are:

(a) 2×3; (b) 2×1; (c) 1×3; (d) 1×1. **Ans.**

TYPES OF MATRICES

(a) Horizontal Matrix: If in a matrix the number of columns is more than the number of rows then it is called a horizontal matrix.

For example $\begin{bmatrix} 1 & 3 & 2 & 3 \\ 2 & 5 & 7 & 9 \end{bmatrix}$ is a horizontal matrix.

(b) Vertical Matrix: If in a matrix the number of rows is more than the number of columns it is called a vertical matrix.

For example $\begin{bmatrix} 2 & 3 \\ 3 & 5 \\ 4 & 6 \\ 5 & 7 \end{bmatrix}$ is a vertical matrix.

(c) Column Matrix: If there if only one column in a matrix, it is called a column matrix.

For example $\begin{bmatrix} 2 \\ 3 \\ 4 \end{bmatrix}$ This is also called column vector.

(d) Square Matrix: If m = n *i.e.,* the number of rows and columns of a matrix are equal, then the matrix is of order $n \times n$ and is called a square matrix of order n.

For example $\begin{bmatrix} 2 & 3 & 1 \\ 1 & 5 & 2 \\ 7 & 6 & 9 \end{bmatrix}$ is a square matrix and $\begin{bmatrix} 1 & 3 & 2 & 3 \\ 2 & 5 & 7 & 9 \end{bmatrix}$ is a rectangular matrix.

(e) Rectangular Matrices: When the number of rows and columns of the array are not equal, then the matrix is known as a rectangular matrix.

(f) Row Matrix: If in a matrix, there is only one row it is called a row matrix. For example [1, 2, 3]. This is also called a row vector.

(g) Null (or zero) Matrix: If all the elements of an m × n matrix are zero, then it is called a null or zero matrix and denoted by $O_{m \times n}$ or simply O.

For example $\begin{bmatrix} 0 & 0 & 0 \\ 0 & 0 & 0 \end{bmatrix}$ is the 2 × 3 mull matrix.

(h) Unit Matrix: A square matrix having unity for its elements in the leading diagonal and all the other elements as zero is called an unit matrix.

For example $\begin{bmatrix} 1 & 0 & 0 & 0 \\ 0 & 0 & 0 & 0 \\ 0 & 0 & 1 & 0 \\ 0 & 0 & 0 & 1 \end{bmatrix}$ is unit matrix of order 4 × 4 denote it byI_4.

∴ an a-rowed square matrix $[a_{ij}]$ is called a unit matrix provided

$$a_{ij} = 1, \text{ whenever } i = j$$
$$= 0, \text{ whenever } i \neq j.$$

(i) Equal Matrix: Two matrices A $[(a_{ij})]$ are said to be equal if (a) they are of the same type *i.e.,* if they have same number of rows and columns and (b) the elements in the corresponding positions of the two matrices are equal.

From the definition given above it is evident that

1. If A = B, then B = A (Symmetry)
2. A = A, where A is any matrix. (Reflexivity
3. If A = B and B = C, then A = C (Transitivity)

i.e., the relation of equality in the set of all matrices is an equivalence relation.

(j) Diagonal Matrix: A square matrix in which all elements except those in the main (or leading) diagonal are zero is know as a diagonal matrix.

For example $\begin{bmatrix} 2 & 0 & 0 \\ 0 & 3 & 0 \\ 0 & 0 & 7 \end{bmatrix}$ is a 3-rowed diagonal matrix.

The um of the diagonal elements of a square matrix A (say) is called the trace of the matrix A.

(k) Sub-Matrix: A matrix which is obtained from a given matrix by deleting any number of rows and number of columns is called a sub-matrix of the given matrix.

For example $\begin{bmatrix} 1 & 2 \\ 3 & 4 \end{bmatrix}$ is a sub-matrix of $\begin{bmatrix} 5 & 3 & 2 \\ 1 & 1 & 2 \\ 7 & 3 & 4 \end{bmatrix}$

(l) Diagonal Element and Orinciple Diagonal: Those elements a_{ij} of any matrix $[a_{ij}]$ are called diagonal elements for which i = j.

The line along which the above elements lie is called the Principal diagonal or the Diagonal of the matrix.

TRIANGULAR MATRICES

It every element above or below the leading diagonal is zero, then the matrix is called a *Triangular Matrix.*

(a) Upper Triangular Matrix: A square matrix A whose elements $a_{ij} = 0$ for $i < j$ is called an upper triangular matrix.

For example $\begin{bmatrix} a_{11} & a_{12} & a_{13} \ldots\ldots a_{1n} \\ 0 & a_{22} & a_{23} \ldots\ldots a_{2n} \\ 0 & 0 & a_{33} \ldots\ldots a_{3n} \\ \ldots & \ldots & \ldots\ldots\ldots\ldots \\ 0 & 0 & 0 \ldots\ldots a_{nn} \end{bmatrix}$

(b) Lower Triangular Matrix: A square matrix A whose element $a_{ij} = 0$ for $i < j$ is called a lower triangular matrix.

For example $\begin{bmatrix} a_{11} & 0 & 0 \ldots\ldots 0 \\ a_{21} & a_{22} & 0 \ldots\ldots 0 \\ a_{31} & a_{32} & a_{33} \ldots\ldots 0 \\ \ldots & \ldots & \ldots \ \ldots\ldots 0 \\ a_{n1} & a_{n2} & a_{n3} \ldots\ldots a_{nn} \end{bmatrix}$

DIAGONAL MATRIX

Defintion: A square matrix in which all element except those element there in the leading diagonal are zero is called a diagonal matrix. Thus, an n-rowed square matrix $[a_{ij}]$ is a diagonal matrix iff $a_{ij} = 0$ wherever $i \neq j$. If A $[a_{ij}]$ is a diagonal matrix of order n. It must be in the following form.

For eample
$$\begin{bmatrix} a_{11} & 0 & 0 \ldots\ldots & 0 \\ 0 & a_{22} & 0 \ldots\ldots & 0 \\ 0 & 0 & a_{33} \ldots\ldots & \ldots \\ \ldots & \ldots & \ldots \ldots\ldots & \ldots \\ 0 & 0 & 0 \ldots\ldots & a_{nn} \end{bmatrix}$$

Theorem 1:

Any two diangonal matrices of the same order commute under mltiplication.

Proof:

Let any two diagonal matrices be

$$A = \begin{bmatrix} a_1 & 0 & 0 & \ldots & 0 \\ 0 & a_2 & 0 & \ldots & 0 \\ \ldots & \ldots & \ldots & \ldots & \ldots \\ 0 & 0 & 0 & \ldots & a_n \end{bmatrix} \text{ and } B = \begin{bmatrix} b_1 & 0 & 0 & \ldots & 0 \\ 0 & b_2 & 0 & \ldots & 0 \\ \ldots & \ldots & \ldots & \ldots & \ldots \\ 0 & 0 & 0 & \ldots & b_n \end{bmatrix}$$

Then we have

$$AB = \begin{bmatrix} a_1 & 0 & 0 & \ldots & 0 \\ 0 & a_2 & 0 & \ldots & 0 \\ \ldots & \ldots & \ldots & \ldots & \ldots \\ 0 & 0 & 0 & \ldots & a_n \end{bmatrix} \times \begin{bmatrix} b_1 & 0 & 0 & \ldots & 0 \\ 0 & b_2 & 0 & \ldots & 0 \\ \ldots & \ldots & \ldots & \ldots & \ldots \\ 0 & 0 & 0 & \ldots & b_n \end{bmatrix}$$

$$\Rightarrow \quad AB = \begin{bmatrix} a_1b_1 & 0 & 0 & \ldots & 0 \\ 0 & a_2b_2 & 0 & \ldots & 0 \\ \ldots & 0 & \ldots & \ldots & \ldots \\ 0 & 0 & 0 & \ldots & a_nb_n \end{bmatrix} \quad \ldots(1)$$

and $BA = \begin{bmatrix} b_1 & 0 & 0 & \dots & 0 \\ 0 & b_2 & 0 & \dots & 0 \\ \dots & \dots & \dots & \dots & 0 \\ 0 & 0 & 0 & \dots & b_n \end{bmatrix} \times \begin{bmatrix} a_1 & 0 & 0 & \dots & 0 \\ 0 & a_2 & 0 & \dots & 0 \\ \dots & \dots & \dots & \dots & \dots \\ 0 & 0 & 0 & \dots & a_n \end{bmatrix}$

$$= \begin{bmatrix} b_1a_1 & 0 & 0 & \dots & 0 \\ 0 & b_2a_1 & 0 & \dots & 0 \\ \dots & \dots & \dots & \dots & 0 \\ 0 & 0 & 0 & \dots & b_na_n \end{bmatrix} \quad \dots\text{(ii)}$$

∴ From (1) and (2), we find that AB = BA and each one of them is a diagonal matrix. **Hence proved.**

Theorem 2:

Sum of any two diagonal matrices of order n is a diagonal matrix of order n and commute under addition.

Proof:

Let any two diagonal matrices be

$A = \begin{bmatrix} a_1 & 0 & 0 & \dots & 0 \\ 0 & a_2 & 0 & \dots & 0 \\ \dots & \dots & \dots & \dots & \dots \\ 0 & 0 & 0 & \dots & a_n \end{bmatrix}$ and $B = \begin{bmatrix} b_1 & 0 & 0 & \dots & 0 \\ 0 & b_2 & 0 & \dots & 0 \\ \dots & \dots & \dots & \dots & \dots \\ 0 & 0 & 0 & \dots & b_n \end{bmatrix}$

$$\therefore \quad A + B = \begin{bmatrix} a_1 + b_1 & 0 & 0 & \dots & 0 \\ 0 & a_2 + b_2 & 0 & \dots & \dots \\ \dots & \dots & \dots & \dots & 0 \\ 0 & 0 & 0 & \dots & a_n + b_n \end{bmatrix} \quad \dots(1)$$

$$\text{And } B + A = \begin{bmatrix} b_1 + a_1 & 0 & 0 & \dots & 0 \\ 0 & b_2 + a_2 & 0 & \dots & 0 \\ \dots & \dots & \dots & \dots & 0 \\ 0 & 0 & 0 & \dots & b_n + a_n \end{bmatrix}$$

...(2)

∴ From (1) and (2), we get A + B = B + A and each one of them is a diagonal matrix of odrer n.

SEALAR MATRIX

Definition: If in a square matrix A all the diagonal elements are equal to a (where $a \neq 0$) and all the remaining elements are equal to zero then it is called a scalar matrix. Thus, $n \times n$ square matrix $[a_{ij}]$ is called a sealar matrix iff for some member a.

For example $\begin{bmatrix} a & 0 & 0 & 0 \\ 0 & a & 0 & 0 \\ 0 & 0 & a & 0 \\ 0 & 0 & 0 & a \end{bmatrix}$ is a scalar matrix of order 4×4.

COMMUTATIVE MATRICES

Definition: If A and B are two square matrices such that AB = BA, then A and B are called *commutative* or are said to *commute.*

If **AB** = – **BA,** the matrices **A** and **B** are said to *anti-commute.*

Example 1:

Show that the matrices A and B anti-commute, where

$$A = \begin{bmatrix} 1 & -1 \\ 2 & -1 \end{bmatrix} \text{ and } B = \begin{bmatrix} 1 & 1 \\ 4 & -1 \end{bmatrix}$$

Solution:

$$\text{Here } AB = \begin{bmatrix} 1 & -1 \\ 2 & -1 \end{bmatrix} \times \begin{bmatrix} 1 & 1 \\ 4 & -1 \end{bmatrix}$$

$$= \begin{bmatrix} 1.1 - 1.4 & 1.1 + (-1).(-1) \\ 2.1 - 1.4 & 2.1 + (-1).(-1) \end{bmatrix} = \begin{bmatrix} -3 & 2 \\ -2 & 3 \end{bmatrix}$$

$$\text{And } BA = \begin{bmatrix} 1 & 1 \\ 4 & -1 \end{bmatrix} \times \begin{bmatrix} 1 & -1 \\ 2 & -1 \end{bmatrix} \qquad ...(1)$$

$$= \begin{bmatrix} 1.1 + 1.2 & 1.(-1) + 1.(-1) \\ 4.1 + (-1).2 & 4.(-1) + (-1).(-1) \end{bmatrix} = \begin{bmatrix} 3 & -2 \\ 2 & -3 \end{bmatrix}$$

$$= - \begin{bmatrix} -3 & 2 \\ -2 & 3 \end{bmatrix} \qquad ...(2)$$

$\therefore$ From (1) and (2) we find that **AB** = – **BA**.

Hence **A** and **B** anti-commute.

Example 2:

If $A = \begin{bmatrix} a & 0 & 0 \\ 0 & a & 0 \\ 0 & 0 & a \end{bmatrix}$ *and* $B = \begin{bmatrix} a_{11} & a_{12} & a_{13} \\ a_{21} & a_{22} & a_{23} \\ a_{31} & a_{32} & a_{33} \end{bmatrix}$

Then prove that AB = BA = aB.

Solution:

$$AB = \begin{bmatrix} a & 0 & 0 \\ 0 & a & 0 \\ 0 & 0 & a \end{bmatrix} \times \begin{bmatrix} a_{11} & a_{12} & a_{13} \\ a_{21} & a_{22} & a_{23} \\ a_{31} & a_{32} & a_{33} \end{bmatrix}$$

$$= \begin{bmatrix} aa_{11} & aa_{12} & aa_{13} \\ aa_{21} & aa_{22} & aa_{23} \\ aa_{31} & aa_{32} & aa_{33} \end{bmatrix} = a \begin{bmatrix} a_{11} & a_{12} & a_{13} \\ a_{21} & a_{22} & a_{23} \\ a_{31} & a_{32} & a_{33} \end{bmatrix}$$

$= aB.$

Similarly $BA = \begin{bmatrix} a_{11} & a_{12} & a_{13} \\ a_{21} & a_{22} & a_{23} \\ a_{31} & a_{32} & a_{33} \end{bmatrix} \times \begin{bmatrix} a & 0 & 0 \\ 0 & a & 0 \\ 0 & 0 & a \end{bmatrix}$

$$= \begin{bmatrix} aa_{11} & aa_{12} & aa_{13} \\ aa_{21} & aa_{22} & aa_{23} \\ aa_{31} & aa_{32} & aa_{33} \end{bmatrix} = a \begin{bmatrix} a_{11} & a_{12} & aa_{13} \\ a_{21} & a_{22} & a_{23} \\ a_{31} & a_{32} & a_{33} \end{bmatrix}$$

$= aB.$

Hence AB = BA = aB.

IDENTITY MATRIX

Definition: *If in a scalar martix the diagonal element a = 1, then the matrix is called the unit matrix or identity matrix and is denoted by* $\boldsymbol{I}_n$ *in the case of n × n matrix. Thus a square matrix* $A = [a_{ij}]$ *n × n is called an identity or unit matrix iff* $a_{ij} \begin{cases} 1 & \text{when } i=j \\ 0 & \text{when } i \neq j \end{cases}$

For example $I_4 = \begin{bmatrix} 1 & 0 & 0 & 0 \\ 0 & 1 & 0 & 0 \\ 0 & 0 & I & 0 \\ 0 & 0 & 0 & 1 \end{bmatrix}$

Example 1:

Prove that $I^m = I^{m-1} = ... = I^2 = I$, where m is any positive integer and I_n is the unit matrix of order $n \times n$.

Solution:

Let A be any n × n matrix and **I** be the nuit martrix of order n × n *i.e.*, $\mathbf{I} = \mathbf{I}_n$.

Now we know that $A\mathbf{I}_n = \mathbf{I}_n A = A$

But $\mathbf{I}_n = \mathbf{I}$. ...(1)

$\therefore \quad A\mathbf{I} = \mathbf{I}A = A$

Taking $A = \mathbf{I}$, we have $\mathbf{I} \cdot \mathbf{I} = \mathbf{I} \Rightarrow \mathbf{I}^2 = \mathbf{I}$ ...(2)

Again form (1), taking $A = \mathbf{I}^2$, where $\mathbf{I}^2 = \mathbf{I}$ (proved), we get

$\mathbf{I}^2 \cdot \mathbf{I} = \mathbf{I}^3 \quad \Rightarrow \quad \mathbf{I}^3 = \mathbf{I}^2 = \mathbf{I}$, from (2).

Proceeding in this way, we can prove that

$\mathbf{I}^m = \mathbf{I}^{m-1} = ... = \mathbf{I}^3 = \mathbf{I}^3 = \mathbf{I}$, where m is any positive integer.

Example 2:

If A be any $n \times n$ matrix and I_n is the identity martix of order $n \times n$, then prove that $AI_n = I_nA = A$.

Solution:

Let us suppose that

$$A = \begin{bmatrix} a_{11} & a_{12} & ... & a_{1n} \\ a_{21} & a_{22} & ... & a_{2n} \\ ... & ... & ... & ... \\ a_{n1} & a_{n2} & ... & a_{nn} \end{bmatrix} \text{ and } I_n = \begin{bmatrix} 1 & 0 & ... & 0 \\ 0 & 1 & ... & 0 \\ ... & ... & ... & 0 \\ 0 & 0 & ... & 1 \end{bmatrix}$$

$$\therefore \quad A \cdot \mathbf{I}_n = \begin{bmatrix} a_{11} & a_{12} & ... & a_{1n} \\ a_{21} & a_{22} & ... & a_{2n} \\ ... & ... & ... & ... \\ a_{n1} & a_{n2} & ... & a_{nn} \end{bmatrix} \times \begin{bmatrix} 1 & 0 & ... & 0 \\ 0 & 1 & ... & 0 \\ ... & ... & ... & 0 \\ 0 & 0 & ... & 1 \end{bmatrix}$$

$$= \begin{bmatrix} a_{11}.1 + a_{12}.0 + ... + a_{1n}.0 & a_{11}.0 + a_{12}.1 + ... + a_{1n}.0 \\ a_{21}.1 + a_{22}.0 + ... + a_{2n}.0 & a_{21}.0 + a_{22}.1 + ... + a_{2n}.0 \\ ... & ... \\ a_{n1}.1 + a_{n2}.0 + ... + a_{nn}.0 & a_{n1}.0 + a_{n2}.1 + ... + a_{nn}.0 \end{bmatrix}$$

$$\begin{array}{ccc} \ldots & \ldots & a_{11}.0 + a_{12}.0 + \ldots + a_{1n}.1 \\ \ldots & \ldots & a_{21}.0 + a_{22}.0 + \ldots + a_{1n}.1 \\ \ldots & \ldots & \ldots \quad \ldots \\ \ldots & \ldots & a_{n1}.0 + a_{n2}.0 + \ldots + a_{nn}.1 \end{array}\Bigg]$$

$$= \begin{bmatrix} a_{11} & a_{12} & \ldots & a_{1n} \\ a_{21} & a_{22} & \ldots & a_{2n} \\ \ldots & \ldots & \ldots & \ldots \\ a_{n1} & a_{n2} & \ldots & a_{nn} \end{bmatrix} = A$$

Similarly we can show that $\mathbf{I}_n \cdot A = A$.

Hence we have $A \cdot \mathbf{I}_n = \mathbf{I}_n \cdot A = A$.

ADJOINT MATRIX

Let A $[a_{ij}]$ n × n be any n × n matrix. The transpose B' of the matrix B = $[A_i]$ n × n, where a_{ij} denoted the cofactor of the element a_{ij} in the determinant |A| is called the adjoint of the matrix A ad is denoted by the symbol ads.

Thus the adjoint o A matrix A is the transpose o the matrix formed by the cofactor of A if

$$A = \begin{bmatrix} a_{11} & a_{12} & a_{1n} \\ a_{21} & a_{22} & a_{2n} \\ \ldots\ldots & & \\ a_n & a_{n2} & a_{nn} \end{bmatrix}$$

Then Adj A = the transpose of the matrix $\begin{bmatrix} A_{11} & A_{22} & A_{1n} \\ A_{21} & A_{22} & A_{2n} \\ \ldots\ldots & \ldots\ldots & \ldots\ldots \\ A_{n1} & A_{n2} & A_{nn} \end{bmatrix}$

THE INVERSE OF A MATRIX

If for a given square matrix **A**, there exists a matrix **B** such that **AB** = **BA** = **I**, where **I** is a nuit matrix, then **A** is called **non-singular** and **B** is called **inverse of A** and we write $\mathbf{B} = \mathbf{A}^{-1}$ (read as **B** equals **A** inverse).

Here **A** is the inverse of **B** and we can write $\mathbf{A} = \mathbf{B}^{-1}$.

If B *i.e.,* $\mathbf{A}^{-1}$ does not exist, then **A** is called **singular.**

Note 1: Non-square matrix has no inverse.

For example: $\begin{bmatrix} 1 & 2 \\ 3 & 4 \end{bmatrix} \begin{bmatrix} -2 & 1 \\ \frac{3}{2} & -\frac{1}{2} \end{bmatrix} = \begin{bmatrix} 1 & 0 \\ 0 & 1 \end{bmatrix} = I$

Each matrix in the product is the inverse of the other.

Note 2: If **AB** and **BA** are both defined and equal then the matrices **A** and **B** should both be square matrices of the same order.

THEOREMS ON INVERSE OF A MATRIX

Theorem 1:

If A and B be two non-singular matrices of the same order then AB is also non-singular and $(AB)^{-1} = B^{-1} A^{-1}$.

Or

The inverse of a product is the product of the inverse taken in the reverse order. This is also known as the Reciprocal lawe for the inverse of a product.

Proof:

A^{-1} and B^{-1} exisxts since A and B are non-singular.

$\therefore \quad (AB)(B^{-1} A^{-1}) = A(BB^{-1})A^{-1}$, by associative law

$= AIA^{-1} = AA^{-1}$

$= I.$

And $(B^{-1} A^{-1})(AB) = B^{-1}(A^{-1} A)B$, by associative law

$= B^{-1}(I)B, \qquad \because A^{-1} A = I$

$= B^{-1}(IB) = B^{-1} B,$

$= I.$

$\therefore \quad (B^{-1} A^{-1})(AB) = (AB)(B^{-1} A^{-1}) = I$

i.e., $\mathbf{B^{-1} A^{-1}}$ is the inverse of **AB** or $\mathbf{(AB)^{-1} = B^{-1} A^{-1}}$, and as such **AB** is also non-singualr.

Theorem 2:

If a given square matrix A has an inverse, then it is unique or there exists one and only one inverse matrix to a given matrix.

Proof:

Let us suppose that **B** and **C** are two possible inverse of **A.** Then we must have

$$AB = BA = I \quad ...(1)$$

and $$AC = CA = I \quad ...(2)$$

$\therefore$ From (1) and (2), we get AB = AC, each being equal to I

$\Rightarrow$ $B (AB) = B (AC)$

$\Rightarrow$ $(BA) B = (BA) C$

$\Rightarrow$ $IB = IC$, from (1)

$\Rightarrow$ $B = C$

Hence there cannot be two inverses of A.

ORTHOGONAL MATRIX

Definition: *A square matrix **A** is called an orthogonal matrix if **AA' = I**, where **I** is an identity matrix and **A'** is the transposed matrix of **A**.*

Theorems on Orthogonal Matrices

Theorem 1:

For any two orthogonal matrices A and B, show that AB is an orthogonal matrix.

Proof:

If A and B are orthogonal matrices, then by definition we have

$$AA' = A'A = I \quad ...(1)$$

and $$BB' = B'B = I \quad ...(2)$$

$\therefore$ $(AB)(AB)' = (AB)(B'A'),$

$= AB\,B'A' = A'(BB')A'$

$= AIA'$, from (2)

$= AA' = I$, from (1).

Similarly we can prove that

$(AB)'(AB) = B'A'\,AB,$

$= B'IB$, from (1)

$= B'B = I$, from (2)

Hence AB is an orthogonal matrix by definition.

Theorem 2:

For any square matrix **A**, *if* **AA' = I**, *then* **A'A = I**.

Proof:

Since AA' = I, so A is invertible (i.e., A possesses an inverse) and there exists another matrix B such that

$$AB = BA = I \qquad ...(1)$$

Now $B = BI = B(AA')$, $\because$ AA' = I (given)

$= (BA)A' = IA'$, from (1)

i.e., $B = A'$

$\therefore$ From (1), we get AA' = A'A = I. Hence proved.

Theorem 3:

If **A** *is an orthogonal matrix, then* **A'** *is also orthogonal.*

Proof:

By definition if A is an orthogonal matrix, then

$$AA' = A'A = I$$

$\Rightarrow$ $(AA')' = (A'A)' = I$, transposing and remembering $I' = I$

$\Rightarrow$ $(A')'A' = A'(A')' = I$.

$\Rightarrow$ A' is othogonal by definition. Hence proved.

i.e., Transpose of an orthogonal matrix is also orthogonal

Theorem 4:

If **A** *is an othogonal matrix, then* $\mathbf{A^{-1}}$ *is also orthogonal*

Poof:

By definition if A is orthogonal, then

$$AA' = A'A = I$$

$\Rightarrow$ $(AA')^{-1} = (A'A)^{-1} = I$,

taking inverse and remembering $I^{-1} = I$

$\Rightarrow$ $(A')^{-1} A^{-1}$

$= A^{-1} (A')^{-1} = I$,

$\Rightarrow$ $(A^{-1})' A^{-1} = A^{-1} (A^{-1})' = I$ (Note)

$\Rightarrow$ A^{-1} is orthogonal by definition. Hence proved.

i.e., Inverse of an orthogonal matrix is also orthogonal matrix is also Lorthogonal.

UNITARY MATRIX

Definition: *A square matrix A is called an unitary matrix if* $A^{\Theta} = I$*, where I is an identity matrix and* A^{Θ} *is the transposed conjugate of A.*

THEORMES ON UNITARY MATRICES

Theorem 1:

For any two unitary matrices A and B show that Ab is an unitary matrix.

Proof:

If **A** and **B** are unitary matrices, then by definition we have

$$AA^{\Theta} = A^{\Theta}\mathbf{A} = \mathbf{I} \qquad \text{...(1)}$$

and $BB^{\Theta} = B^{\Theta}\mathbf{B} = \mathbf{I}$ ...(2)

$$\therefore \quad (AB)\,(AB)^{\Theta} = (AB)\,(B^{\Theta}A^{\Theta}),$$

$$= \mathbf{A}\,(BB^{\Theta})\,A^{\Theta} = AIA^{\Theta}, \text{ from (2)}$$

$$= AA^{\Theta} = \mathbf{I}, \text{ from (1)}$$

Similarly $(AB)^{\Theta}(AB) = B^{\Theta}A^{\Theta}\mathbf{AB}$,

$$= B^{\Theta}\,\mathbf{IB}, \text{ from (1)}$$

$$= B^{\Theta}\,\mathbf{B} = \mathbf{I}, \text{ from (2)}$$

Hence AB is an Unitary Matrix. **Hence proved.**

Theorem 2:

If A is an unitary matrix, then A^{-1} *is also unitary.*

Proof:

By definition if **A** is an unitary matrix, then

$$AA^{\Theta} = \mathbf{A}^{\varnothing}\mathbf{A} = \mathbf{I}$$

$$\Rightarrow \quad (AA^{\Theta})^{-1} = (A^{\Theta}A)^{-1} = \mathbf{I}, \text{ taking inverse}$$

$$\Rightarrow \quad (A^{\Theta})^{-1}\mathbf{A}^{-1} = \mathbf{A}^{-1}(A^{\Theta})^{-1} = \mathbf{I}$$

$\Rightarrow \quad (A^{-1})^{\Theta} A^{-1} = A^{-1} (A^{-1})^{\Theta} = I$ **(Note)**

$\Rightarrow$ A^{-1} is an unitary matrix by definition. **Hence proved.**

Theorem 3:

If **A** *is an unitary matrix, then* **A'** *is also unitary.*

Proof:

By definition if A is an unitary matrix, then

$$AA^{\Theta} = A^{\Theta}A = I.$$

$\Rightarrow \quad (AA^{\Theta})^{\Theta} = (A^{\Theta}A)^{\Theta} = I$ taking transposed conjugate and remembering that $I^{\Theta} = I$ **(Note)**

$\Rightarrow \quad (A^{\Theta})^{\Theta} A^{\Theta} = A^{\Theta} (A^{\Theta})^{\Theta} = I,$

$\Rightarrow \quad AA^{\Theta} = A^{\Theta} A = I$, since $(A^{\Theta})^{\Theta} = A$

$\Rightarrow \quad (AA^{\Theta})' = (A^{\Theta}A)' = I$, taking transpose of each side

$\Rightarrow \quad (A^{\Theta})' A' = A' (A^{\Theta})' = I$ using

$\Rightarrow \quad (A')^{\Theta} A' = A' (A')^{\Theta} = I$ **(Note)**

$\Rightarrow$ **A'** is an uniary matrix. **Hence proved.**

Theorem 4:

For any square matrix **A,** *if* $AA^{\Theta} = I$, *then* $A^{\Theta} A = I$.

Proof:

Since $AA^{\Theta} = I$, where **I** is the unit matrix, so we find that **A** is invertible and there exists another matrix **B** such that

$$AB = BA = I \quad ...(1)$$

Now $B = BI = B\,(AA^{\Theta})$, $\because AA^{\Theta} = I$ (given)

$= (BA)\,A^{\Theta} = IA^{\Theta}$ form (1)

i.e ., $B = A^{\Theta}$

$\therefore$ From (1), we get $AA^{\Theta} = A^{\Theta} A = I$. **Hence proved.**

Example 1:

Prove that the matrix $\frac{1}{\sqrt{3}}\begin{bmatrix} 1 & 1+i \\ 1-i & -1 \end{bmatrix}$ *is unitary.*

Solution:

$$\text{Let } A = \frac{1}{\sqrt{3}}\begin{bmatrix} 1 & 1+i \\ 1-i & -1 \end{bmatrix}$$

$$\text{Then } A^{\Theta} = \frac{1}{\sqrt{3}}\begin{bmatrix} 1 & 1+i \\ 1-i & -1 \end{bmatrix}$$

$$\therefore \quad A^{\Theta} A = \frac{1}{\sqrt{3}}\begin{bmatrix} 1 & 1+i \\ 1-i & -1 \end{bmatrix} \times \frac{1}{\sqrt{3}}\begin{bmatrix} 1 & 1+i \\ 1-i & -1 \end{bmatrix}$$

$$= \frac{1}{3}\begin{bmatrix} 1.1 + (1+i).(1-i) & 1.(1+i) + (1+i)(-1) \\ (1-i).1 + (-1)(1-i) & (1-i)(1+i) + (-1)(-1) \end{bmatrix}$$

$$= \frac{1}{3}\begin{bmatrix} 1 + i - i^2 & 0 \\ 0 & 1 - i^2 + 1 \end{bmatrix} = \frac{1}{3}\begin{bmatrix} 3 & 0 \\ 0 & 3 \end{bmatrix} = \begin{bmatrix} 1 & 0 \\ 0 & 1 \end{bmatrix} = I.$$

Example 2:

Show that the matrix $A = \begin{bmatrix} \cos\alpha & \sin\alpha \\ -\sin\alpha & \cos\alpha \end{bmatrix}$ *is orthogonal.*

Solution:

$$A' = \begin{bmatrix} \cos\alpha & -\sin\alpha \\ \sin\alpha & \cos\alpha \end{bmatrix}$$

$$\therefore \quad A'A = \begin{bmatrix} \cos\alpha & -\sin\alpha \\ \sin\alpha & \cos\alpha \end{bmatrix}\begin{bmatrix} \cos\alpha & \sin\alpha \\ -\sin\alpha & \cos\alpha \end{bmatrix}$$

$$= \begin{bmatrix} \cos^2\alpha + \sin^2\alpha & \cos\alpha\sin\alpha - \sin\alpha\cos\alpha \\ \sin\alpha\cos\alpha - \cos\alpha\sin\alpha & \sin^2\alpha \cos^2\alpha \end{bmatrix}$$

$$= \begin{bmatrix} 1 & 0 \\ 0 & 1 \end{bmatrix} = I. \text{ Hence A orthogonal.}$$

PARTITIONING OF MATRICES

Submatrix

Definition: *A matrix obtained by striking off some of the rows and columns of another matrix A is defined as a* **sub-matrix** *of A.*

For example if A = $\begin{bmatrix} 2 & 3 & 1 \\ 3 & 5 & 7 \end{bmatrix}$, then [2], [3], [5] etc.

$\begin{bmatrix} 2 & 3 \\ 3 & 5 \end{bmatrix}, \begin{bmatrix} 3 & 1 \\ 5 & 7 \end{bmatrix}$ etc. are all sub-martices of A.

It is sometimes found useful to subdivide a matrix into submatrices by drawing lines parallel to its rows and columns and to consider these sub-matrices as the elements of the original matrix.

Consider the matrix,

$$A = \begin{bmatrix} x_1 & y_1 & z_1 & : & \alpha_1 & \beta_1 \\ x_2 & y_2 & z_2 & : & \alpha_2 & \beta_2 \\ x_3 & y_3 & z_3 & : & \alpha_3 & \beta_3 \\ \cdots & \cdots & \cdots & \cdots & \cdots & \cdots \\ p_2 & q_1 & r_1 & : & a_1 & b_1 \\ p_2 & q_2 & r_2 & : & a_2 & b_2 \end{bmatrix}$$

Let $A_{11} = \begin{bmatrix} x_1 & y_1 & z_1 \\ x_2 & y_2 & z_2 \\ x_3 & y_3 & z_3 \end{bmatrix}$; $A_{12} = \begin{bmatrix} \alpha_1 & \beta_1 \\ \alpha_2 & \beta_2 \\ \alpha_3 & \beta_3 \end{bmatrix}$;

$A_{21} = \begin{bmatrix} p_1 & q_1 & r_1 \\ p_2 & q_2 & r_2 \end{bmatrix}$; $A_{22} = \begin{bmatrix} a_1 & b_1 \\ a_2 & b_2 \end{bmatrix}$

Then we may write A = $\begin{bmatrix} A_{11} & A_{12} \\ A_{21} & A_{22} \end{bmatrix}$.

The matrix A is then said to have been *partitioned* and the dotted lies indicate the partitions. Here it is obvious that a matrix can be partitioned in several ways. The elements A_{11}, A_{12}, A_{21} and A_{22} are themeselves matrices and are the sub-matrices of A.

IDENTICALLY PARTITIONED MATRICES

Two matrices of the same size know as identically partitioned matrices if when expressed as matrices of matrices (*i.e.,* when partitioned) they are of the same order and the corresponding sub-matrices (or elements) are also of the same size. Such matrices are said to be *additively coherent.*

For exmaple:

$$\begin{pmatrix} 1 & 2 & 3 & : & 7 & 8 \\ 4 & 5 & 6 & : & 9 & 5 \\ \cdots & \cdots & \cdots & \cdots & \cdots & \cdots \\ 2 & 3 & 4 & : & 2 & 3 \\ 5 & 6 & 7 & : & 4 & 5 \\ 4 & 5 & 8 & : & 6 & 7 \end{pmatrix} \text{ and } \begin{pmatrix} 1 & 2 & 4 & : & 3 & 0 \\ 2 & 0 & 5 & : & 4 & 6 \\ \cdots & \cdots & \cdots & \cdots & \cdots & \cdots \\ 1 & 0 & 2 & : & 1 & 2 \\ 2 & 5 & 4 & : & 3 & 4 \\ 2 & 6 & 2 & : & 5 & 6 \end{pmatrix}$$

Two matrices A and B, which are conformble to the procuct AB, are called *multiplicately coherent* if A and B are parittioned in such a way that the columns of A are partitioned in the same way as the rows of B are partitioned. Here the rows of A and columns of B can be partitioned in any way.

$$\text{Let } A = \begin{bmatrix} 1 & 0 & 0 & 0 \\ 2 & 0 & 0 & 0 \\ 2 & 3 & 1 & 1 \end{bmatrix} \text{ and } B = \begin{bmatrix} 2 & 3 & 5 \\ 3 & 7 & 1 \\ 4 & 0 & 2 \\ 2 & 5 & 1 \end{bmatrix}$$

Here A is a 3×4 matrix and B is a 4×3 matrix, so these are conformable to the product AB (*i.e.*, the product AB exists), Now if we write

$$A = \begin{bmatrix} 1 & 0 & 0 & : & 0 \\ 2 & 1 & 0 & : & 0 \\ \cdots & \cdots & \cdots & \cdots & \cdots \\ 2 & 3 & 1 & : & 1 \end{bmatrix} \text{ and } B = \begin{bmatrix} 2 & : & 3 & 5 \\ 3 & : & 7 & 1 \\ 4 & : & 0 & 2 \\ \cdots & \cdots & \cdots & \cdots \\ 2 & : & 5 & 1 \end{bmatrix}$$

then the partitioning of the columns of A is in the same way as the partitioning of the rows of **B**.(Here we note that after third column in A the partitioning has been done and in **B** the partitioning has been done after thired row). Thus, according to definition given above the matrices A and B are called multiplicative coherent.

PERIODIC MATRIX

Definition: *A square matrix A is called periodic, if* $A^{k+1} = A$, *where k is a positive integer.*

If k is the least positive integer for which $A^{k+1} = A$, *then A is said to be of* **period k.**

Idempotent Matrix

Definition: A square matrix A is called idempotent provided it satisfies the relation $A^2 = A$.

SYMMETRIC IDEMPOTENT MATRIX

Definition: *A square matrix A is called symmetric idempotent if* $A = A'$ *and* $A^2 = A$*, where* A' *is the transposed matrix of A.*

Example 1:

Show that if A and B are matrices of order $n \times n$ *and such that* $AB = A$ *and* $BA = B$*, then A and B are idempotent martices.*

Solution:

We have $ABA = (AB)\,A = (A)\,A$, $\quad \because AB = A$ (given)

$\Rightarrow \quad ABA = A^2 \qquad$ ...(1)

Also $\quad ABA = A\,(BA) = A\,(B) \quad \because BA = B$ (given)

$= AB = A, \quad \because AB = A$ (given)

$\Rightarrow \quad ABA = A \qquad$ ...(2)

From (1) and (2), we have $A^2 = A$ i.e., A is idempotent.

In a similar manner, we can prove that

$BAB = B\,(AB) = B\,(A), \quad \because AB = A$ (given)

$= BA = B, \quad \because BA = B$ (given)

$\Rightarrow \quad BAB = B \qquad$ ...(3)

Also $\quad BAB = (BA)\,B = (B)\,B, \quad \because BA = B$ (given)

$\Rightarrow \quad BAB = B^2 \qquad$ ...(4)

$\therefore$ From (3) and (4), we have $B^2 = B$ *i.e.,* B is idempotent.

Hence proved.

Example 2:

If a is an idempotent matrix, then the matrix $B = I - A$ *is idempotent and* $AB = O = BA$.

Solution:

We know $\mathbf{IA = AI = A}$. $\qquad$ ...(1)

Also **A** being an idempotent matrix, we have $\mathbf{A^2 = A}$. $\qquad$ ...(2)

Since **I** and A are square matrices, so **I – A** is also a square matrix and therefore we have

$$(I - A)^2 = (I - A)(I - A)$$
$$= (I - A)I - (I - A)A, \text{ by distributive law}$$
$$= I^2 - AI - IA + A^2$$
$$= I - A - A + A, \text{ from (1), (2) and } I^2 = I$$

$\Rightarrow$ $(I - A)^2 = I - A$, i.e., $I - A$ or B is an idempotent matrix by definition.

Again $AB = A(I - A) = AI - A^2$, by distributive law

$= A - A$, from (1) and (2)

i.e., $AB = O$.

And $BA = (I - A)A = IA - A^2$, by distributive law

$= A - A$, from (1) and (2)

$\Rightarrow$ $BA = O$. **Hence proved.**

Example 3:

If A and B are idempotent martices, then show that AB is idempotent if A and B commute.

Solution:

If A is idempotent, then $A^2 = A$ and if B is idempotent, then $B^2 = B$. ...(1)

And if A and B commute, then $AB = BA$ ...(2)

Now $(AB)^2 = (AB).(AB)$

$= A(BA)B$, by associative law

$= A(AB)B$, from (2)

$= (AA)(BB)$, by associative law

$= A^2B^2$

$= AB$, by (1)

Hence AB is idempotent.

Example 4:

Show that the matrix $A = \begin{bmatrix} 2 & -2 & -4 \\ -1 & 3 & 4 \\ 1 & -2 & -3 \end{bmatrix}$ *is idempotent.*

Solution:

$$A^2 = A \cdot A = \begin{bmatrix} 2 & -2 & -4 \\ -1 & 3 & 4 \\ 1 & -2 & -3 \end{bmatrix} \times \begin{bmatrix} 2 & -2 & -4 \\ -1 & 3 & 4 \\ 1 & -2 & -3 \end{bmatrix}$$

$$= \begin{bmatrix} 2.2 - 2(-1) - 4.1 & 2(-2) - 2.3 - 4(-2) & 2(-4) - 2.4 - 4(-3) \\ -1.2 + 3(-1) + 4.1 & -1(-2) + 3.3 + 4(-2) & -1(-4) + 3.4 + 4(-3) \\ 1.2 - 2(-1) - 3.1 & 1(-2) - 2.3 - 3(-2) & 1(-4) - 2.4 - 3(-3) \end{bmatrix}$$

$$= \begin{bmatrix} 2 & -2 & -4 \\ -1 & 3 & 4 \\ 1 & -2 & -3 \end{bmatrix} = A$$

Hence the matrix A is idempotent.

INVOLUTORY MATRIX

Definition: *A square matrix A is called Involutory provided it satisfies the relation $A^2 = \boldsymbol{I}$, where $\boldsymbol{I}$ is the identity matrix.*

For example, the matrix $A = \begin{bmatrix} 1 & 0 \\ 1 & -1 \end{bmatrix}$ is involutory matrix,

since $A^2 = \begin{bmatrix} 1 & 0 \\ 1 & -1 \end{bmatrix} \times \begin{bmatrix} 1 & 0 \\ 1 & -1 \end{bmatrix}$

$$= \begin{bmatrix} 1.1 + 0.0 & 1.0 + 0.(-1) \\ 0.1 + (-1).0 & 0.0 + (-1).(-1) \end{bmatrix} = \begin{bmatrix} 1 & 0 \\ 0 & 1 \end{bmatrix} = I$$

Example 1:

If A is any square martix of order n and I_n is the identity martix of order n, such that $(I_n - A) = O$, then show that A is an involutory martix.

Solution:

Given that $(I_n - A)(I_n + A) = O$

$\Rightarrow \quad I_n^2 + I_n \cdot A - A \cdot I_n - A^2 = O$

$$\Rightarrow \quad I_n + A - A - A^2 = O, \qquad \because I_n^2 = I_n, I_n \cdot A = A = A \cdot I_n.$$

$$\Rightarrow \quad I_n - A^2 = O$$

$$\Rightarrow \quad A^2 = I_n$$

i.e., **A** is involutory by definition.

Example 2:

Show that the matrix $A = \begin{bmatrix} -5 & -8 & 0 \\ 3 & 5 & 0 \\ 1 & 2 & -1 \end{bmatrix}$ *is involutory,*

Solution:

$$A^2 = \begin{bmatrix} -5 & -8 & 0 \\ 3 & 5 & 0 \\ 1 & 2 & -1 \end{bmatrix} \times \begin{bmatrix} -5 & -8 & 0 \\ 3 & 5 & 0 \\ 1 & 2 & -1 \end{bmatrix}$$

$$= \begin{bmatrix} (-5).(-5) + (-8).3 + 0.1 & (-5).(-8) + (-8).5 + 0.2 & (-5).0 + (-8).0 + 0\,(-1) \\ 3.(-5) + 5.3 + 0.1 & 3.(-8) + 5.5 + 0.2 & 3.0 + 5.0 + 0.(-1) \\ 1.(-5) + 2.3 + (-1).1 & 1.(-8) + 2.5 + (-1).2 & 1.0 + 2.0 + (-1)\,(-1) \end{bmatrix}$$

$$= \begin{bmatrix} 25 - 24 + 0 & -40 - 40 + 0 & 0 + 0 + 0 \\ -15 + 15 + 0 & -24 + 25 + 0 & 0 + 0 + 0 \\ -5 + 6 - 1 & -8 + 10 - 2 & 0 + 0 + 1 \end{bmatrix} = \begin{bmatrix} 1 & 0 & 0 \\ 0 & 1 & 0 \\ 0 & 0 & 1 \end{bmatrix} = I$$

Hence the given matrix a is involutory.

NILPOTENT MATRIX

Definition: *A square matrix **A** si called Nilpotent matrix of order **m**, provided it satisfies the relation* $A^m = O$ *and* $A^{m-1} \neq O$*, where m is a positive integer and **O** is the null matrix of order m.*

For example, the matrix $A = \begin{bmatrix} 0 & 1 \\ 0 & 0 \end{bmatrix}$ is a nilpotent matrix, since

$$A = \begin{bmatrix} 0 & 1 \\ 0 & 0 \end{bmatrix} \neq \mathbf{O},$$

$$A^2 = \begin{bmatrix} 0 & 1 \\ 0 & 0 \end{bmatrix} + \begin{bmatrix} 0 & 1 \\ 0 & 0 \end{bmatrix} = \begin{bmatrix} 0.0 + 1.0 & 0.1 + 1.0 \\ 0.0 + 0.0 & 0.1 + 0.0 \end{bmatrix}$$

$$= \begin{bmatrix} 0 & 0 \\ 0 & 0 \end{bmatrix} = \mathbf{O},$$

$$A^3 = \lambda^2 \cdot A = O \cdot A = O.$$

i.e., A is a matrix which is not itself a zero matrix though its powers are zero matrices and so it is a nilpotent matrix *(Another definition of nilpotent matrix).*

Example:

Show that $A = \begin{bmatrix} 1 & 2 & 3 \\ 1 & 2 & 3 \\ -1 & -2 & -3 \end{bmatrix}$ *is a nilpotent matrix of order 2.*

Solution:

Given $A = \begin{bmatrix} 1 & 2 & 3 \\ 1 & 2 & 3 \\ -1 & -2 & -3 \end{bmatrix} \neq O$

$$\therefore \quad A^2 = \begin{bmatrix} 1 & 2 & 3 \\ 1 & 2 & 3 \\ -1 & -2 & -3 \end{bmatrix} \times \begin{bmatrix} 1 & 2 & 3 \\ 1 & 2 & 3 \\ -1 & -2 & -3 \end{bmatrix}$$

$$= \begin{bmatrix} 1.1 + 2.2 + 3(-1) & 1.2 + 2.2 + 3(-2) & 1.3 + 2.3 + 3(-3) \\ 1.1 + 2.1 + 3(-1) & 1.2 + 2.2 + 3(-2) & 1.3 + 2.3 + 3(-3) \\ -1.1 - 2.1 - 3(-1) & -1.2 - 2.2 - 3(-2) & -1.3 - 2.3 - 3(-3) \end{bmatrix}$$

$$= \begin{bmatrix} 0 & 0 & 0 \\ 0 & 0 & 0 \\ 0 & 0 & 0 \end{bmatrix} = \mathbf{O},$$ where **O** is the null matrix of order 3.

i.e., $\mathbf{A^2 = O}$ $\mathbf{A \neq O}$. Hence **A** is a nilpotent matrix of order 2.

SCALAR MULTIPLE OF A MATRIX

If A is a matrix and λ is a number then λA is defined as the matrix each element of which λ times the corresponding element of the matrix. A.

For example: $2\begin{bmatrix} 3 & 5 & 7 \\ 2 & 3 & 4 \end{bmatrix} = \begin{bmatrix} 6 & 10 & 14 \\ 4 & 6 & 8 \end{bmatrix}$

⇒ if A $[a_{ij}]$, then $\lambda A = [\lambda a_{ij}]$,

where λ is a number.

ADDITION OF MATRICES

If there be two m × n matrices given by $A = [a_{ij}]$ and $B = [b_{ij}]$, then the matrix A + B is defined as the matrix each element of which is the sum of the corresponding elements of A and B *i.e.,*

$$A + B = [a_{ij} + b_{ij}],$$

where i = 1, 2, 3,..., m

and j = 1, 2, 3,..., n.

For example: If $A = \begin{bmatrix} a_1 & b_1 & c_1 \\ a_2 & b_2 & c_2 \end{bmatrix}$ and $B = \begin{bmatrix} a_3 & b_3 & c_3 \\ a_4 & b_4 & c_4 \end{bmatrix}$

then $A = B = \begin{bmatrix} a_1 + a_3 & b_2 + b_3 & c_1 + c_3 \\ a_2 + a_4 & b_2 + b_4 & c_2 + c_4 \end{bmatrix}$

SUBTRACTION OF MATRICES

It there be two m × n matrices given by $A = [a_{ij}]$ and $B = [b_{ij}]$, then the matrix A – B is defined as the matrix each element of which is obtained by subtracting the element of B from the corresponding element A *i.e.,* $A - B = [a_{ij} - b_{ij}]$,

where i = 1, 2,...,m and j = 1, 2,...,n.

For example: If $A = \begin{bmatrix} a_1 & b_1 & c_1 \\ a_2 & b_2 & c_2 \end{bmatrix}$ and $B = \begin{bmatrix} a_3 & b_3 & c_3 \\ a_4 & b_4 & c_4 \end{bmatrix}$

then $A - B = \begin{bmatrix} a_1 - a_3 & b_2 - b_3 & c_1 - c_3 \\ a_2 - a_4 & b_2 - b_4 & c_2 - c_4 \end{bmatrix}$

Note: If the two matrices A and B are of the same order, then only their addition and subtraction is possible and these matrices are said to be conformable for addition or subtraction.

On the other hand if the matrices A and B are of different orders, then their addition and subtraction is not possible and these matrices are called non-conformable for addition and subtraction.

PROPERTIES OF MATRIX ADDITION

Property I: Addition of Matrices is Commutative

$(A + B) = (B + A)$

i.e., $[a_{ij}] + [b_{ij}] = [b_{ij}] + [a_{ij}]$,

where $[a_{ij}]$ and $[b_{ij}]$ are any two $m \times n$ matrices *i.e.,* matrices of the same order.

Proof:

$[a_{ij}] + [b_{ij}] = [a_{ij} + b_{ij}]$, by definition of addition

$= [b_{ij} + a_{ij}]$, $\because$ addition of numbers (elements) is commutative

$= [b_{ij} + a_{ij}]$

i.e., $[a_{ij}] + [b_{ij}] = [b_{ij}] + [a_{ij}]$. Hence the theorem.

Property II: Addition of Matrices is Associative

$[(A + B) + C] = A + [B + C]$

i.e., $\{[a_{ij}] + [b_{ij}]\} + [c_{ij}] = [a_{ij}] + [b_{ij}] + [c_{ij}]\}$,

where $[a_{ij}]$, $[b_{ij}]$ and $[c_{ij}]$ are any three matrices of the same order $m \times n$, say.

Proof:

$\{[a_{ij}] + [b_{ij}]\} + [c_{ij}]$

$= [a_{ij} + b_{ij}] + [c_{ij}]$, by law of addition for matrices

$= [(a_{ij} + b_{ij}) + c_{ij}]$, by law of addition for matrices

$= [a_{ij} + (b_{ij} + c_{ij})]$, $\because$ addition of numbers is associative

$= [a_{ij}] + [b_{ij} + c_{ij}]$

$= [a_{ij}] + \{[b_{ij}] + [c_{ij}]\}$. Hence the theorem.

Property III: Addition of Matrices Obey the Distributive Law

$k [A + B] = kA + kB$

i.e., $k ([a_{ij}] + [b_{ij}]) = k [a_{ij}] + 0 [b_{ij}]$

where $[a_{ij}]$ and $[b_{ij}]$ are any two matrices of the same order $m \times n$, say.

Proof:

$k ([a_{ij}] + [b_{ij}] = k [a_{ij} + b_{ij}]$, by law of addition

$= [k (a_{ij} + b_{ij})]$, by law of scalar multiplication

$= (ka_{ij} + kb_{ij})]$, by distributive law for numbers

$= [ka_{ij}] + [kb_{ij}]$

$= k\,[a_{ij}] + k\,[b_{ij}]$. **Hence the theorem.**

Property IV: Existence of Additive Identity:

If $A = [a_{ij}]$ be any $m \times n$ matrix and O be the $m \times n$ null matrix, then

$$A + O = A = O + A.$$

Proof:

Is obvious, since $a_{ij} + 0 = a_{ij} = 0 + a_{ij}$

Property V: Existence of Additive Inverse

If A $[a_{ij}]$ be any $m \times n$ matrix, then there exists another $m \times n$ matrix B such that

$$A + B + O = B = A,$$

where O is the $m \times n$ null matrix.

Here the matrix B is called the additive inverse of the matrix A or the negative of A.

Also the (i, j) the element of b is $-a_{ij}$ if $A = [a_{ij}]$

SOLVED EXAMPLES

Example 1:

$$A = \begin{bmatrix} 1 & 1 & -1 \\ 2 & -3 & 4 \\ 3 & -2 & 3 \end{bmatrix};\ B = \begin{bmatrix} -1 & -2 & -1 \\ 6 & 12 & 6 \\ 5 & 10 & 5 \end{bmatrix},\ C = \begin{bmatrix} -1 & -1 & 1 \\ 2 & 2 & -2 \\ -3 & -3 & 3 \end{bmatrix}$$

show that AB and CA are null matrices but $AB \neq O$, $AC \neq O$.

Solution:

$$AB = \begin{bmatrix} 1 & 1 & -1 \\ 2 & -3 & 4 \\ 3 & -2 & 3 \end{bmatrix} \times \begin{bmatrix} -1 & -2 & -1 \\ 6 & 12 & 6 \\ 5 & 10 & 5 \end{bmatrix}$$

$$= \begin{bmatrix} 1(-1) + 1.6 + (-1).5 & 1(-2) + 1.12 + (-1).10 & 1(-1) + 1.6 + (-1).5 \\ 2(-1) - 3.6 + 4.5 & 2(-2) - 3.12 + 4.10 & 2(-1) - 3.6 + 4.5 \\ 3(-1) - 2.6 + 3.5 & 3(-2) - 2.12 + 3.10 & 3(-1) - 2.6 + 3.5 \end{bmatrix}$$

$$= \begin{bmatrix} 0 & 0 & 0 \\ 0 & 0 & 0 \\ 0 & 0 & 0 \end{bmatrix}, \text{ which is null matrix.}$$

This is known as 'unusual property' of matrix multiplication

$$CA = \begin{bmatrix} -1 & -1 & 1 \\ 2 & 2 & -2 \\ -3 & -3 & 3 \end{bmatrix} \times \begin{bmatrix} 1 & 1 & -1 \\ 2 & -3 & 4 \\ 3 & -2 & 3 \end{bmatrix}$$

$$= \begin{bmatrix} 1.1 - 1.2 + 1.3 & -1.1 - 1(-3) + 1(-2) & -1(-1) - 1.4 + 1.3 \\ 2.1 + 2.2 - 2.3 & 2.1 + 2(-3) - 2(-2) & 2(-1) + 2.4 - 2.3 \\ -3.1 - 3.2 + 3.3 & -3.1 - 3(-3) + 3(-2) & -3(-1) - 3.4 + 3.3 \end{bmatrix}$$

$$= \begin{bmatrix} 0 & 0 & 0 \\ 0 & 0 & 0 \\ 0 & 0 & 0 \end{bmatrix}, \text{ which is a null matrix.}$$

Hence proved.

We can prove in a similar way that BA $\neq$ O and AC $\neq$ O.

Example 2(a):

If $A = \begin{bmatrix} 1 & -2 & 3 \\ -4 & 2 & 5 \end{bmatrix}$ and $B = \begin{bmatrix} 2 & 3 \\ 4 & 5 \\ 2 & 1 \end{bmatrix}$ *find AB and show that* $AB \neq BA$

Solution:

$$AB = \begin{bmatrix} 1 & -2 & 3 \\ -4 & 2 & 5 \end{bmatrix} \times \begin{bmatrix} 2 & 3 \\ 4 & 5 \\ 2 & 1 \end{bmatrix}$$

$$= \begin{bmatrix} 1.2 + (-2).4 + 3.2 & 1.3 + (-2).5 + 3.1 \\ -4.2 + 2.4 + 5.2 & -4,3 + 2.5 + 5.1 \end{bmatrix}$$

$$= \begin{bmatrix} 0 & -4 \\ 10 & 3 \end{bmatrix}$$

and $$BA = \begin{bmatrix} 2 & 3 \\ 4 & 5 \\ 2 & 1 \end{bmatrix} \times \begin{bmatrix} 1 & -2 & 3 \\ -4 & 2 & 5 \end{bmatrix}$$

$$= \begin{bmatrix} 2.1 + 3(-4) & 2(-2) + 3(2) & 2(3) + 3(5) \\ 4.1 + 5(-4) & 4(-2) + 5(2) & 4(3) + 5(5) \\ 2.1 + 1(-4) & 2(-2) + 1(2) & 2(3) + 1(5) \end{bmatrix}$$

$$= \begin{bmatrix} -10 & 2 & 21 \\ -16 & 2 & 37 \\ -2 & -2 & 11 \end{bmatrix}$$

Hence AB $\neq$ BA.

Example 3:

If $A = \begin{bmatrix} 1 & 2 \\ 3 & 0 \\ 4 & 1 \end{bmatrix}$ and $B = \begin{bmatrix} 0 & 1 & 0 \\ 0 & 2 & 1 \\ 2 & 3 & 0 \end{bmatrix}$, *find BA.*

Solution:

$$BA = \begin{bmatrix} 0 & 1 & 0 \\ 0 & 2 & 1 \\ 2 & 3 & 0 \end{bmatrix} \times \begin{bmatrix} 1 & 2 \\ 3 & 0 \\ 4 & 1 \end{bmatrix}$$

$$= \begin{bmatrix} 0.1 + 1.3 + 0.4 & 0.2 + 1.0 + 0.1 \\ 0.1 + 2.3 + 1.4 & 0.2 + 2.0 + 1.1 \\ 2.1 + 3.3 + 0.4 & 2.2 + 3.0 + 0.1 \end{bmatrix}$$

$$= \begin{bmatrix} 3 & 0 \\ 10 & 1 \\ 11 & 4 \end{bmatrix}$$

Example 4:

Multiply [3 – 1 4] and $\begin{bmatrix} -2 \\ 6 \\ 3 \end{bmatrix}$

Solution:

$$[3 \; -1 \; 4] \times \begin{bmatrix} -2 \\ 6 \\ 3 \end{bmatrix}$$

$$= [3(-2) + (-1) \, . \, 6 + 4.3] = [0]$$ **Ans.**

Example 5(a):

If $A = \begin{bmatrix} 0 & 1 \\ 1 & 2 \end{bmatrix}$ and $B = \begin{bmatrix} 1 \\ 2 \end{bmatrix}$, *find AB and BA, if they exist.*

Solution:

$$AB = \begin{bmatrix} 0.1 + 1.2 \\ 1.1 + 2.2 \end{bmatrix} = \begin{bmatrix} 2 \\ 5 \end{bmatrix}$$ **Ans.**

Also BA does not exist as the number of columns in B and number of rows in A are not equal.

Example 5(b):

If $A = \begin{bmatrix} 1 & 4 & 0 \\ 2 & 5 & 0 \\ 3 & 6 & 0 \end{bmatrix}$, $B = \begin{bmatrix} 3 & 2 & 1 \\ 1 & 2 & 3 \\ 4 & 5 & 6 \end{bmatrix}$ and $C = \begin{bmatrix} 3 & 2 & 1 \\ 1 & 2 & 3 \\ 7 & 8 & 9 \end{bmatrix}$

then evaluate AB – AC.

Solution:

$$AB = \begin{bmatrix} 1 & 4 & 0 \\ 2 & 5 & 0 \\ 3 & 6 & 0 \end{bmatrix} \times \begin{bmatrix} 3 & 2 & 1 \\ 1 & 2 & 3 \\ 4 & 5 & 6 \end{bmatrix}$$

$$= \begin{bmatrix} 1.3 + 4.1 + 0.4 & 1.2 + 4.2 + 0.5 & 1.1 + 4.3 + 0.6 \\ 2.3 + 5.1 + 0.4 & 2.2 + 5.2 + 0.5 & 2.1 + 5.3 + 0.6 \\ 3.3 + 6.1 + 0.4 & 3.2 + 6.2 + 0.5 & 3.1 + 6.3 + 0.6 \end{bmatrix}$$

$$= \begin{bmatrix} 7 & 10 & 13 \\ 11 & 14 & 17 \\ 15 & 18 & 21 \end{bmatrix} \quad ...(1)$$

And
$$AC = \begin{bmatrix} 1 & 4 & 0 \\ 2 & 5 & 0 \\ 3 & 6 & 0 \end{bmatrix} \times \begin{bmatrix} 3 & 2 & 1 \\ 1 & 2 & 3 \\ 7 & 8 & 9 \end{bmatrix}$$

$$= \begin{bmatrix} 1.3 + 4.1 + 0.7 & 1.2 + 4.2 + 0.8 & 1.1 + 4.3 + 0.9 \\ 2.3 + 5.1 + 0.7 & 2.2 + 5.2 + 0.8 & 2.1 + 5.3 + 0.2 \\ 3.3 + 6.1 + 0.7 & 3.2 + 6.2 + 0.8 & 3.1 + 6.3 + 0.9 \end{bmatrix}$$

$$= \begin{bmatrix} 7 & 10 & 13 \\ 11 & 14 & 17 \\ 15 & 18 & 21 \end{bmatrix} \quad ...(2)$$

$\therefore$ From (1) and (2) we get AB – AC

$$= \begin{bmatrix} 7 & 10 & 13 \\ 11 & 14 & 17 \\ 15 & 18 & 21 \end{bmatrix} - \begin{bmatrix} 7 & 10 & 13 \\ 11 & 14 & 17 \\ 15 & 18 & 21 \end{bmatrix} = \begin{bmatrix} 0 & 0 & 0 \\ 0 & 0 & 0 \\ 0 & 0 & 0 \end{bmatrix}$$ **Ans.**

Example 6:

Find the square of the matrix

$$\begin{bmatrix} -1 & 1 & 1 & 1 \\ 1 & -1 & 1 & 1 \\ 1 & 1 & -1 & 1 \\ 1 & 1 & 1 & -1 \end{bmatrix}$$

Solution:

$$\begin{bmatrix} -1 & 1 & 1 & 1 \\ 1 & -1 & 1 & 1 \\ 1 & 1 & -1 & 1 \\ 1 & 1 & 1 & -1 \end{bmatrix}^2$$

$$= \begin{bmatrix} -1 & 1 & 1 & 1 \\ 1 & -1 & 1 & 1 \\ 1 & 1 & -1 & 1 \\ 1 & 1 & 1 & -1 \end{bmatrix} \times \begin{bmatrix} -1 & 1 & 1 & 1 \\ 1 & -1 & 1 & 1 \\ 1 & 1 & -1 & 1 \\ 1 & 1 & 1 & -1 \end{bmatrix}$$

$$= \begin{bmatrix} (-1)(-1) + 1.1 + 1.1 + 1.1 & (-1).1 + 1(-1) + 1,1 + 1.1 & (-1).1 + 1.1 + 1(-1) + 1.1 & (-1).1 + 1.1 + 1.1 + 1(-1) \\ 1.(-1) + (-1).1 + 1.1 + 1.1 & 1.1 + (-1)(-1) + 1.1 + 1.1 & 1.1 + (-1).1 + 1(-1) + 1.1 & 1.1 + (-1).1 + 1.1 + 1(-1) \\ 1.(-1) + 1.1 + (-1).1 + 1.1 & 1.1 + 1(-1) + (-1).1 + 1.1 & 1.1 + 1.1 + (-1).(-1) + 1.1 & 1.1 + 1.1 + (-1).1 + 1(-1) \\ 1.(-1) + 1.1 + 1.1 + (-1).1 & 1.1 + 1.(-1) + 1.1 + (-1).1 & 1.1 + 1.1 + 1,(-1) + (-1).1 & 1.1 + 1.1 + 1.1 + (-1)(-1) \end{bmatrix}$$

$$= \begin{bmatrix} 4 & 0 & 0 & 0 \\ 0 & 4 & 0 & 0 \\ 0 & 0 & 4 & 0 \\ 0 & 0 & 0 & 4 \end{bmatrix} = 4 \begin{bmatrix} 1 & 0 & 0 & 0 \\ 0 & 1 & 0 & 0 \\ 0 & 0 & 1 & 0 \\ 0 & 0 & 0 & 1 \end{bmatrix}$$ **Ans.**

Example 7(a):

I*if* $A = \begin{bmatrix} 1 & 1 & 3 \\ 2 & 2 & 6 \\ -1 & -1 & -3 \end{bmatrix}$, *show that* $A^2 = O$.

Solution:

$$A^2 = \begin{bmatrix} 1 & 1 & 3 \\ 2 & 2 & 6 \\ -1 & -1 & -3 \end{bmatrix} \times \begin{bmatrix} 1 & 1 & 3 \\ 2 & 2 & 6 \\ -1 & -1 & -3 \end{bmatrix}$$

$$= \begin{bmatrix} 1.1 + 1.2 + 3.(-1) & 1.1 + 1.2 + 3.(-1) & 1.3 + 1.6 + 3.(-3) \\ 2.1 + 2.2 + 6.(-1) & 2.1 + 2.2 + 6.(-1) & 2.3 + 2.6 + 6.(-3) \\ -1.1 - 1.2 - 3.(-1) & -1.1 + 1.2 - 3(-1) & -1.3 + 1.6 - 3.(-3) \end{bmatrix}$$

$$= \begin{bmatrix} 0 & 0 & 0 & 0 \\ 0 & 0 & 0 & 0 \\ 0 & 0 & 0 & 0 \end{bmatrix} = O, \text{ where O is } 3 \times 3 \text{ null matrix.}$$ **Hence proved**

Example 7(b):

Evaluate A^3 if $A = \begin{bmatrix} \cosh\theta & \sinh\theta \\ \sinh\theta & \cosh\theta \end{bmatrix}$

Solution:

$$A^2 = \begin{bmatrix} \cosh\theta & \sinh\theta \\ \sinh\theta & \cosh\theta \end{bmatrix} \times \begin{bmatrix} \cosh\theta & \sinh\theta \\ \sinh\theta & \cosh\theta \end{bmatrix}$$

$$= \begin{bmatrix} \cosh^2\theta + \sinh^2\theta & \cosh\theta\sinh\theta + \sinh\theta\cosh\theta \\ \sinh\theta\cosh\theta + \cosh\theta\sinh\theta & \sinh^2\theta + \cosh\theta \end{bmatrix}$$

$$= \begin{bmatrix} \cosh 2\theta & \sinh 2\theta \\ \sinh 2\theta & \cosh 2\theta \end{bmatrix} \because \cosh^2\theta + \sinh^2\theta = \cosh 2\theta, \quad 2\cosh\theta\cosh\theta = \sinh 2\theta$$

$$\therefore \quad A^3 = A^2A = \begin{bmatrix} \cosh 2\theta & \sinh 2\theta \\ \sinh 2\theta & \cosh 2\theta \end{bmatrix} \begin{bmatrix} \cosh 2\theta & \sinh 2\theta \\ \sinh 2\theta & \cosh 2\theta \end{bmatrix}$$

$$= \begin{bmatrix} \cosh 2\theta\cosh\theta + \sinh 2\theta\sinh\theta & \cosh 2\theta\sinh\theta + \sinh 2\theta\cosh\theta \\ \sinh 2\theta\cosh\theta + \cosh 2\theta\sinh\theta & \sinh 2\theta\sinh\theta + \cosh 2\theta\cosh\theta \end{bmatrix}$$

$$= \begin{bmatrix} \cosh(2\theta + \theta) & \sinh(2\theta + \theta) \\ \sinh(2\theta + \theta) & \cosh(2\theta + \theta) \end{bmatrix},$$

$\because$ sinh (A + B) = sinh A cosh B + cosh A sin B

sinh (A + B) = cosh A cosh B + sinh A sin B

$$= \begin{bmatrix} \cosh 3\theta & \sinh 2\theta \\ \sinh 3\theta & \cosh 2\theta \end{bmatrix}$$

Example 7(c):

If A, B, C are three matrices such that

$A = [x, y, z]$, $B = \begin{bmatrix} a & h & g \\ h & b & f \\ g & f & c \end{bmatrix}$, $C = \begin{bmatrix} x \\ y \\ z \end{bmatrix}$, *evaluate ABC.*

Solution:

$$B = [x, y, z] \times \begin{bmatrix} a & h & g \\ h & b & f \\ g & f & c \end{bmatrix}$$

$$= [x.a + y.h + z.g \quad x.h + y.b + z.f \quad x.g + y.f + z.c]$$

$$\Rightarrow ABC = [ax + hy + gz \quad hx + by + fz \quad gx + fy + cz \times \begin{bmatrix} x \\ y \\ z \end{bmatrix}$$

$$= [x\,(ax + hy + gz) + y\,(hx + by + fz) + z\,(gx + fy + cz)]$$ **(Note)**

$$= [ax^2 + by^2 + cz^2 + 2hxy + 2gzx + 2fyz].$$ **Ans.**

Example 7(d):

Find the product of the following two matrices

$$\begin{bmatrix} 0 & c & -b \\ -c & 0 & a \\ b & -a & 0 \end{bmatrix} \times \begin{bmatrix} a^2 & ab & ac \\ ab & b^2 & bc \\ ac & bc & c^2 \end{bmatrix}$$

Solution:

The required product

$$= \begin{bmatrix} 0 & c & -b \\ -c & 0 & a \\ b & -a & 0 \end{bmatrix} \times \begin{bmatrix} a^2 & ab & ac \\ ab & b^2 & bc \\ ac & bc & c^2 \end{bmatrix}$$

$$= \begin{bmatrix} 0.a^2 + c.ab - b.ac & 0.ab + c.b^2 - b.bc & 0ac + c.bc - b.c^2 \\ -ca^2 + 0.ab\ a.ac & -c.ab + 0b^2 + a.bc & -c.ac + 0.bc + a.c^2 \\ b.a^2 - a.ab + 0.ac & b.ab - a.b^2 + 0.bc & b.ac - a.bc + 0.c^2 \end{bmatrix}$$

$$= \begin{bmatrix} 0 & 0 & 0 \\ 0 & 0 & 0 \\ 0 & 0 & 0 \end{bmatrix}$$

Example 8:

Find the values of x, y, z in the following equation

$$\begin{bmatrix} 1 & 2 & 3 \\ 3 & 1 & 2 \\ 2 & 3 & 1 \end{bmatrix} \times \begin{bmatrix} x \\ y \\ z \end{bmatrix} = \begin{bmatrix} 4 & -2 \\ 0 & -6 \\ -1 & 2 \end{bmatrix} \times \begin{bmatrix} 2 \\ 1 \end{bmatrix}$$

Solution:

$$\begin{bmatrix} 1 & 2 & 3 \\ 3 & 1 & 2 \\ 2 & 3 & 1 \end{bmatrix} \times \begin{bmatrix} x \\ y \\ z \end{bmatrix} = \begin{bmatrix} 1.x + 2y + 2z \\ 3x + 1.y + 2z \\ 2x + 3y + 1.z \end{bmatrix} \quad ...(1)$$

And $\begin{bmatrix} 4 & -2 \\ 0 & -6 \\ -1 & 2 \end{bmatrix} \times \begin{bmatrix} 2 \\ 1 \end{bmatrix} = \begin{bmatrix} 4.2 + (-2).1 \\ 0.2 + (-6).1 \\ -1.2 + \quad 2.1 \end{bmatrix} = \begin{bmatrix} 6 \\ -6 \\ 0 \end{bmatrix}$.(2)

With the help of (1) and (2), the given equation reduces to

$$\begin{bmatrix} x + 2y + 3z \\ 3x + \ y + 2z \\ 2x + 3y + \ z \end{bmatrix} = \begin{bmatrix} 6 \\ -6 \\ 0 \end{bmatrix}$$

From this on comparing the corresponding elements on both sides we get $x + 2y + 3z = 6$; $3x + y + 2z = -6$ and $2x + 3y + z = 0$.

Solving these we get $x = -4$, $y = 2$, $z = 2$. **Ans.**

Example 9:

If $A = \begin{bmatrix} 1 & 2 & 3 \\ 0 & 1 & 2 \\ 0 & 0 & 1 \end{bmatrix}$; $B = \begin{bmatrix} x \\ y \\ z \end{bmatrix}$ and $AB = \begin{bmatrix} 6 \\ 3 \\ 1 \end{bmatrix}$ *find the values of x, y, z.*

Solution:

$$AB = \begin{bmatrix} 1 & 2 & 3 \\ 0 & 1 & 2 \\ 0 & 0 & 1 \end{bmatrix} \times \begin{bmatrix} x \\ y \\ z \end{bmatrix}$$

$$\Rightarrow \quad \begin{bmatrix} 6 \\ 3 \\ 1 \end{bmatrix} = \begin{bmatrix} 1.x + 2.y + 3z \\ 0.x + 1.y + 2z \\ 0.x + 0.y + 1.z \end{bmatrix}$$

$\Rightarrow \quad 6 = x + 2y + 3z,\ 3 = y + 2z,\ 1 = z,$ **(Note)**

comparing the corresponding elements of the matrices on both sides.

Solving these we get $x = 1,\ y = 1,\ z = 1.$ **Ans.**

Example 10(a):

If $A = \begin{bmatrix} i & 0 \\ 0 & -i \end{bmatrix}, B = \begin{bmatrix} 0 & -1 \\ 1 & 0 \end{bmatrix}, C = \begin{bmatrix} 0 & i \\ i & 0 \end{bmatrix}$, *prove that*

$A^2 = B^2 = C^2 = -I$ *and* $AB = -C = -BA$, *where* $I = \begin{bmatrix} 1 & 1 \\ 0 & 1 \end{bmatrix}$

Solution:

$$A^2 = \begin{bmatrix} i & 0 \\ 0 & -i \end{bmatrix} \times \begin{bmatrix} i & 0 \\ 0 & -i \end{bmatrix}$$

$$= \begin{bmatrix} i.i + 0.0 & i.0 + (-1) \\ 0.i - i.0 & 0.0 + (-i)(-i) \end{bmatrix} = \begin{bmatrix} -1 & 0 \\ 0 & -1 \end{bmatrix}$$

$$= \begin{bmatrix} 1 & 0 \\ 0 & 1 \end{bmatrix}$$

$$= -I.$$

and $B^2 = \begin{bmatrix} 1 & -1 \\ 0 & 0 \end{bmatrix} \times \begin{bmatrix} 1 & -1 \\ 0 & 0 \end{bmatrix} = \begin{bmatrix} 0.0 - 1.1 & 0.(-1) + (-1).0 \\ 1.0 + 0.1 & 1(-1) + 0.0 \end{bmatrix}$

$$= \begin{bmatrix} -1 & 0 \\ 0 & -1 \end{bmatrix} = -\begin{bmatrix} 1 & 0 \\ 0 & 1 \end{bmatrix} = -1$$

Similarly we can prove that $C^2 = -I$. Hence $A^2 = B^2 = C^2 = -I$.

Again $AB = \begin{bmatrix} i & 0 \\ 0 & -i \end{bmatrix} \times \begin{bmatrix} 0 & -1 \\ 1 & 0 \end{bmatrix}$

$$= \begin{bmatrix} i.0 + 0.1 & i(-1) + 0.0 \\ 0.0 - i(1) & 0(-1) - i.0 \end{bmatrix}$$

$$= \begin{bmatrix} 0 & -i \\ -i & 0 \end{bmatrix} = - \begin{bmatrix} 0 & i \\ i & 0 \end{bmatrix} = -C$$

and $$BA = \begin{bmatrix} 0 & -1 \\ 1 & 0 \end{bmatrix} \times \begin{bmatrix} i & 0 \\ 0 & -i \end{bmatrix}$$

$$= \begin{bmatrix} 0.i - i.o & 0.0 - 1(-i) \\ 1.i + 0.0 & 1.0 + 0(-1) \end{bmatrix}$$

$$= \begin{bmatrix} 0 & i \\ i & 0 \end{bmatrix} = C$$

Hence AB = – C = – BA.

Example 10(b):

Given $A_i = \begin{bmatrix} 0 & 0 & 0 & 1 \\ 0 & 0 & 1 & 0 \\ 0 & 1 & 0 & 0 \\ 1 & 0 & 0 & 0 \end{bmatrix}$; $A_2 = \begin{bmatrix} 0 & 0 & 0 & i \\ 0 & 0 & -i & 0 \\ 0 & i & 0 & 0 \\ -i & 0 & 0 & 0 \end{bmatrix}$

$A_2 = \begin{bmatrix} 0 & 0 & 1 & 0 \\ 0 & 0 & 0 & -1 \\ 1 & 0 & 0 & 0 \\ 0 & -1 & 0 & 0 \end{bmatrix}$ and $A_4 = \begin{bmatrix} 1 & 0 & 0 & 0 \\ 0 & 1 & 0 & 0 \\ 0 & 0 & -1 & 0 \\ 0 & 0 & 0 & -1 \end{bmatrix}$

Show that $A_i A_k + A_k A_i = 2I$ or O according as $i = k$ or $i \neq k$ and I is the matrix of order 4 and i and k take the values 1, 2, 3 and 4.

Solution:

Let i = k = 1 (say). Then

$A_iA_k = A_1A_2 = A_kA_i$

$\therefore \quad A_iA_k = A_1A_1 = \begin{bmatrix} 0 & 0 & 0 & 1 \\ 0 & 0 & 1 & 0 \\ 0 & 1 & 0 & 0 \\ 1 & 0 & 0 & 0 \end{bmatrix} \times \begin{bmatrix} 0 & 0 & 0 & 1 \\ 0 & 0 & 1 & 0 \\ 0 & 1 & 0 & 0 \\ 1 & 0 & 0 & 0 \end{bmatrix}$

$$= \begin{bmatrix} 0+0+0+1 & 0+0+0+0 & 0+0+0+0 & 0+0+0+0 \\ 0+0+0+0 & 0+0+1+0 & 0+0+0+0 & 0+0+0+0 \\ 0+0+0+0 & 0+0+0+0 & 0+1+0+0 & 0+0+0+0 \\ 0+0+0+0 & 0+0+0+0 & 0+0+0+0 & 1+0+0+0 \end{bmatrix}$$

$$= \begin{bmatrix} 1 & 0 & 0 & 0 \\ 0 & 1 & 0 & 0 \\ 0 & 0 & 1 & 0 \\ 0 & 0 & 0 & 1 \end{bmatrix} = I$$

$\therefore \quad A_iA_k + A_kA_i = I + I = 2I$ **Hence proved.**

If $i \neq k$, let $i = 3$ and $k = 2$

Then $A_iA_k = A_3A_2 = \begin{bmatrix} 0 & 0 & 1 & 0 \\ 0 & 1 & 0 & -1 \\ 1 & 0 & 0 & 0 \\ 0 & -1 & 0 & 0 \end{bmatrix} \times \begin{bmatrix} 0 & 0 & 0 & i \\ 0 & 0 & -i & 0 \\ 0 & i & 0 & 0 \\ -i & 0 & 0 & 0 \end{bmatrix}$

$$= \begin{bmatrix} 0+0+0+0 & 0+0+i+0 & 0+0+0+0 & 0+0+0+0 \\ 0+0+0+i & 0+0+0+0 & 0+0+0+0 & 0+0+0+0 \\ 0+0+0+0 & 0+0+0+0 & 0+0+0+0 & i+0+0+0 \\ 0+0+0+0 & 0+0+0+0 & 0+i+0+0 & 0+0+0+0 \end{bmatrix}$$

$$= \begin{bmatrix} 0 & i & 0 & 0 \\ i & 0 & 0 & 0 \\ 0 & 0 & 0 & i \\ 0 & 0 & i & 0 \end{bmatrix} = i \begin{bmatrix} 0 & 1 & 0 & 0 \\ 1 & 0 & 0 & 0 \\ 0 & 0 & 0 & 1 \\ 0 & 0 & 1 & 0 \end{bmatrix}$$

And $A_kA_i = A_2A_3 = \begin{bmatrix} 0 & 0 & 0 & i \\ 0 & 0 & -i & 0 \\ 0 & i & 0 & 0 \\ -i & 0 & 0 & 0 \end{bmatrix} \times \begin{bmatrix} 0 & 0 & 1 & 0 \\ 0 & 0 & 0 & -1 \\ 1 & 0 & 0 & 0 \\ 0 & -1 & 0 & 0 \end{bmatrix}$

$= \begin{bmatrix} 0 & -i & 0 & 0 \\ -i & 0 & 0 & 0 \\ 0 & 0 & 0 & -i \\ 0 & 0 & -i & 0 \end{bmatrix}$ Multiplying in the usual way

$$= -i\begin{bmatrix} 0 & 1 & 0 & 0 \\ 1 & 0 & 0 & 0 \\ 0 & 0 & 0 & 1 \\ 0 & 0 & 1 & 0 \end{bmatrix}$$

$$\therefore \quad A_iA_k + A_kA_i = i\begin{bmatrix} 0 & 1 & 0 & 0 \\ 1 & 0 & 0 & 0 \\ 0 & 0 & 0 & 1 \\ 0 & 0 & 1 & 0 \end{bmatrix} - i\begin{bmatrix} 0 & 1 & 0 & 0 \\ 1 & 0 & 0 & 0 \\ 0 & 0 & 0 & 1 \\ 0 & 0 & 1 & 0 \end{bmatrix}$$

$$= O$$ **Hence proved.**

We can in a similar way prove the above result by giving i and k other values also.

Example 11:

If $A = \begin{bmatrix} 2 & 3 & 1 \\ 0 & -1 & 5 \end{bmatrix}$ and $B = \begin{bmatrix} 1 & 2 & -6 \\ 0 & -1 & 3 \end{bmatrix}$ *evaluate (a)* $A^2 - B^2$ *and (b) AB and BA.*

Solution:

(a) $A^2 = \begin{bmatrix} 2 & 3 & 1 \\ 0 & -1 & 5 \end{bmatrix} \times \begin{bmatrix} 2 & 3 & 1 \\ 0 & -1 & 5 \end{bmatrix}$, which does not exist as number of columns in the first matrix is not equal to number of rows in the second matrix.

Similarly B^2 does not exist.

(b) AB and BA both do not exist, the reason being the same as in part (a) above.

Example 12:

If $A = \begin{bmatrix} \cos\theta & -\sin\theta \\ \sin\theta & \cos\theta \end{bmatrix}$, $B = \begin{bmatrix} \cos\phi & -\sin\phi \\ \sin\phi & \cos\phi \end{bmatrix}$ *show that AB = BA.*

Solution:

$$AB = \begin{bmatrix} \cos\theta & -\sin\theta \\ \sin\theta & \cos\theta \end{bmatrix} \times \begin{bmatrix} \cos\phi & -\sin\phi \\ \sin\phi & \cos\phi \end{bmatrix}$$

$$= \begin{bmatrix} \cos\theta\cos\phi - \sin\theta\sin\phi & -\cos\theta\sin\phi - \sin\theta\cos\phi \\ \sin\theta\cos\phi + \cos\theta\sin\phi & -\sin\theta\sin\phi + \cos\theta\cos\phi \end{bmatrix}$$

$$= \begin{bmatrix} \cos(\theta+\phi) & -\sin(\theta+\phi) \\ \sin(\theta+\phi) & \cos(\theta+\phi) \end{bmatrix} \qquad ...(1)$$

And $BA = \begin{bmatrix} \cos\phi & -\sin\phi \\ \sin\phi & \cos\phi \end{bmatrix} \times \begin{bmatrix} \cos\theta & -\sin\theta \\ \sin\theta & \cos\theta \end{bmatrix}$

$$= \begin{bmatrix} \cos\phi\cos\theta - \sin\phi\sin\theta & -\cos\phi\sin\theta - \sin\phi\cos\theta \\ \sin\phi\cos\theta + \cos\phi\sin\theta & -\sin\phi\sin\theta + \cos\phi\cos\theta \end{bmatrix}$$

$$= \begin{bmatrix} \cos(\theta+\phi) & -\sin(\theta+\phi) \\ \sin(\theta+\phi) & \cos(\theta+\phi) \end{bmatrix}$$

$\therefore$ From (1) and (2) we get AB = BA. **Hence proved.**

Example 13(a):

If $A = \begin{bmatrix} 1 & 5 & 6 \\ -6 & 7 & 0 \end{bmatrix}$ and $B = \begin{bmatrix} 1 & -5 & 7 \\ 8 & -7 & 7 \end{bmatrix}$ *then find A + B and A – B.*

Solution:

$$A + B = \begin{bmatrix} 1 & 5 & 6 \\ -6 & 7 & 0 \end{bmatrix} + \begin{bmatrix} 1 & -5 & 7 \\ 8 & -7 & 7 \end{bmatrix}$$

$$= \begin{bmatrix} 1+1 & 5-5 & 6+7 \\ -6+8 & 7-7 & 0+7 \end{bmatrix} = \begin{bmatrix} 2 & 0 & 13 \\ 2 & 0 & 7 \end{bmatrix} \qquad \textbf{Ans.}$$

$$\text{and } A - B = \begin{bmatrix} 1 & 5 & 6 \\ -6 & 7 & 0 \end{bmatrix} - \begin{bmatrix} 1 & -5 & 7 \\ 8 & -7 & 7 \end{bmatrix}$$

$$= \begin{bmatrix} 1-1 & 5-(-5) & 6-7 \\ -6-7 & 8-7-(-7) & 0-7 \end{bmatrix} = \begin{bmatrix} 0 & 10 & -1 \\ -14 & 14 & -7 \end{bmatrix} \qquad \textbf{Ans.}$$

Example 13(b):

Solve the following equations for A and B:

$2A - B = \begin{bmatrix} 3 & -3 & 0 \\ 3 & 3 & 2 \end{bmatrix}$, $2B + A = \begin{bmatrix} 4 & 1 & 5 \\ -1 & 4 & -4 \end{bmatrix}$

Solution:

Given $2A - B = \begin{bmatrix} 3 & -3 & 0 \\ 3 & 3 & 2 \end{bmatrix}$

Multiplying both sides by 2, we get

$$4A - 2B = 2\begin{bmatrix} 3 & -3 & 0 \\ 3 & 3 & 2 \end{bmatrix} = \begin{bmatrix} 6 & -6 & 0 \\ 6 & 6 & 4 \end{bmatrix} \quad ...(1)$$

Also given that $2B + A = \begin{bmatrix} 4 & 1 & 5 \\ -1 & 4 & -4 \end{bmatrix}$ ...(2)

Adding (1) and (2) we get

$$5A = \begin{bmatrix} 6 & -6 & 0 \\ 6 & 6 & 4 \end{bmatrix} + \begin{bmatrix} 4 & 1 & 5 \\ -1 & 4 & -4 \end{bmatrix}$$

$$= \begin{bmatrix} 6+4 & -6+1 & 0+5 \\ 6-1 & 6+4 & 4+4 \end{bmatrix} = \begin{bmatrix} 10 & -5 & 5 \\ 5 & 10 & 0 \end{bmatrix}$$

$$\Rightarrow \quad A = \frac{1}{2}\begin{bmatrix} 10 & -5 & 5 \\ 5 & 10 & 0 \end{bmatrix} = \begin{bmatrix} 2 & -1 & 1 \\ 1 & 2 & 0 \end{bmatrix}$$

Again from (2) we get

$$2B = \begin{bmatrix} 4 & 1 & 5 \\ -1 & 4 & -4 \end{bmatrix} - A$$

$$\Rightarrow \quad 2B = \begin{bmatrix} 4 & 1 & 5 \\ -1 & 4 & -4 \end{bmatrix} - \begin{bmatrix} 2 & -1 & 1 \\ 1 & 2 & 0 \end{bmatrix}$$

$$= \begin{bmatrix} 4-2 & 1+1 & 5-1 \\ -1-1 & 4-2 & -4-0 \end{bmatrix} = \begin{bmatrix} 2 & 2 & 4 \\ -2 & 2 & -4 \end{bmatrix}$$

$$\Rightarrow \quad B = \frac{1}{2}\begin{bmatrix} 2 & 2 & 4 \\ -2 & 2 & -4 \end{bmatrix} = \begin{bmatrix} 1 & 1 & 2 \\ -1 & 1 & -2 \end{bmatrix}.$$

Ans.

Example 13(c):

Given $A = \begin{bmatrix} 1 & 2 & -3 \\ 5 & 0 & 2 \\ 1 & -1 & 1 \end{bmatrix}$ and $B = \begin{bmatrix} 3 & -1 & 2 \\ 4 & 2 & 5 \\ 2 & 0 & 3 \end{bmatrix}$,

find the matrix C such that $A + 2C = B$.

Solution:

Given that A + 2C = B or 2= B – A

$$\Rightarrow \quad 2C = \begin{bmatrix} 3 & -1 & 2 \\ 4 & 2 & 5 \\ 2 & 0 & 3 \end{bmatrix} - \begin{bmatrix} 1 & 2 & -3 \\ 5 & 0 & 2 \\ 1 & -1 & 1 \end{bmatrix}$$

$$= \begin{bmatrix} 3-1 & -1-2 & 2-(-3) \\ 4-5 & 2-0 & 5-2 \\ 2-1 & 0-(-1) & 3-1 \end{bmatrix} = \begin{bmatrix} 2 & -3 & 5 \\ -1 & 2 & 3 \\ 1 & 1 & 2 \end{bmatrix}$$

$$\Rightarrow \quad C = \frac{1}{2}\begin{bmatrix} 2 & -3 & 0 \\ -1 & 2 & 3 \\ 1 & 1 & 2 \end{bmatrix} = \begin{bmatrix} 1 & -\frac{3}{2} & \frac{5}{2} \\ -\frac{1}{2} & 1 & \frac{3}{2} \\ \frac{1}{2} & \frac{1}{2} & 1 \end{bmatrix}$$

Example 13(d):

If $A = \begin{bmatrix} 2 & 3 & 1 \\ 0 & -1 & 5 \end{bmatrix}$ and $B = \begin{bmatrix} 1 & 2 & -6 \\ 0 & -1 & 3 \end{bmatrix}$ *evaluate 3A – 4B.*

Solution:

$$2A - 4B = 3\begin{bmatrix} 2 & 3 & 1 \\ 0 & -1 & 5 \end{bmatrix} - 4\begin{bmatrix} 1 & 2 & -6 \\ 0 & -1 & 3 \end{bmatrix}$$

$$= \begin{bmatrix} 6 & 9 & 3 \\ 0 & -3 & 15 \end{bmatrix} - \begin{bmatrix} 4 & 8 & -24 \\ 0 & -4 & 12 \end{bmatrix}$$

$$= \begin{bmatrix} 6-4 & 9-8 & 3-(-24) \\ 0-0 & -3-(-4) & 15-12 \end{bmatrix}$$

$$= \begin{bmatrix} 2 & 1 & 27 \\ 0 & 1 & 3 \end{bmatrix}.$$

Example 13(e):

Calculate AB and BA if

$$A = \begin{bmatrix} 1 & -1 & 1 \\ -3 & 2 & -1 \\ -2 & 1 & 0 \end{bmatrix}, B = \begin{bmatrix} 1 & 2 & 3 \\ 2 & 4 & 6 \\ 1 & 2 & 3 \end{bmatrix}$$

Solution:

$$AB = \begin{bmatrix} 1 & -1 & 1 \\ -3 & 2 & -1 \\ -2 & 1 & 0 \end{bmatrix} \times \begin{bmatrix} 1 & 2 & 3 \\ 2 & 4 & 6 \\ 1 & 2 & 3 \end{bmatrix}$$

$$= \begin{bmatrix} 1.1 + (-1).2 + 1.1 & 1.2 \ (-1).4 + 1.2 & 1.3 + (-1).6 + 1.3 \\ (-3).1 + 2.2 + (-1).1 & (-3).2 + 2.4 + (-1).2 & (-3).3 + 2.6 + (-1).3 \\ (-2).1 + 1.2 + 0.1 & (-2).2 + 1.4 + 0.2 & (-2).3 + 1.6 + 0.3 \end{bmatrix}$$

$$= \begin{bmatrix} 0 & 0 & 0 \\ 0 & 0 & 0 \\ 0 & 0 & 0 \end{bmatrix} = O, \text{ where } O \text{ is } 3 \times 3 \text{ null matrix.}$$ **Ans.**

$$\text{And } BA = \begin{bmatrix} 1 & 2 & 3 \\ 2 & 4 & 6 \\ 1 & 2 & 3 \end{bmatrix} \times \begin{bmatrix} 1 & -1 & 1 \\ -3 & 2 & -1 \\ -2 & 1 & 0 \end{bmatrix}$$

$$= \begin{bmatrix} 1.1 + 2.(-3) + 3(-2) & 1(-1) + 2.2 + 3.1 & 1.1 + 2.(-1) + 3.0 \\ 2.1 + 4.(-3) + 6(-2) & 2(-1) + 4.2 + 6.1 & 2.1 + 4.(-1) + 6.0 \\ 1.1 + 2.(-3) + 3(-2) & 1(-1) + 2.2 + 3.1 & 1.1 + 2.(-1) + 30 \end{bmatrix}$$

$$= \begin{bmatrix} -11 & 6 & -1 \\ -22 & 12 & -2 \\ -11 & 6 & -1 \end{bmatrix}$$ **Ans.**

[**Note:** $AB \neq BA$].

Example 14:

If $A = \begin{bmatrix} 2 & 3 & 4 \\ 1 & 2 & 3 \\ -1 & 1 & 2 \end{bmatrix}$ and $B = \begin{bmatrix} 1 & 3 & 0 \\ -1 & 2 & 1 \\ 0 & 0 & 2 \end{bmatrix}$ *evaluate AB, BA or whichever exists.*

Solution:

$$AB = \begin{bmatrix} 2 & 3 & 4 \\ 1 & 2 & 3 \\ -1 & 1 & 2 \end{bmatrix} \times \begin{bmatrix} 1 & 3 & 0 \\ -1 & 2 & 1 \\ 0 & 0 & 2 \end{bmatrix}$$

$$= \begin{bmatrix} 2.1 + 3(-1) + 4.0 & 2.3 + 3.2 + 4.0 & 2.0 + 3.1 + 4.2 \\ 1.1 + 2(-1) + 3.0 & 1.3 + 2.2 + 3.0 & 1.0 + 2.1 + 3.2 \\ -1.1 + 1(-1) + 2.0 & 1.3 + 1.2 + 2.0 & -1.0 + 1.1 + 2.2 \end{bmatrix}$$

$$= \begin{bmatrix} -1 & 12 & 11 \\ -1 & 7 & 8 \\ -2 & -1 & 5 \end{bmatrix}$$

And $BA = \begin{bmatrix} 1 & 3 & 0 \\ -1 & 2 & 1 \\ 0 & 0 & 2 \end{bmatrix} \times \begin{bmatrix} 2 & 3 & 4 \\ 1 & 2 & 3 \\ -1 & 1 & 2 \end{bmatrix}$

$$= \begin{bmatrix} 1.2 + 3.1 + 0(-1) & 1.3 + 3.2 + 0.1 & 1.4 + 3.3 + 0.2 \\ -1.2 + 2.1 + 1(-1) & -1.3 + 2.2 + 1.1 & -1.4 + 2.3 + 1.2 \\ 0.2 + 0.1 + 2(-1) & 0.3 + 0.2 + 2.1 & 0.4 + 0.3 + 2.2 \end{bmatrix}$$

$$= \begin{bmatrix} 5 & 9 & 13 \\ -1 & 2 & 4 \\ -2 & 2 & 4 \end{bmatrix}$$

Example 15:

Find the product matrix of the matrices

$$\begin{bmatrix} 2 & 1 & 2 & 1 \\ 1 & 1 & 1 & 1 \end{bmatrix} \text{ and } \begin{bmatrix} 2 & -1 & 0 \\ 0 & 4 & 1 \\ -2 & 1 & 0 \\ 1 & -3 & 2 \end{bmatrix}$$

Solution:

The required matrix

$$= \begin{bmatrix} 2 & 1 & 2 & 1 \\ 1 & 1 & 1 & 1 \end{bmatrix} \times \begin{bmatrix} 2 & -1 & 0 \\ 0 & 4 & 1 \\ -2 & 1 & 0 \\ 1 & -3 & 2 \end{bmatrix}$$

$$= \begin{bmatrix} 2.2 + 1.0 + 2(-2) + 1.1 & 2(-1) + 1.4 + 2.1 + 1(-3) & 2.0 + 1.1 + 2.0 + 1.2 \\ 1.2 + 1.0 + 1(-2) + 1.1 & 1(-1) + 1.4 + 1.1 + 1(-3) & 1.0 + 1.1 + 1.0 + 1.2 \end{bmatrix}$$

$$= \begin{bmatrix} 1 & 1 & 3 \\ 1 & 1 & 3 \end{bmatrix}.$$ **Ans.**

Example 16:

If $\begin{bmatrix} 4 \\ 1 \\ 3 \end{bmatrix} A = \begin{bmatrix} -4 & 8 & 4 \\ -1 & 2 & 1 \\ -3 & 6 & 3 \end{bmatrix}$, *find A.*

Solution:

We know that if X is an m × n matrix, Y is an n × k matrix, then the product XY is an m × k matrix.

Hence $\begin{bmatrix} 4 \\ 1 \\ 3 \end{bmatrix}$ is 3 × 1 matrix and $\begin{bmatrix} -4 & 8 & 4 \\ -1 & 2 & 1 \\ -3 & 6 & 3 \end{bmatrix}$

is 3 × 3 matrix, os A must be a 1 × 3 matrix *i.e.,* a row matrix. (**Note**)

∴ Let A = [a, b, c]

Then $\begin{bmatrix} 4 \\ 1 \\ 3 \end{bmatrix} \times [a\ b\ c] = \begin{bmatrix} -4 & 8 & 4 \\ -1 & 2 & 1 \\ -3 & 6 & 3 \end{bmatrix}$

Which gives $\begin{bmatrix} 4a & 4b & 4c \\ a & b & c \\ 3a & 3b & 3c \end{bmatrix} = \begin{bmatrix} -4 & 8 & 4 \\ -1 & 2 & 1 \\ -3 & 6 & 3 \end{bmatrix}$

Comparing corresponding elements we have

$4a = -4, a = -1,$

$3a = -3,$

$4b = 8, b = 2,$

$3b = 6$

and $4c = 4, c = 1, 3c = 3.$

All these are satisfied by $a = -1, b = 2, c = 1.$

Hence from (1) we have

$A = [a\ b\ c] = [-1, 2, 1].$ **Ans.**

Example 17:

Evaluate $\begin{bmatrix} \cos\theta + \sin\theta & \sqrt{2}\sin\theta \\ -\sqrt{2}\sin\theta & \cos\theta - \sin\theta \end{bmatrix}^n$

Solution:

Let $A = \begin{bmatrix} \cos\theta + \sin\theta & \sqrt{2}\sin\theta \\ -\sqrt{2}\sin\theta & \cos\theta - \sin\theta \end{bmatrix}$ ...(1)

Then $A^2 = A.\ A$

$$= \begin{bmatrix} \cos\theta + \sin\theta & \sqrt{2}\sin\theta \\ -\sqrt{2}\sin\theta & \cos\theta - \sin\theta \end{bmatrix}\begin{bmatrix} \cos\theta + \sin\theta & \sqrt{2}\sin\theta \\ -\sqrt{2}\sin\theta & \cos\theta - \sin\theta \end{bmatrix}$$

$$= \begin{bmatrix} (\cos\theta + \sin\theta)^2 - 2\sin^2\theta & (\cos + \sin\theta)\sqrt{2}\sin\theta \\ & + \sqrt{2}\sin\theta(\cos\theta - \sin\theta \\ -\sqrt{2}\sin\theta(\cos\theta + \sin\theta) & -\sqrt{2}\sin\theta\sqrt{2}\sin\theta \\ -\sqrt{2}\sin\theta(\cos\theta - \sin\theta) & + (\cos\theta - \sin\theta)^2 \end{bmatrix}$$

$$= \begin{bmatrix} (\cos^2\theta - \sin^2\theta) + 2\sin\theta\cos\theta & 2\sqrt{2}\sin\theta\cos\theta \\ -2\sqrt{2}\sin\theta\cos\theta & (\cos^2\theta - \sin^2\theta) - 2\cos\theta\sin\theta \end{bmatrix}$$

$$\Rightarrow \quad A^2 = \begin{bmatrix} \cos 2\theta + \sin 2\theta & \sqrt{2}\sin 2\theta \\ -\sqrt{2}\sin 2\theta & (\cos 2\theta - \sin 2\theta \end{bmatrix} \quad ...(2)$$

Looking at (1) and (2), let us assume that

$$A^n = \begin{bmatrix} \cos n\theta + \sin n\theta & \sqrt{2}\sin n\theta \\ -\sqrt{2}\sin n\theta & (\cos n\theta - \sin n\theta \end{bmatrix} \quad ...(3)$$

Let (3) be true for an = k

i.e., $A^k = \begin{bmatrix} \cos k\theta + \sin k\theta & \sqrt{2}\sin k\theta \\ -\sqrt{2}\sin k\theta & (\cos k\theta - \sin k\theta \end{bmatrix}$ ...(4)

$$\therefore \quad A^{k+1} = A^k \cdot A$$

$$= \begin{bmatrix} \cos k\theta + \sin k\theta & \sqrt{2}\sin k\theta \\ -\sqrt{2}\sin k\theta & \cos k\theta - \sin k\theta \end{bmatrix} \times \begin{bmatrix} \cos\theta + \sin\theta & \sqrt{2}\sin\theta \\ -\sqrt{2}\sin\theta & \cos\theta - \sin\theta \end{bmatrix}$$

$$= \begin{bmatrix} (\cos k\theta + \sin k\theta)(\cos\theta + \sin\theta) & (\cos k\theta + \sin k\theta)(\sqrt{2}\sin\theta) \\ +(\sqrt{2}\sin k\theta)(-\sqrt{2}\sin\theta) & +(\sqrt{2}\sin k\theta)(\cos\theta - \sin\theta) \\ -\sqrt{2}\sin k\theta(\cos\theta + \sin\theta) & (-\sqrt{2}\sin k\theta)(\sqrt{2}\sin\theta) \\ +(\cos k\theta - \sin k\theta)(-\sqrt{2}\sin\theta) & +(\cos k\theta - \sin k\theta(\cos\theta - \sin\theta) \end{bmatrix}$$

$$= \begin{bmatrix} \cos k\theta\cos\theta + \cos k\theta\sin\theta + \sin k\theta\cos\theta & \sqrt{2}(\sin k\theta\cos\theta \\ +\sin k\theta\sin\theta - 2\sin k\theta\sin\theta & +\cos k\theta\sin\theta) \\ & -2\sin k\theta + \sin\theta + \cos k\theta\cos\theta \\ -\sqrt{2}(\sin k\theta\cos\theta + \cos k\theta\sin\theta) & -\cos k\theta\sin\theta - \sin k\theta\cos\theta \\ & +\sin k\theta\sin\theta \end{bmatrix}$$

$$= \begin{bmatrix} \cos(k\theta + \theta) + \sin(k\theta + \theta) & \sqrt{2}\sin(k\theta + \theta) \\ -\sqrt{2}\sin(k\theta + \theta) & \cos(k\theta + \theta) - \sin(k\theta + \theta) \end{bmatrix}$$

$$= \begin{bmatrix} \cos(k+1)\theta + \sin(k+1)\theta & \sqrt{2}\sin(k+1)\theta \\ -\sqrt{2}\sin(k+1)\theta & \cos(k+1)\theta - \sin(k+1)\theta \end{bmatrix}$$

$\therefore$ (3) is true for n = k + 1 provided (4) is true.

Also we have shown in (2) that (3) is true for n = 2.

Hence it is true for n = 2 + 1 *i.e.,* 3 and so on.

Hence (3) is true for all positive integral values of n.

Hence $A^n = \begin{bmatrix} \cos n\theta + \sin n\theta & \sqrt{2}\sin n\theta \\ -\sqrt{2}\sin n\theta & \cos n\theta - \sin n\theta \end{bmatrix}$ **Ans.**

Example 18(a):

If $A = \begin{bmatrix} 1 & -3 & 2 \\ 2 & 1 & -3 \\ 4 & -3 & -1 \end{bmatrix}$; $B = \begin{bmatrix} 1 & 4 & 1 & 0 \\ 2 & 1 & 1 & 1 \\ 1 & -2 & 1 & 2 \end{bmatrix}$ *and* $C = \begin{bmatrix} 2 & 1 & -1 & -2 \\ 3 & -2 & -1 & -1 \\ 2 & -5 & -1 & 0 \end{bmatrix}$,

show that AB = AC

Solution:

$$AB = \begin{bmatrix} 1 & -3 & 2 \\ 2 & 1 & -3 \\ 4 & -3 & -1 \end{bmatrix} \times \begin{bmatrix} 1 & 4 & 1 & 0 \\ 2 & 1 & 1 & 1 \\ 1 & -2 & 1 & 0 \end{bmatrix}$$

$$= \begin{bmatrix} 1.1 - 3.2 + 2.1 & 1.4 - 3.1 + 2\,(-2) & 1.1 - 3.1 + 2.1 & 1.0 - 3.1 + 2.2 \\ 2.1 + 1.2 - 3.1 & 2.4 + 1.1 - 3\,(-2) & 2.1 + 1.1 - 3.1 & 2.0 + 1.1 - 3.2 \\ 4.1 - 3.2 - 1.1 & 4.4 - 3.1 - 1\,(-2) & 4.1 - 3.2 - 1.1 & 4.0 - 3.1 - 1.2 \end{bmatrix}$$

$$= \begin{bmatrix} -3 & -3 & 0 & 1 \\ 1 & 15 & 0 & -5 \\ -3 & 15 & 0 & -5 \end{bmatrix}$$

Ans.

$$AC = \begin{bmatrix} 1 & -3 & 2 \\ 2 & 1 & -3 \\ 4 & -3 & -1 \end{bmatrix} \times \begin{bmatrix} 2 & 1 & -1 & -2 \\ 3 & -2 & -1 & -1 \\ 2 & -5 & -1 & 0 \end{bmatrix}$$

$$= \begin{bmatrix} 1.2 - 3.3 + 2.2 & 1.1 - 3\,(-2) + 2\,(-5) & 1(-1) - 3(-1) + 2(-1) & 1(-2) - 3(-1) + 2.0 \\ 2.2 + 1.3 - 3.2 & 2.1 + 1\,(-2) - 3\,(-5) & 2(-1) + 1(-1) - 3(-1) & 2(-2) + 1(-1) - 3.0 \\ 4.2 - 3.3 - 1.2 & 4.1 - 3\,(-2) - 1\,(-5) & 4(-1) - 3(-1) - 1(-1) & 4(-2) - 3(-1) - 1.0 \end{bmatrix}$$

$$= \begin{bmatrix} -3 & -3 & 0 & 1 \\ 1 & 15 & 0 & -5 \\ -3 & 15 & 0 & -5 \end{bmatrix}$$

$\therefore$ From (1) and (2), we get AB = AC.

[**Note:** Students should see that AB = AC does not necessarily simply that B = C.

Example 18(b):

If $A = \begin{bmatrix} 1 & 2 & 3 \\ 3 & -2 & 1 \\ 4 & 2 & 1 \end{bmatrix}$, *find the matrix X such that* $A + X + I = 0$, *where*

I and O are unit and zero 3 × 3 matrices respectively.

Solution:

Given that A + X + I = O ⇒ X = O – A – I

$$= \begin{bmatrix} 0 & 0 & 0 \\ 0 & 0 & 0 \\ 0 & 0 & 0 \end{bmatrix} - \begin{bmatrix} 1 & 2 & 3 \\ 3 & -2 & 1 \\ 4 & 2 & 1 \end{bmatrix} - \begin{bmatrix} 1 & 0 & 0 \\ 0 & 1 & 0 \\ 0 & 0 & 1 \end{bmatrix}$$

subtituting values of A, I and O

$$= \begin{bmatrix} 0-1-1 & 0-2-0 & 0-3-0 \\ 0-3-0 & 0+2-1 & 0-1-0 \\ 0-4-0 & 0-2-0 & 0-1-1 \end{bmatrix} = \begin{bmatrix} -2 & -2 & -3 \\ -3 & 1 & -1 \\ -4 & -2 & -2 \end{bmatrix}$$

Example 18(c):

Given $A = \begin{bmatrix} 1 & 2 & -3 \\ 5 & 0 & 2 \\ 1 & -1 & 1 \end{bmatrix}$ *and* $B = \begin{bmatrix} 3 & -1 & 2 \\ 4 & 2 & 5 \\ 2 & 0 & 3 \end{bmatrix}$ *find the matrix C such that A + C = B.*

Solution:

A + C = B

or C = B – A

$$\Rightarrow \quad C = \begin{bmatrix} 3 & -1 & 2 \\ 4 & 2 & 5 \\ 2 & 0 & 3 \end{bmatrix} - \begin{bmatrix} 1 & 2 & -3 \\ 5 & 0 & 2 \\ 1 & -1 & 1 \end{bmatrix}$$

$$= \begin{bmatrix} 3-1 & -1-2 & 2+3 \\ 4-5 & 2-0 & 5-2 \\ 2-1 & 0-(0-1) & 3-1 \end{bmatrix} = \begin{bmatrix} 2 & -3 & 5 \\ -1 & 2 & 3 \\ 1 & 1 & 2 \end{bmatrix}$$

Ans.

Example 19:

If $A = \begin{bmatrix} 1 & 2 \\ 3 & 4 \\ 5 & 6 \end{bmatrix}$ *and* $B = \begin{bmatrix} -3 & -2 \\ 1 & -5 \\ 4 & 3 \end{bmatrix}$ *find* $D = \begin{bmatrix} p & q \\ r & s \\ t & u \end{bmatrix}$, *such that A + B – D = 0*

Solution:

$$A + B - D = O \text{ or } D = A + B$$

$$\Rightarrow \quad D = \begin{bmatrix} 1 & 2 \\ 3 & 4 \\ 5 & 6 \end{bmatrix} + \begin{bmatrix} -3 & -2 \\ 1 & -5 \\ 4 & 3 \end{bmatrix}$$

$$= \begin{bmatrix} 1-3 & 2-2 \\ 3+1 & 4-5 \\ 5+4 & 6+3 \end{bmatrix} = \begin{bmatrix} -2 & 0 \\ 4 & -1 \\ 9 & 9 \end{bmatrix} = \begin{bmatrix} p & q \\ r & s \\ t & u \end{bmatrix} \text{ given}$$

∴ We have p = – 2, q = 0, r = 4, s = – 1, t = 9, u = 9 which gives D.

Ans.

Example 20:

If $A = \begin{bmatrix} 1 & -1 \\ 2 & -1 \end{bmatrix}$, $B = \begin{bmatrix} a & 1 \\ b & -1 \end{bmatrix}$ *and* $(A + B)^2 = A^2 + B^2$, *find a and b.*

Solution:

Here we have

$$A^2 = \begin{bmatrix} 1 & -1 \\ 2 & - \end{bmatrix} \times \begin{bmatrix} 1 & -1 \\ 2 & -1 \end{bmatrix}$$

$$= \begin{bmatrix} 1-2 & -1+1 \\ 2-2 & -2+1 \end{bmatrix} = \begin{bmatrix} -1 & 0 \\ 0 & -1 \end{bmatrix}$$

$$B^2 = \begin{bmatrix} a & 1 \\ b & -1 \end{bmatrix} \times \begin{bmatrix} a & 1 \\ b & -1 \end{bmatrix} = \begin{bmatrix} a^2+b & a-1 \\ ab-b & b+1 \end{bmatrix}$$

$$\therefore \quad A^2 + B^2 = \begin{bmatrix} -1 & 0 \\ 0 & -1 \end{bmatrix} + \begin{bmatrix} a^2+b & a-1 \\ ab-b & b+1 \end{bmatrix}$$

$$= \begin{bmatrix} -1+a^2+b & 0+a-1 \\ 0+ab-b & -1+b+1 \end{bmatrix} = \begin{bmatrix} a^2+b-1 & a-1 \\ ab-b & b \end{bmatrix} \quad \text{...(1)}$$

$$\text{Also } A + B = \begin{bmatrix} 1 & -1 \\ 2 & -1 \end{bmatrix} + \begin{bmatrix} a & 1 \\ b & -1 \end{bmatrix}$$

$$= \begin{bmatrix} 1+a & -1+1 \\ 2+b & -1-1 \end{bmatrix} = \begin{bmatrix} 1+a & 0 \\ 2+b & -2 \end{bmatrix}$$

$$\therefore \quad (A + B)^2 = \begin{bmatrix} 1+a & 0 \\ 2+b & -2 \end{bmatrix} \times \begin{bmatrix} 1+a & 0 \\ 2+b & -2 \end{bmatrix}$$

$$= \begin{bmatrix} (1+a)^2 + 0 & 0+0 \\ (2+b)(1+a) - 2(2+b) & 0+4 \end{bmatrix}$$

$$= \begin{bmatrix} (1+a)^2 & 0 \\ (2+b)(1+a) & 4 \end{bmatrix} \qquad ...(2)$$

Now it is given that $(A + B)^2 = A^2 + B^2$.

$$\Rightarrow \quad \begin{bmatrix} (1+a)^2 & 0 \\ (2+b)(1+a) & 4 \end{bmatrix} = \begin{bmatrix} a^2+b-1 & a-1 \\ ab-b & b \end{bmatrix}, \text{ from (1) and (2)}$$

$\Rightarrow$ $0 = a - 1$ and $4 = b$, comparing the elements of second column on both sides

$\Rightarrow$ $a = 1$ and $b = 4$. **Ans.**

Example 21(a):

If A, B are two $n \times n$ matrices and if

$$C = A + B, \; AB = BA, \; B^2 = O,$$

then show that for every integer m,

$$C^{m+1} = A^m [A + (m + 1) B].$$

Solution:

We hall prove that $C^{m+1} = A^m [A + (m + 1) B]$, ...(1)

by mathematical induction.

For $m = 1$, from (1) we get $C^2 = A [A + 2B]$ ...(2)

Also $C = A + B$, given

$\therefore$ $C^2 = (A + B)^2 = (A + B)(A + B)$

$= A^2 + BA + AB + B^2$,

$= A^2 + 2\,AB$, since $AB = BA$, $B^2 = O$ (given)

$\Rightarrow$ $C^2 = A(A + 2B)$, which is the same as (2).

Hence (1) is true for $m = 1$.

Let us now assume that (1) holds when $m = k$

i.e., $C^{k+1} = A^k [A + (k + 1) B]$ ...(3)

Now $C^{k+1} = C^{k+1}\, C$, by

$= A^k [A + (k + 1) B].(A + B)$, from (3) and $C = A + B$ (given)

$$\Rightarrow \quad C^k = A^k[A(A + B) + (k + 1) B(A + B)]$$

$$= A^k[A^2 + AB + (k + 1) BA + (k + 1) B^2]$$

$$= A^k[A^2 + AB + (k + 1) AB], \quad \because BA = AB,\ B^2 = O$$

$$= A^k[A^2 + (1 + k + 1) AB],$$

$$= A^k.A [A + \{(k + 1) + 1\} B]$$

$$\Rightarrow \quad C^{k+2} = A^{k+1} [A + \{(k + 1) + 1\} B].$$

Hence (1) is true for $m = k + 1$ provided (3) is true *i.e.,* for $m = k$. Also we have shown that (1) true for $m = 1$, so it is true for $m = 1 +$ *i.e.,* $m = 2$ and so on. Hence (1) is true for all positive integral values of m.

Hence proved.

Example 21(b):

Show that

$$\begin{bmatrix} \cos\theta & -\sin\theta \\ \sin\theta & \cos\theta \end{bmatrix} = \begin{bmatrix} 1 & -\tan\frac{1}{2}\theta \\ \tan\frac{1}{2}\theta & 1 \end{bmatrix} \begin{bmatrix} 1 & \tan\frac{1}{2}\theta \\ -\tan\frac{1}{2}\theta & 1 \end{bmatrix}^{-1}$$

Solution:

We have

$$\begin{bmatrix} \cos\theta & -\sin\theta \\ \sin\theta & \cos\theta \end{bmatrix} \times \begin{bmatrix} 1 & \tan\frac{1}{2}\theta \\ -\tan\frac{1}{2}\theta & 1 \end{bmatrix}$$

$$= \begin{bmatrix} \cos\theta + \sin\theta\tan\frac{1}{2}\theta & \cos\theta\tan\frac{1}{2}\theta - \sin\theta \\ \sin\theta - \cos\theta\tan\frac{1}{2}\theta & \sin\theta\tan\frac{1}{2}\theta + \cos\theta \end{bmatrix}$$

$$= \begin{bmatrix} \dfrac{\cos\theta\cos\frac{1}{2}\theta + \sin\theta\sin\frac{1}{2}\theta}{\cos\frac{1}{2}\theta} & \dfrac{\cos\theta\sin\frac{1}{2}\theta - \sin\theta\cos\frac{1}{2}\theta}{\cos\frac{1}{2}\theta} \\ \dfrac{\sin\theta\cos\frac{1}{2}\theta - \cos\theta\sin\frac{1}{2}\theta}{\cos\frac{1}{2}\theta} & \dfrac{\cos\theta\sin\frac{1}{2}\theta + \cos\theta\cos\frac{1}{2}\theta}{\cos\frac{1}{2}\theta} \end{bmatrix}$$

$$= \frac{1}{\cos\frac{1}{2}\theta}\begin{bmatrix} \cos\theta\cos\frac{1}{2}\theta + \sin\theta\sin\frac{1}{2}\theta & \cos\theta\sin\frac{1}{2}\theta - \sin\theta\cos\frac{1}{2}\theta \\ \sin\theta\cos\frac{1}{2}\theta - \cos\theta\sin\frac{1}{2}\theta & \cos\theta\cos\frac{1}{2}\theta + \sin\theta\sin\frac{1}{2}\theta \end{bmatrix}$$

$$= \left(\sec\frac{1}{2}\theta\right)\begin{bmatrix} \cos\left(\theta - \frac{1}{2}\theta\right) & -\sin\left(\theta - \frac{1}{2}\theta\right) \\ \sin\left(\theta - \frac{1}{2}\theta\right) & \cos\left(\theta - \frac{1}{2}\theta\right) \end{bmatrix}$$

$$= \left(\sec\frac{1}{2}\theta\right)\begin{bmatrix} \cos\frac{1}{2}\theta & -\sin\frac{1}{2}\theta \\ \sin\frac{1}{2}\theta & \cos\frac{1}{2}\theta \end{bmatrix}$$

$$= \begin{bmatrix} \cos\frac{1}{2}\theta\sec\frac{1}{2}\theta & -\sin\frac{1}{2}\theta\sec\frac{1}{2}\theta \\ \sin\frac{1}{2}\theta\sec\frac{1}{2}\theta & \cos\frac{1}{2}\theta\sec\frac{1}{2}\theta \end{bmatrix}.$$

$$= \begin{bmatrix} 1 & -\tan\frac{1}{2}\theta \\ \tan\frac{1}{2}\theta & 1 \end{bmatrix}$$

$$\Rightarrow \begin{bmatrix} \cos\theta & -\sin\theta \\ \sin\theta & \cos\theta \end{bmatrix} = \begin{bmatrix} 1 & -\tan\frac{1}{2}\theta \\ \tan\frac{1}{2}\theta & 1 \end{bmatrix}\begin{bmatrix} 1 & \tan\frac{1}{2}\theta \\ -\tan\frac{1}{2}\theta & 1 \end{bmatrix}^{-1}$$

Hence proved

Example 21(c):

If A and B be n-rowed square matrices, then show that

1. $(A + B)^2 = A^2 + AB + BA + B^2$;
2. $(A + B)(A - B) = A^2 - AB + BA - B^2$;
3. $(A - B)(A + B) = A^2 + AB - BA - B^2$; *and*
4. $(A - B)^2 = A^2 - AB - BA + B^2$.

Solution:

As A and B are n-rowed square matrices therefore A + B and A – B are also n-rowed square matrices and as such distributive law is true.

1. $(A + B)^2 = (A + B) \times (A + B)$
$= (A + B) A + (A + B) B$, by distributive law
$= AA + BA + AB + BB$, by distributive law
$= A^2 + BA + AB + B^2$.

2. $(A + B) (A - B)$
$= (A + B) A + (A + B) (-B)$, by distributive law
$= AA + BA + A (-B) + B (-B)$, by distributive law
$= A^2 + BA - AB - B^2$.

3. $(A - B) (A + B) = (A - B) A + (A - B) B$, by distributive law
$= AA + BA + AB - BB$, by distributive law
$= A^2 - BA + AB - B^2$.

4. $(A - B)^2 = (A - B). (A - B)$
$= AA + A (-B) + (-B) A + (-B) (-B)$, by distributive law
$= A^2 - AB - BA + B^2$. **Hence proved.**

Example 21(d):

Show that if A, B, C are matrices, such that A (BC) is defined, then (AB) C is also defined and A (BC) = (AB) C.

Solution:

Since A (BC) is defined so the matrices A, B, C are conformable to multiplication and we can take $A = [a_{ij}]$, $B = [b_{jk}]$ and $C = [c_{kl}]$, where A, B, C are $m \times n$, $n \times p$, $p \times q$ matrices.

Then $AB = [a_{ij}] [b_{jk}]$ is a $m \times p$ matrix

i.e., (i, k)th element of the product $AB = \sum_{j=1}^{n} a_{ij} b_{jk}$ **(Note)**

Similarly (j, *l*)th element of the product $BC = \sum_{k=1}^{p} b_{jk} c_{kl}$ **(Note)**

Also (AB) C is the product of an $m \times p$ and a $p \times q$ matrices and so is conformable to multiplication, hence defined.

$\therefore$ (i, *l*)th element in the product of (AB) and C

= sum of products of corresponding elements in the ith row of AB and *l*th column of C with k common

$$= \sum_{k=1}^{p} \left[\left(\sum_{j=1}^{n} a_{ij} b_{jk} \right) c_{kl} \right]$$ **(Note)**

$$= \sum_{k=1}^{p}\sum_{j=1}^{n} a_{ij}\, b_{jk}\, c_{kl} \qquad ...(1)$$

Again (i, l)th element in the product of A and (BC)

= sum of products of cor corresponding elements in the ith row of A and lth column of (BC)

$$= \sum_{j=1}^{n} a_{ij} \sum_{k=1}^{p} b_{jk}\, c_{kl} \qquad \textbf{(Note)}$$

$$= \sum_{k=1}^{p}\ \sum_{j=1}^{n} a_{ij}\, b_{jk}\, c_{kl} \qquad ...(2)$$

∴ From (1) and (2) we conclude that (AB) C = A (BC).

Example 22:

If $A = \begin{bmatrix} 2 & 0 & 0 \\ 0 & 2 & 0 \\ 0 & 0 & 2 \end{bmatrix}$ *and* $B = \begin{bmatrix} x_1 & y_1 & z_1 \\ x_2 & y_2 & z_2 \\ x_3 & y_3 & z_3 \end{bmatrix}$ *then prove that AB = 2B.*

Solution:

$$AB = \begin{bmatrix} 2 & 0 & 0 \\ 0 & 2 & 0 \\ 0 & 0 & 2 \end{bmatrix} \times \begin{bmatrix} x_1 & y_1 & z_1 \\ x_2 & y_2 & z_2 \\ x_3 & y_3 & z_3 \end{bmatrix}$$

$$= \begin{bmatrix} 2x_1 + 0 + 0 & 2y_1 + 0 + 0 & 2z_1 + 0 + 0 \\ 0 + 2x_2 + 0 & 0 + 2y_2 + 0 & 0 + 2z_2 + 0 \\ 0 + 0 + 2x_3 & 0 + 0 + 2y_3 & 0 + 0 + 2z_3 \end{bmatrix}$$

$$= \begin{bmatrix} 2x_1 & 2y_1 & 2z_1 \\ 2x_2 & 2y_2 & 2z_2 \\ 2x_3 & 2y_3 & 2z_3 \end{bmatrix} = 2 \begin{bmatrix} x_1 & y_1 & z_1 \\ x_2 & y_2 & z_2 \\ x_3 & y_3 & z_3 \end{bmatrix} = 2B$$

Example 23:

Prove that the product of two matrices

$$\begin{bmatrix} \cos^2\theta & \cos\theta\sin\theta \\ \cos\theta\sin\theta & \sin^2\theta \end{bmatrix} \text{ and } \begin{bmatrix} \cos^2\phi & \cos\phi\sin\phi \\ \cos\phi\sin\phi & \sin^2\phi \end{bmatrix}$$

is zero when θ aznd ϕ differ by an odd multiple of $\frac{1}{2}\pi$.

Solution:

The required product

$$= \begin{bmatrix} \cos^2\theta\cos^2\phi + \cos\theta\sin\theta\cos\phi\sin\phi & \cos^2\theta\cos\phi + \cos\theta\sin\theta\cos\phi\sin^2\phi \\ \cos\theta\sin\theta\cos^2\phi + \sin^2\theta\cos\phi\sin\phi & \cos\theta\sin\theta\cos\phi\sin\phi + \sin^2\theta\sin\phi \end{bmatrix}$$

$$= \begin{bmatrix} \cos\theta\cos\phi(\cos\theta\cos\phi + \sin\theta\sin\phi) & \cos\theta\sin\phi(\cos\theta\cos\phi + \sin\theta\sin\phi) \\ \sin\theta\cos\phi(\cos\theta\cos\phi + \sin\theta\sin\phi) & \sin\theta\sin\phi(\cos\theta\cos\phi + \sin\theta\sin\phi) \end{bmatrix}$$

$$= \begin{bmatrix} \cos\theta\cos\phi\cos(\theta\sim\phi) & \cos\theta\sin\phi\cos(\theta\sim\phi) \\ \sin\theta\cos\phi\cos(\theta\sim\phi) & \sin\theta\sin\phi\cos(\theta\sim\phi) \end{bmatrix}$$

If $\theta \sim \phi$ = an odd multiple of $\frac{1}{2}\pi$, then $\cos(\theta \sim \phi) = 0$ and consequently the above product is zero (*i.e.*, the null matrix of order 2×2).

Example 24(a):

If $I = \begin{bmatrix} 1 & 0 \\ 0 & 1 \end{bmatrix}$ *and* $E = \begin{bmatrix} 0 & 1 \\ 0 & 0 \end{bmatrix}$, *Prove that*

$$(aI + bE)^3 = a^3I + 3a^2bE.$$

Solution:

$$aI + bE = a\begin{bmatrix} 1 & 0 \\ 0 & 1 \end{bmatrix} + b\begin{bmatrix} 0 & 1 \\ 0 & 0 \end{bmatrix}$$

$$= \begin{bmatrix} a & 0 \\ 0 & a \end{bmatrix} + \begin{bmatrix} 0 & b \\ 0 & 0 \end{bmatrix} = \begin{bmatrix} a+0 & 0+b \\ 0+0 & a+0 \end{bmatrix}$$

$$= \begin{bmatrix} a & b \\ 0 & a \end{bmatrix} = B \text{ (say)}$$

$\therefore \quad (aI + bE)^2 = B^2 = \begin{bmatrix} a & b \\ 0 & a \end{bmatrix} \times \begin{bmatrix} a & b \\ 0 & a \end{bmatrix}$

$$= \begin{bmatrix} a.a + b.0 & a.b + b.a \\ 0.a + a.0 & 0.b + a.a \end{bmatrix} = \begin{bmatrix} a^2 & 2ab \\ 0 & a^2 \end{bmatrix}$$

$\therefore \quad (aI + BE)^3 = B^3 = B^2B$ **(Note)**

$$= \begin{bmatrix} a^2 & 2ab \\ 0 & a^2 \end{bmatrix} \times \begin{bmatrix} a & b \\ 0 & a \end{bmatrix}$$

$$= \begin{bmatrix} a^2.a + 2ab.0 & a^2.b + 2ab.a \\ 0.a + a^2.0 & 0.b + a^2.a \end{bmatrix} = \begin{bmatrix} a^3 & 3a^2b \\ 0 & a^3 \end{bmatrix} \quad ...(1)$$

Now $a^3I + 3a^2bE = a^2 \begin{bmatrix} 1 & 0 \\ 0 & 1 \end{bmatrix} + 3a^2b \begin{bmatrix} 0 & 1 \\ 0 & 0 \end{bmatrix}$

$$= \begin{bmatrix} a^3 & 0 \\ 0 & a^3 \end{bmatrix} + \begin{bmatrix} 0 & 3a^2b \\ 0 & 0 \end{bmatrix}$$

$$= \begin{bmatrix} a^3 + 0 & 0 + 3a^2b \\ 0 + 0 & a^3 + 0 \end{bmatrix} = \begin{bmatrix} a^3 & 3a^2b \\ 0 & a^2 \end{bmatrix} \quad ...(2)$$

$\therefore$ From (1) and (2) we get $(aI + bE)^3 = a^3I + 3a^2bE$.

Example 24(b):

If $A = \begin{bmatrix} 1 & -1 \\ -1 & 1 \end{bmatrix}$, *then show that* $A^2 = 2A$ *and* $A^3 = 4A$.

Solution:

Given $A = \begin{bmatrix} 1 & -1 \\ -1 & 1 \end{bmatrix}$ ...(1)

$\therefore \quad A^2 = A \cdot A = \begin{bmatrix} 1 & -1 \\ -1 & 1 \end{bmatrix} \times \begin{bmatrix} 1 & -1 \\ -1 & 1 \end{bmatrix}$

$$= \begin{bmatrix} 1.1 + (-1).(-1) & 1.(-1) + (-1).1 \\ (-1).1 + 1.(-1) & (-1).(-1) + 1.1 \end{bmatrix} = \begin{bmatrix} 2 & -2 \\ -2 & 2 \end{bmatrix}$$

$$= 2 \begin{bmatrix} 1 & -1 \\ -1 & 1 \end{bmatrix} = 2A, \text{ from (1)} \qquad ...(2)$$

Again $\quad A^3 = A \cdot A^2 = A \cdot (2A)$, from (2)

$= 2A \cdot A = 2A^2 = 2\ (2A)$, from (2)

$= 4A.$ **Hence proved.**

Example 24(c):

If A denoteds the matrix $\begin{bmatrix} a & b \\ c & d \end{bmatrix}$, *prove that* $A^2-(a+d)A+(ad-bc)\,I = O.$

Solution:

$$A^2 = \begin{bmatrix} a & b \\ c & d \end{bmatrix} \times \begin{bmatrix} a & b \\ c & d \end{bmatrix}$$

$$= \begin{bmatrix} a.a + b.c & a.b + b.d \\ c.a + d.c & c.d + d.d \end{bmatrix} = \begin{bmatrix} a^2 + bc & b(a + d) \\ c(a + d) & cb + d^2 \end{bmatrix}$$

$\therefore \quad A^2 - (a + d)\ A + (ad - dc)\ I$

$$= \begin{bmatrix} a^2 + bc & b(a+d) \\ c(a+d) & cb + d^2 \end{bmatrix} - (a + d) \begin{bmatrix} a & b \\ c & d \end{bmatrix} + (ad - bc) \begin{bmatrix} 1 & 0 \\ 0 & 1 \end{bmatrix}$$

$$= \begin{bmatrix} a^2 + bc & b(a+d) \\ c(a+d) & cb + d^2 \end{bmatrix} + \begin{bmatrix} -a(a+d) & -b(a+d) \\ -c(a+d) & -d(a+d) \end{bmatrix}$$

$$+ \begin{bmatrix} ad - bc & 0 \\ 0 & ad - bc \end{bmatrix}$$

$$= \begin{bmatrix} a^2 + bc - a(a+d) + ad - bc & b(a+d) - b(a+d) + 0 \\ c(c+d) - c(a+d) + 0 & cb + d^2 - d(a+d) + ad - bc \end{bmatrix}$$

$$= \begin{bmatrix} 0 & 0 \\ 0 & 0 \end{bmatrix} = O, \text{ where O is the } 2 \times 2 \text{ null matrix.}$$

Example 24(d):

Show that $\begin{bmatrix} \cos\theta & -\sin\theta \\ \sin\theta & \cos\theta \end{bmatrix}^n = \begin{bmatrix} \cos n\theta & -\sin n\theta \\ \sin n\theta & \cos n\theta \end{bmatrix}$ *where n is a positive integer.*

Solution:

Let $A = \begin{bmatrix} \cos\theta & -\sin\theta \\ \sin\theta & \cos\theta \end{bmatrix}$ ...(1)

Then $(A^2) = A.A = \begin{bmatrix} \cos\theta & -\sin\theta \\ \sin\theta & \cos\theta \end{bmatrix} \times \begin{bmatrix} \cos\theta & -\sin\theta \\ \sin\theta & \cos\theta \end{bmatrix}$

$$= \begin{bmatrix} \cos^2\theta - \sin^2\theta & -\sin\theta\cos\theta - \sin\theta\cos\theta \\ \sin\theta\cos\theta + \sin\theta\cos\theta & -\sin^2\theta + \cos^2\theta \end{bmatrix}$$

$\Rightarrow \quad A^2 = \begin{bmatrix} \cos 2\theta & -\sin 2\theta \\ \sin 2\theta & \cos 2\theta \end{bmatrix}$ ...(2)

Similarly $(A)^3 = (A)^2.\ A$

$$= \begin{bmatrix} \cos 2\theta & -\sin 2\theta \\ \sin 2\theta & \cos 2\theta \end{bmatrix} \times \begin{bmatrix} \cos\theta & -\sin\theta \\ \sin\theta & \cos\theta \end{bmatrix}, \text{ from (1)}$$

$$= \begin{bmatrix} \cos 2\theta\cos\theta - \sin 2\theta\sin\theta & -\cos 2\theta\sin\theta - \sin 2\theta\cos\theta \\ \sin 2\theta\cos\theta + \cos 2\theta\sin\theta & -\sin 2\theta\sin\theta + \cos 2\theta\cos\theta \end{bmatrix}$$

$$= \begin{bmatrix} \cos(2\theta+\theta) & -\sin(2\theta+\theta) \\ \sin(2\theta+\theta) & \cos(2\theta+\theta) \end{bmatrix}$$

$\Rightarrow \quad (A)^3 = \begin{bmatrix} \cos 3\theta & -\sin 3\theta \\ \sin 3\theta & \cos 3\theta \end{bmatrix}$ ...(3)

In the light of (1), (2) and (3) let us assume that

$(A)^n = \begin{bmatrix} \cos n\theta & -\sin n\theta \\ \sin n\theta & \cos n\theta \end{bmatrix}$ ...(4)

Now $(A)^{n+1} = (A)^n \cdot (A)$

$$= \begin{bmatrix} \cos n\theta & -\sin n\theta \\ \sin n\theta & \cos n\theta \end{bmatrix} \times \begin{bmatrix} \cos\theta & -\sin\theta \\ \sin\theta & \cos\theta \end{bmatrix}$$

$$= \begin{bmatrix} \cos n\theta \cos\theta - \sin n\theta \sin\theta & -\cos n\theta \sin\theta - \sin n\theta \cos\theta \\ \sin n\theta \cos\theta + \cos n\theta \sin\theta & -\sin n\theta \sin\theta + \cos n\theta \cos\theta \end{bmatrix}$$

$$= \begin{bmatrix} \cos(n\theta+\theta) & -\sin(n\theta+\theta) \\ \sin(n\theta+\theta) & \cos(n\theta+\theta) \end{bmatrix} = \begin{bmatrix} \cos(n+1)\theta & -\sin(n+1)\theta \\ \sin(n+1)\theta & \sin(n+1)\theta \end{bmatrix}$$

i.e., (4) holds for n + 1 if it is true for n.

We have already proved in (2) and (3) that (4) holds for n = 2 and 3. Hence (4) holds for all positive integral values of n.

i.e., $(A)^n = \begin{bmatrix} \cos\theta & -\sin\theta \\ \sin\theta & \cos\theta \end{bmatrix}^n \begin{bmatrix} \cos n\theta & -\sin n\theta \\ \sin n\theta & \sin n\theta \end{bmatrix}$ **Hence proved.**

Example 25:

If $A = \begin{bmatrix} 3 & -4 \\ 1 & -1 \end{bmatrix}$, *show that* $A^n = \begin{bmatrix} 1+2n & -4n \\ n & 1-2n \end{bmatrix}$

Solution:

$$A^2 = A\cdot A = \begin{bmatrix} 3 & -4 \\ 1 & -1 \end{bmatrix} \times \begin{bmatrix} 3 & -4 \\ 1 & -1 \end{bmatrix}$$

$$= \begin{bmatrix} 3.3 - 4(1) & 3.(-4) - 4.(-1) \\ 1.3 - 1.(1) & 1.(-4) - 1.(-1) \end{bmatrix} = \begin{bmatrix} 5 & -8 \\ 2 & -3 \end{bmatrix}$$

$$= \begin{bmatrix} 1+2(2) & -4n(2) \\ (2) & 1-2(2) \end{bmatrix}$$

$= A^n$, when n = 2 **(Note)**

$\therefore \quad A^n = \begin{bmatrix} 1+2n & -4n \\ n & 1-2n \end{bmatrix}$ holds when n = 2

Now $A^{n+1} = A^n.A$

$$= \begin{bmatrix} 1+2n & -4n \\ n & 1-2n \end{bmatrix} . \begin{bmatrix} 3 & -4 \\ 1 & -1 \end{bmatrix}$$

$$= \begin{bmatrix} (1+2n).3 - 4n(1) & (1+2n)(-4) - 4n(-1) \\ n.3 + (1-2n)(1) & n(-4) + (1-2n)(-1) \end{bmatrix}$$

$$= \begin{bmatrix} 3+2n & -4-4n \\ 1+n & -1-2n \end{bmatrix} = \begin{bmatrix} 1+2(n+1) & -4(n+1) \\ (n+1) & 1-2(n+1) \end{bmatrix}$$

i.e., $A^n = \begin{bmatrix} 1+2n & -4n \\ n & 1-2n \end{bmatrix}$ holds for n + 1.

Also we have shown above that it holds for n = 2.

Hence it is true for all positive integral values of n.

Example 26:

Let $A = \begin{bmatrix} a & b \\ 0 & 1 \end{bmatrix}$*, where* $a \neq 0$*. Show that for* $n \geq 0$*,*

$$A^n = \begin{bmatrix} a^n & \dfrac{b(a^n-1)}{(a-1)} \\ 0 & 1 \end{bmatrix}$$

Solution:

$$A^2 = A.A = \begin{bmatrix} a & b \\ 0 & 1 \end{bmatrix} \times \begin{bmatrix} a & b \\ 0 & 1 \end{bmatrix}$$

$$= \begin{bmatrix} a.a + b.0 & a.b + b.1 \\ 0.a + 1.0 & 0.b + 1.1 \end{bmatrix} = \begin{bmatrix} a^2 & b(a+1) \\ 0 & 1 \end{bmatrix}$$

$$= \begin{bmatrix} a^2 & \dfrac{b(a^2-1)}{(a-1)} \\ 0 & 1 \end{bmatrix} = A^n, \text{ when } n = 2.$$ **(Note)**

$\therefore$ $A^n = \begin{bmatrix} a^n & b(a^n-1)/(a-1) \\ 0 & 1 \end{bmatrix}$ holds when n = 2.

Now $A^{n+1} = A^n. A$.

$$= \begin{bmatrix} a^n & b(a^n - 1)/(a-1) \\ 0 & 1 \end{bmatrix} \times \begin{bmatrix} a & b \\ 0 & 1 \end{bmatrix}$$

$$= \begin{bmatrix} a^n.a+0 & a^n.b+1.\{b(a^n - 1)/(a-1) \\ 0+0 & 0+1 \end{bmatrix}$$

$$= \begin{bmatrix} a^{n+1} & b\{a^n(a-1) + (a-1)\}/(a-1) \\ 0 & 1 \end{bmatrix}$$

$$= \begin{bmatrix} a^{n+1} & b(a^{n+1} - 1)/(a-1) \\ 0 & 1 \end{bmatrix}$$

$\therefore$ $A^n = \begin{bmatrix} a^n & b(a^n - 1)/(a-1) \\ 0 & 1 \end{bmatrix}$ holds for n = 1.

Also we have shown above that it holds for n = 2.

Hence it is true for all positive integral values of $n \geq 0$.

Hence proved.

Example 27(a):

Show that if $A = \begin{bmatrix} \cosh\theta & \sinh\theta \\ \sinh\theta & \cosh\theta \end{bmatrix}$, *then* $A^n = \begin{bmatrix} \cosh n\theta & \sinh n\theta \\ \sinh n\theta & \cosh n\theta \end{bmatrix}$

Solution:

Here $A^2 = A \cdot A$

$$= \begin{bmatrix} \cosh\theta & \sinh\theta \\ \sinh\theta & \cosh\theta \end{bmatrix} \times \begin{bmatrix} \cosh\theta & \sinh\theta \\ \sinh\theta & \cosh\theta \end{bmatrix}$$

$$= \begin{bmatrix} \cosh^2\theta + \sinh^2\theta & \cosh\theta\sinh\theta + \sinh\theta\cosh\theta \\ \text{Sinh}\,\theta\cosh\theta + \cosh\theta\sinh\theta & \sinh^2\theta + \cosh^2\theta \end{bmatrix}$$

$\Rightarrow$ $A^2 = \begin{bmatrix} \cosh 2\theta & \sinh 2\theta \\ \sinh 2\theta & \cosh 2\theta \end{bmatrix}$ ...(1)

Similarly $A^3 = A^2 \cdot A$

$$= \begin{bmatrix} \cosh 2\theta & \sinh 2\theta \\ \sinh 2\theta & \cosh 2\theta \end{bmatrix} \times \begin{bmatrix} \cosh \theta & \sinh \theta \\ \sinh \theta & \cosh \theta \end{bmatrix}, \text{ from (1)}$$

$$= \begin{bmatrix} \cosh 2\theta \cosh \theta + \sinh 2\theta \sinh \theta & \cosh 2\theta \sinh \theta + \sinh 2\theta \cosh \theta \\ \sinh 2\theta \cosh \theta + \cosh 2\theta \sinh \theta & \sinh 2\theta \sinh \theta + \cosh 2\theta \cosh \theta \end{bmatrix}$$

$$= \begin{bmatrix} \cosh (2\theta+\theta) & \sinh (2\theta+\theta) \\ \sinh (2\theta+\theta) & \cosh (2\theta+\theta) \end{bmatrix}$$

$$\Rightarrow \quad A^3 = \begin{bmatrix} \cosh 3\theta & \sinh 3\theta \\ \sinh 3\theta & \cosh 3\theta \end{bmatrix} \qquad \text{....(1)}$$

In the light of (1), (2) and the given value of A, let us assume that

$$A^n = \begin{bmatrix} \cosh n\theta & \sinh n\theta \\ \sinh n\theta & \cosh n\theta \end{bmatrix} \qquad \text{...(2)}$$

Now $A^{n+1} = A^n \cdot A$

$$= \begin{bmatrix} \cosh n\theta & \sinh n\theta \\ \sinh n\theta & \cosh n\theta \end{bmatrix} \times \begin{bmatrix} \cosh \theta & \sinh \theta \\ \sinh \theta & \cosh \theta \end{bmatrix}$$

$$= \begin{bmatrix} \text{csoh } n\theta \cosh \theta + \sinh n\theta \sinh \theta & \cosh n\theta \sinh \theta + \sinh n\theta \cosh \theta \\ \sinh n\theta \cosh \theta + \cosh n\theta \sinh \theta & \sinh n\theta \sinh \theta + \cosh n\theta \cosh \theta \end{bmatrix}$$

$$= \begin{bmatrix} \cosh (n\theta+\theta) & \sinh (n\theta+\theta) \\ \sinh (n\theta+\theta) & \cosh (n\theta+\theta) \end{bmatrix}$$

$$= \begin{bmatrix} \cosh (n+1)\theta & \sinh (n+1)\theta \\ \sinh (n+1)\theta & \cosh (n+1)\theta \end{bmatrix}$$

i.e., (3) holds for n + 1 if it is true for n.

Also from (1) and (2) we know that (3) holds for n = 2 and n = 3. Hence (3) holds for all positive integral values of n.

$$i.e.,\ A^n = \begin{bmatrix} \cosh n\theta & \sinh n\theta \\ \sinh n\theta & \cosh n\theta \end{bmatrix}$$

Hence proved

Example 27(b):

If $A = \begin{bmatrix} 2 & -1 \\ 0 & 1 \end{bmatrix}$ and $B = \begin{bmatrix} 1 & 0 \\ -1 & -1 \end{bmatrix}$, *show that*

$$(A + B)^2 = A^2 + AB + BA + B^2 \ ' \ A^2 + 2AB + B^2$$

Solution:

$$A^2 = \begin{bmatrix} 2 & -1 \\ 0 & 1 \end{bmatrix} \times \begin{bmatrix} 2 & -1 \\ 0 & 1 \end{bmatrix}$$

$$= \begin{bmatrix} 2.2 + (-1).0 & 2.(-1) - 1.1 \\ 0.2 + 1.0 & 0.(-1) + 1.1 \end{bmatrix} = \begin{bmatrix} 4 & -3 \\ 0 & -1 \end{bmatrix}$$

$$AB = \begin{bmatrix} 2 & -1 \\ 0 & 1 \end{bmatrix} \times \begin{bmatrix} 1 & 0 \\ -1 & -1 \end{bmatrix}$$

$$= \begin{bmatrix} 2.2 - 1(-1) & 2.0 - 1\,(-\,1) \\ 0.1 + 1(-1) & 0.0 + 1\,(-1) \end{bmatrix} = \begin{bmatrix} 3 & 1 \\ -1 & -1 \end{bmatrix};$$

$$BA = \begin{bmatrix} 1 & 0 \\ -1 & -1 \end{bmatrix} \times \begin{bmatrix} 0 & -1 \\ 0 & 1 \end{bmatrix}$$

$$= \begin{bmatrix} 1.2 + 0.0 & 1\,(-\,1) + 0.1 \\ -1.2 - 1.0 & -1\,(-1) - 1.1 \end{bmatrix} = \begin{bmatrix} 2 & -1 \\ -2 & 0 \end{bmatrix}$$

$$B^2 = \begin{bmatrix} 1 & 0 \\ -1 & -1 \end{bmatrix} \times \begin{bmatrix} 1 & 0 \\ -1 & -1 \end{bmatrix}$$

$$= \begin{bmatrix} 1.1 + 0\,(-1) & 1.0 + 0\,(-\,1) \\ -1.1 - 1\,(-1) & -1.0 - 1\,(-1) \end{bmatrix} = \begin{bmatrix} 1 & 0 \\ 0 & 1 \end{bmatrix}$$

$$A + B = \begin{bmatrix} 2 & -1 \\ 0 & 1 \end{bmatrix} + \begin{bmatrix} 1 & 0 \\ -1 & -1 \end{bmatrix}$$

$$= \begin{bmatrix} 2+1 & -1+0 \\ 0-1 & 1-1 \end{bmatrix} = \begin{bmatrix} 3 & -1 \\ -1 & 0 \end{bmatrix}$$

$$(A+B)^2 = \begin{bmatrix} 3 & -1 \\ -1 & 0 \end{bmatrix} \times \begin{bmatrix} 3 & -1 \\ -1 & 0 \end{bmatrix}$$

$$= \begin{bmatrix} 3.3-1(-1) & 3(-1)-1.0 \\ -1.3+0(-1) & -1(-1)+0.0 \end{bmatrix} = \begin{bmatrix} 10 & -3 \\ -3 & 1 \end{bmatrix} \quad ...(1)$$

Now $A^2 + AB + BA + B^2$

$$= \begin{bmatrix} 4 & -3 \\ 0 & 1 \end{bmatrix} + \begin{bmatrix} 3 & 1 \\ -1 & -1 \end{bmatrix} + \begin{bmatrix} 2 & -1 \\ -2 & 0 \end{bmatrix} + \begin{bmatrix} 1 & 0 \\ 0 & 1 \end{bmatrix}$$

$$= \begin{bmatrix} 4+3+2+1 & -3+1-1+0 \\ 0-1-2+0 & 1-1+0+1 \end{bmatrix} = \begin{bmatrix} 10 & -3 \\ -3 & 1 \end{bmatrix}$$

$= (A+B)^2$, from (i) **Hence proved.**

Also $A^2 + 2AB + B^2$

$$= \begin{bmatrix} 4 & -3 \\ 0 & 1 \end{bmatrix} + 2\begin{bmatrix} 3 & 1 \\ -1 & -1 \end{bmatrix} + \begin{bmatrix} 1 & 0 \\ 0 & 1 \end{bmatrix}$$

$$= \begin{bmatrix} 4 & -3 \\ 0 & 1 \end{bmatrix} + \begin{bmatrix} 6 & 2 \\ -2 & -2 \end{bmatrix} + \begin{bmatrix} 1 & 0 \\ 0 & 1 \end{bmatrix}$$

$$= \begin{bmatrix} 4+6+1 & -3+2+0 \\ 0-2+0 & 1-2+1 \end{bmatrix} = \begin{bmatrix} 11 & -1 \\ -2 & 0 \end{bmatrix} \neq (A+B)^2$$

Hence proved

Example 27(c):

Evaluate $A^2 - 4A - 5I$, where

$$A = \begin{bmatrix} 1 & 2 & 2 \\ 2 & 1 & 2 \\ 2 & 2 & 1 \end{bmatrix} \text{ and } I = \begin{bmatrix} 1 & 0 & 0 \\ 0 & 1 & 0 \\ 0 & 0 & 1 \end{bmatrix}$$

Solution:

$$A^2 = \begin{bmatrix} 1 & 2 & 2 \\ 2 & 1 & 2 \\ 2 & 2 & 1 \end{bmatrix} \times \begin{bmatrix} 1 & 2 & 2 \\ 2 & 1 & 2 \\ 2 & 2 & 1 \end{bmatrix}$$

$$= \begin{bmatrix} 1.1 + 2.2 + 2.2 & 1.2 + 2.1 + 2.2. & 1.2 + 2.2 + 2.1 \\ 2.1 + 1.2 + 2.2 & 2.2 + 1.1 + 2.2 & 2.2 + 1.2 + 2.1 \\ 2.1 + 2.2 + 1.2 & 2.2 + 2.1 + 1.2 & 2.2 + 2.2 + 1.1 \end{bmatrix}$$

$$= \begin{bmatrix} 1+4+4 & 2+2+4 & 2+4+2 \\ 2+2+4 & 4+1+4 & 4+2+2 \\ 2+4+2 & 4+2+2 & 4+4+2 \end{bmatrix} = \begin{bmatrix} 9 & 8 & 8 \\ 8 & 9 & 8 \\ 8 & 8 & 9 \end{bmatrix}$$

$\therefore$ $A^2 - 4A - 5I$

$$= \begin{bmatrix} 9 & 8 & 8 \\ 8 & 9 & 8 \\ 8 & 8 & 9 \end{bmatrix} - 4\begin{bmatrix} 1 & 2 & 2 \\ 2 & 1 & 2 \\ 2 & 2 & 1 \end{bmatrix} - 5\begin{bmatrix} 1 & 0 & 0 \\ 0 & 1 & 0 \\ 0 & 0 & 1 \end{bmatrix}$$

$$= \begin{bmatrix} 9 & 8 & 8 \\ 8 & 9 & 8 \\ 8 & 8 & 9 \end{bmatrix} + \begin{bmatrix} -4 & -8 & -8 \\ -8 & -4 & -8 \\ -8 & -8 & -4 \end{bmatrix} + \begin{bmatrix} -5 & 0 & 0 \\ 0 & -5 & 0 \\ 0 & 0 & -5 \end{bmatrix}$$

$$= \begin{bmatrix} 9-4-5 & 8-8+0 & 8-8+0 \\ 8-8+0 & 9-4-5 & 8-8+0 \\ 8-8+0 & 8-8+0 & 9-4-5 \end{bmatrix} = \begin{bmatrix} 0 & 0 & 0 \\ 0 & 0 & 0 \\ 0 & 0 & 0 \end{bmatrix} = O,$$

where O is the null matrix. **Ans.**

Example 27(d):

Let $f(x) = x^2 - 5x + 6$, find $f(A)$ if

$$A = \begin{bmatrix} 2 & 0 & 1 \\ 2 & 1 & 3 \\ 1 & -1 & 0 \end{bmatrix}$$

Solution:

$f(A) = A^2 - 5A + 6$

$= A^2 - 5A + 6I$, where $I = \begin{bmatrix} 1 & 0 & 0 \\ 0 & 1 & 0 \\ 0 & 0 & 1 \end{bmatrix}$

Now proceed $\begin{bmatrix} 1 & -1 & -3 \\ -1 & -1 & -10 \\ -5 & 4 & 4 \end{bmatrix}$ **Ans.**

Example 28(a):

Show that if A is the matrix $\begin{bmatrix} 2 & -1 & 1 \\ -1 & 2 & -1 \\ 1 & -1 & 2 \end{bmatrix}$, *then* $A^3 - 6A^2 + 9A - 4I$ *= O, I being the 3 × 3 unit matrix and O being the 3 × 3 null matrix.*

Solution:

$$A^2 = \begin{bmatrix} 2 & -1 & 1 \\ -1 & 2 & -1 \\ 1 & -1 & 2 \end{bmatrix} \times \begin{bmatrix} 2 & -1 & 1 \\ -1 & 2 & -1 \\ 1 & -1 & 2 \end{bmatrix}$$

$$= \begin{bmatrix} 4+1+1 & -2-2-1 & 2+1+2 \\ -2-2-1 & 1+4+1 & -1-2-2 \\ 2+1+2 & -1-2-2 & 1+1+4 \end{bmatrix} = \begin{bmatrix} 6 & -5 & 5 \\ -5 & 6 & -5 \\ 5 & -5 & 6 \end{bmatrix}$$

$$\therefore \quad A^3 = A^2 \times A = \begin{bmatrix} 6 & -5 & 5 \\ -5 & 6 & -5 \\ 5 & -5 & 6 \end{bmatrix} \times \begin{bmatrix} 2 & -1 & 1 \\ -1 & 2 & -1 \\ 1 & -1 & 2 \end{bmatrix}$$

$$= \begin{bmatrix} 12+5+5 & -6-10-5 & 6+5+10 \\ -10-6-5 & 5+12+5 & -5-6-10 \\ 10+5+6 & -5-10-6 & 5+5+12 \end{bmatrix}$$

$$= \begin{bmatrix} 22 & -21 & 21 \\ -21 & 22 & -21 \\ 21 & -21 & 22 \end{bmatrix}$$

$$\therefore \quad A^3 - 6A^2 + 9A - 4I$$

$$= \begin{bmatrix} 22 & -21 & 21 \\ -21 & 22 & -21 \\ 21 & -21 & 22 \end{bmatrix} - 6 \begin{bmatrix} 6 & -5 & 5 \\ -5 & 6 & -5 \\ 5 & -5 & 6 \end{bmatrix}$$

$$+ 9 \begin{bmatrix} 2 & -1 & 1 \\ -1 & 2 & -1 \end{bmatrix} - 4 \begin{bmatrix} 1 & 0 & 0 \\ 0 & 0 & 0 \end{bmatrix}$$

$$= \begin{bmatrix} 22 & -21 & 21 \\ -21 & 22 & -21 \\ 21 & -21 & 22 \end{bmatrix} + \begin{bmatrix} -36 & 30 & -30 \\ 30 & -36 & 30 \\ -30 & 30 & -36 \end{bmatrix}$$

$$+ \begin{bmatrix} 18 & -9 & 9 \\ -9 & 18 & -9 \\ 9 & -9 & 18 \end{bmatrix} + \begin{bmatrix} -4 & 0 & 0 \\ 0 & -4 & 0 \\ 0 & 0 & -4 \end{bmatrix}$$

$$= \begin{bmatrix} 22+36+18-4 & -21+30-9+0 & 21-30+9+0 \\ -21+30-9+0 & 22-36+18-4 & -21+30-9+0 \\ 21-30+9+0 & -21+30-9+0 & 22+36+18-4 \end{bmatrix}$$

$$= \begin{bmatrix} 0 & 0 & 0 \\ 0 & 0 & 0 \\ 0 & 0 & 0 \end{bmatrix} = O$$

Hence proved

Example 28(b):

If $A = \begin{bmatrix} 0 & 1 \\ 1 & 1 \end{bmatrix}$ *and* $B = \begin{bmatrix} 0 & -1 \\ 1 & 0 \end{bmatrix}$, *show that* $(A + B)(A - B) \neq A^2 - B^2$

Solution:

$$A + B = \begin{bmatrix} 0 & 1 \\ 1 & 1 \end{bmatrix} - \begin{bmatrix} 0 & -1 \\ 1 & 0 \end{bmatrix} = \begin{bmatrix} 0+0 & 1-1 \\ 1+1 & 1+0 \end{bmatrix}$$

$$= \begin{bmatrix} 0 & 0 \\ 2 & 1 \end{bmatrix}$$

$$A - B = \begin{bmatrix} 0 & 1 \\ 1 & 1 \end{bmatrix} - \begin{bmatrix} 0 & -1 \\ 1 & 0 \end{bmatrix} = \begin{bmatrix} 0-0 & 1-(-1) \\ 1-1 & 1-0 \end{bmatrix}$$

$$= \begin{bmatrix} 0 & 2 \\ 0 & 1 \end{bmatrix}$$

$$\therefore \quad (A + B)(A - B) = \begin{bmatrix} 0 & 0 \\ 2 & 1 \end{bmatrix} \times \begin{bmatrix} 0 & 2 \\ 0 & 1 \end{bmatrix}$$

$$= \begin{bmatrix} 0.0 + 0.0 & 0.2 + 0.1 \\ 2.0 + 1.0 & 2.2 + 1.1 \end{bmatrix} = \begin{bmatrix} 0 & 0 \\ 0 & 5 \end{bmatrix}$$

$$A^2 = \begin{bmatrix} 0 & 1 \\ 1 & 1 \end{bmatrix} \times \begin{bmatrix} 0 & 1 \\ 1 & 1 \end{bmatrix} = \begin{bmatrix} 0.0 + 1.1 & 0.1 + 1.1 \\ 1.0 + 1.1 & 1.1 + 1.1 \end{bmatrix}$$

$$= \begin{bmatrix} 1 & 1 \\ 1 & 2 \end{bmatrix}$$

$$B^2 = \begin{bmatrix} 0 & -1 \\ 1 & 0 \end{bmatrix} \times \begin{bmatrix} 0 & -1 \\ 1 & 0 \end{bmatrix} = \begin{bmatrix} 0.0 - 1.1 & 0.1 - 1.0 \\ 1.0 - 0.1 & 1.1 + 0.0 \end{bmatrix}$$

$$= \begin{bmatrix} -1 & 0 \\ 0 & -1 \end{bmatrix}$$

$$\therefore \quad A^2 - B^2 = \begin{bmatrix} 1 & 1 \\ 1 & 2 \end{bmatrix} - \begin{bmatrix} -1 & 0 \\ 0 & -1 \end{bmatrix} = \begin{bmatrix} 1+1 & 1-0 \\ 1-0 & 2+1 \end{bmatrix}$$

$$= \begin{bmatrix} 2 & 1 \\ 1 & 3 \end{bmatrix}$$

Hence $(A + B)(A - B) \neq A^2 - B^2$.

Example 28(c):

If $A = \begin{bmatrix} 0 & -\tan \frac{1}{2}\alpha \\ \tan \frac{1}{2}\alpha & 0 \end{bmatrix}$ *and I is a unit matrix, then prove that I* $+ A = (I - A) \begin{bmatrix} \cos\alpha & -\sin\alpha \\ \sin\alpha & \cos\alpha \end{bmatrix}$

Solution:

$$I + A = \begin{bmatrix} 1 & 0 \\ 0 & 1 \end{bmatrix} + \begin{bmatrix} 0 & -\tan \frac{1}{2}\alpha \\ \tan \frac{1}{2}\alpha & 0 \end{bmatrix}$$

$$= \begin{bmatrix} 1+0 & 0-\tan \frac{1}{2}\alpha \\ 0+\tan \frac{1}{2}\alpha & 1+0 \end{bmatrix} = \begin{bmatrix} 1 & -\tan \frac{1}{2}\alpha \\ \tan \frac{1}{2}\alpha & 1 \end{bmatrix} \quad ...(1)$$

$$I - A = \begin{bmatrix} 1 & 0 \\ 0 & 2 \end{bmatrix} - \begin{bmatrix} 0 & -\tan\frac{1}{2}\alpha \\ \tan\frac{1}{2}\alpha & 0 \end{bmatrix}$$

$$= \begin{bmatrix} 1-0 & 0+\tan\frac{1}{2}\alpha \\ 0-\tan\frac{1}{2}\alpha & 1-0 \end{bmatrix} = \begin{bmatrix} 1 & \tan\frac{1}{2}\alpha \\ -\tan\frac{1}{2}\alpha & 1 \end{bmatrix}$$

$$\therefore \quad (I - A)\begin{bmatrix} \cos\alpha & -\sin\alpha \\ \sin\alpha & \cos\alpha \end{bmatrix}$$

$$= \begin{bmatrix} 1 & \tan\frac{1}{2}\alpha \\ -\tan\frac{1}{2}\alpha & 1 \end{bmatrix}\begin{bmatrix} \cos\alpha & -\sin\alpha \\ \sin\alpha & \cos\alpha \end{bmatrix}$$

$$= \begin{bmatrix} 1.\cos\alpha + \tan\frac{1}{2}\alpha.\sin\alpha & 1.(-\sin\alpha) + \tan\frac{1}{2}\alpha\cos\alpha \\ -\tan\frac{1}{2}\alpha\cos\alpha + 1.\sin\alpha & (\sin\alpha).\tan\frac{1}{2}\alpha + 1.\cos\alpha \end{bmatrix},$$

$$= \begin{bmatrix} \left(1-2\sin^2\frac{1}{2}a\right) + 2\sin^2\frac{1}{2}\alpha & -2\sin\frac{1}{2}\alpha\cos\frac{1}{2}\alpha + \tan\frac{1}{2}\alpha\cos\alpha \\ -\tan\frac{1}{2}\alpha\cos\alpha + 2\sin\frac{1}{2}\alpha\cos\frac{1}{2}\alpha & 2\sin^2\frac{1}{2}\alpha + \left(1+2\sin^2\frac{1}{2}\alpha\right) \end{bmatrix}$$

writing $\cos\alpha = 1 - 2\sin^2\frac{1}{2}\alpha$

$$= \begin{bmatrix} 1 & -2\tan\frac{1}{2}\alpha\cos^2\frac{1}{2}\alpha + \tan\frac{1}{2}\alpha\cos\alpha \\ -\tan\frac{1}{2}\alpha\cos\alpha + 2\tan\frac{1}{2}\alpha\cos^2\frac{1}{2}\alpha & 1 \end{bmatrix},$$

writiting $\sin \frac{1}{2}\alpha$ as $\tan \frac{1}{2}\alpha \cos \frac{1}{2}\alpha$

$$= \begin{bmatrix} 1 & -\tan \frac{1}{2}\alpha \, 2\cos^2 \frac{1}{2}\alpha - \left(2\cos^2 \frac{1}{2}\alpha - 1\right) \\ \tan \frac{1}{2}\alpha \left\{-\left(2\cos^2 \frac{1}{2}\alpha - 1\right) + 2\cos^2 \frac{1}{2}\alpha\right\} & 1 \end{bmatrix}$$

writing $\cos \alpha = 2 \cos^2 \frac{1}{2}\alpha - 1$

$$= \begin{bmatrix} 1 & -\tan \frac{1}{2}\alpha \\ \tan \frac{1}{2}\alpha & 1 \end{bmatrix} = I + A, \text{ from (1)}$$

Hence proved

Example 28(d):

In a development plan of a city, a contractor has taken a contract to construct certain houses for which he needs building materials like stones, sand etc. There are three firms A, B, C that can supply him these materials. At one time these firms A, B, C supplied him 40, 35 and 25 truck loads of stones and 10, 5 and 8 truck loads of sand respectively. If the cost of one truck load of stone and sand are Rs. 1,200 and Rs. 500 respectively, then find the total amount paid by the contractor to each of these firms A, B, C separately.

Solution:

The truck-loads of stone and sand supplied by the firms A, B, and C can be written in the form of a matrix A (say) given by

$$A = \begin{matrix} \text{Stone} \\ \text{Sand} \end{matrix} \overset{\begin{matrix} A & B & C \end{matrix}}{\begin{bmatrix} 40 & 35 & 25 \\ 10 & 5 & 8 \end{bmatrix}}, \text{ which is a } 2 \times 3 \text{ matrix.}$$

And the cost per truck of stone and sand can be given in the form of a matrix B (say), given by

$$B = \overset{\begin{matrix} \text{Stone} & \text{Sand} \end{matrix}}{[1200 \quad 500]}$$

The required total amount paid to each of the firms A, B and C are given by the product matrix BA. (Note here AB can not be calculated).

$$\text{Now BA} = [1200 \quad 500] \times \begin{bmatrix} 40 & 35 & 25 \\ 10 & 5 & 8 \end{bmatrix}$$

$$= [(1200 \times 40) + (500 \times 10) \ (1200 \times 35) + (500 \times 5) \ (1200 \times 25) + (500 \times 8)]$$

$$= [48000 + 5000 \quad 42000 + 2500 \quad 30000 + 4000]$$

$$= [53{,}000 \quad 44{,}500 \quad 34{,}000]$$

$\therefore$ The amounts paid to the firms A, B and C by the contractor are Rs. 53,000, Rs. 44,500 and Rs. 34,000 respectively. **Ans.**

Example 28(e):

A man buys 8 dozens of mangoes, 10 dozens of ampples and 4 dozens of bananas. Mangoes cost Rs. 18 per dozen, apples Rs. 9 per dozen and bananas Rs. 6 per dozen. Represent the quantities bought by a row matrix and the prices by a column matrix and hence obtain the total cost.

Solution:

The quantities bought are represented by 3 × 1 row matrix [8 10 4] and the prices are represented by 3 × 1 column matrix

$$\begin{bmatrix} 18 \\ 9 \\ 6 \end{bmatrix}$$

$\therefore$ The cost of fruits is a single number *i.e.*, 1 × 1 matrix given by the product matrix

$$[8 \quad 10 \quad 4] \times \begin{bmatrix} 18 \\ 9 \\ 6 \end{bmatrix}$$

i.e., $[(8 \times 18) + (10 \times 9) + (4 \times 6)$ *i.e.*, $[144 + 90 + 24]$ *i.e.*, $[258]$

$\therefore$ The required total cost = Rs. 258. **Ans.**

Example 28(f):

A manufacturer produces three products A, B, C which he sells in the market. Annual sale volumes are indicated as follows:

	Markets	*Products*	
	A	*B*	*C*
I	*8,000*	*10,000*	*15,000*
II	*10,000*	*2,000*	*20,000*

(1) If unit sale prices of A, B and C are Rs. 2.25, Rs. 1.50 and Rs. 1.25 respectively find the total revenue in each market with the help os matrices, (2) if the unit costs of the above three products are Rs. 1.60, Rs 1.20 and Rs. 0. 90 respectively, find the gross profit with the help of matrices.

Solution:

(1) The total revenue in each market is given by the product matrix.

$$[2.25 \quad 1.50 \quad 1.25] \times \begin{bmatrix} 8,00 & 10,000 \\ 10,000 & 2,000 \\ 15,000 & 20,000 \end{bmatrix}$$ **(Note)**

$= [(2, 25 \times 8,000) + (1.50 \times 10,000) + (1.25 \times 15,000)$

$(2.25 \times 10,000) + (1.50 \times 2,000) + (1.25 \times 20,000)]$

$= [18,000 + 15,000 + 18,750 \qquad 22,500 + 3,000 + 25,000]$

$= [51750 \quad 50500]$

$\therefore$ Total revenue from the market I = Rs. 51.750 and total revenue from the market II = Rs. 50,500. **Ans.**

(2) similarly the total cost of products which the manufacturer sells in the markets are given by the product matrix.

$$[1.60 \quad 1.20 \quad 0.90] \times \begin{bmatrix} 8,00 & 10,000 \\ 10,000 & 2,000 \\ 15,000 & 20,000 \end{bmatrix}$$

$= [(1.60 \times 8,000) + (1.20 \times 10,000) + (0.90 \times 15,000)$

$(1.60 \times 10,000) + (1,20 \times 2,000) + (0.90 \times 20,000)]$

$= [12,800 + 12,000 + 13,500 \qquad 16,500 + 2,400 + 18,000]$

$= [38,300 \quad 36,400]$

$\therefore$ Total cost of products which the manufacturer sells in the markets I and I are Rs. 38,300 and Rs. 36,400 respectively.

∴ Requited gross profit = (Total revenue received from both the markets) – (Total cost of products which the manufacturer sold in both the markets)

= (Rs. 51,750 + Rs. 50,500) – (Rs. 38,300 + Rs. 36,400)

= Rs. 102,250 – Rs. 74,700 = Rs. 27, 550. **Ans.**

Example 29:

A store has in stock 20 dozen shirts, 15 dozen trousers and 25 dozen pairs of socks. If the selling prices are Rs. 50 per shirt, Rs. 90 per trouser and Rs. 12 per pair of socks, then find the total amount the stove owner will get after selling all the items in the stock.

Solution:

The stock in the store can be written in the form of a row matrix A given by $A = [20 \times 12 \quad 15 \times 12 \quad 25 \times 12]$

⇒ $A = [240 \quad 180 \quad 300]$, which is a 1×3 matrix.

The prices can be written in the form of a column matrix B given by

$B = \begin{bmatrix} 50 \\ 90 \\ 12 \end{bmatrix}$, which is a 3×1 matrix.

The required amount is a single number *i.e.*, a matrix of order 1×1 and so the same can be obtained by multiplying the matrices A and B, since their product would be a 1×1 matrix.

$$\text{Now } AB = [240 \quad 180 \quad 300] \times \begin{bmatrix} 50 \\ 90 \\ 12 \end{bmatrix}$$

$$= [(240 \times 50) + (180 \times 90) + (300 \times 12)]$$

$$= [12000 + 16200 + 3600] = [31800]$$

∴ The required amount received by the store owner

= Rs. 31800.

Example 30:

A finance company has offices located in every division, every district and every taluka in a certain state in India. Assume that there are five divisions, thirty districts and 200 talukas in the state. Each office has one heeadclerk, one cashier, one clerk and one person. A divisional office has, in addition, one office superintendent, two clerks, one typist and one peon. A district office, has in addition, one clerk and one peon. The basic monthly salaries are as follows: office superintendent Rs. 500, Head clerk Rs. 200, cashier Rs. 175, clerks and typists Rs. 150 and person Rs 100. Using matrix notation find:

(1) the total number of posts of each kind in all the offices taken together,

(2) the total basic monthly salary bill of all the offices taken together.

Solution:

Let us use the symbols Div, Dis, Tal for division, district, taluka respectively and O, H, C, CL, T and P for office superintendent, Head clerk, cashier, clerk, typist and peon respectively.

Then the number of offices can be arranged as elements of a row matrix A (say) given by

$$\begin{array}{cccc} & \text{Div.} & \text{Dis.} & \text{Tal.} \\ A = (& 5 & 30 & 200) \end{array}$$

The composition of staff in various offices can be arranged in a 3 × 6 matrix B (say) given by

$$\begin{array}{cc} & \begin{array}{cccccc} O & H & C & Cl & T & P \end{array} \\ B = & \begin{bmatrix} 1 & 1 & 1 & 2+1 & 1 & 1+1 \\ 0 & 1 & 1 & 1+1 & 0 & 1+1 \\ 0 & 1 & 1 & 1 & 0 & 1 \end{bmatrix} \end{array}$$

The basic monthly salaries of various types of employees of these offices correspond to the elements of the column matrix C (say) given by

$$C = \begin{array}{cc} O & 500 \\ H & 200 \\ C & 175 \\ Cl & 150 \\ T & 150 \\ P & 100 \end{array}$$

(1) Total number of posts of each kind in all the offices are the elements of the product matrix AB

i.e., $[5 \quad 30 \quad 200] \times \begin{bmatrix} 1 & 1 & 1 & 3 & 1 & 2 \\ 0 & 1 & 1 & 2 & 0 & 2 \\ 0 & 1 & 1 & 1 & 0 & 1 \end{bmatrix}$ **(Note)**

i.e., $[5 + 0 + 0, \quad 5 + 30 + 200, \quad 5 + 30 + 200, \quad 15 + 60 + 200,$

$5 + 0 + 0, \quad 10 + 60 + 200]$

	O	H	C	Cl	T	P	
i.e., [5	235	235	275	5	270		**Ans.**

[*i.e.,* required number of posts in all the offices taken together are 5 office supdts, 235. Head clerks, 235 cashiers, 275 cashiers, 275 clerks, 5 typists and 270 persons.

(2) Total basic monthly salary bill of each kind of office are the elements of the product matrix BC

$$i.e., \begin{bmatrix} 1 & 1 & 1 & 3 & 1 & 2 \\ 0 & 1 & 1 & 2 & 0 & 2 \\ 0 & 1 & 1 & 1 & 0 & 1 \end{bmatrix} \times \begin{pmatrix} 500 \\ 200 \\ 175 \\ 150 \\ 150 \\ 100 \end{pmatrix}$$

$$= \begin{bmatrix} (1\times 500)+(1\times 200)+(1\times 175)+(3\times 150)+(1\times 150)+(2\times 150) \\ (0\times 500)+(1\times 200)+(1\times 175)+(2\times 150)+(0\times 150)+(2\times 150) \\ (0\times 500)+(1\times 200)+(1\times 175)+(1\times 150)+(0\times 150)+(1\times 150) \end{bmatrix}$$

$$= \begin{bmatrix} 500+200+175+450+150+200 \\ 0+200+175+300+0+200 \\ 0+200+175+150+0+100 \end{bmatrix} = \begin{bmatrix} 1675 \\ 875 \\ 625 \end{bmatrix}$$

[*i.e.,* The total basic monthly salary bill of each divisional, district and taluka offices are Rs. 1675, Rs. 875 and Rs. 625 respectively].

(3) Total basic monthly salary bill of all the offices (*i.e.,* of five divisional, 30 district and 200 taluka offices) is the element of the product matrix ABC

$$i.e., \quad [5 \quad 30 \quad 200] \times \begin{bmatrix} 1675 \\ 875 \\ 625 \end{bmatrix}$$ **(Note)**

i.e., [(5 × 1675) + (30 × 875) + (200 × 625)

i.e., [8375 + 26250 + 125000] *i.e.,* [159625] **Ans.**

i.e., [total basic monthly salary bill of all the offices taken together is Rs. 159,625].
Ans.

Example 31(a):

A trust fund has Rs. 50.000 that is to be invested into two types of bonds. The first bond pays 5% interest per year and the second bond pays 6% interest per year. Using matrix multiplication, determine how to divide Rs. 50.000 among the two types of bonds so as to obtain an annual total interest of Rs. 2780.

Solution:

Let Rs. 50,000 be divided into two parts Rs. x and Rs. (50,000 – x) out of which first part is invested in first type of bonds and the second part is invested in second type of bonds.

The values of these bonds can be written in the form of a row matrix A given by A = [x 50,000 – x], which is a 1 × 2 matrix.

And the amounts received as interest per rupee annually fıom these two types of bonds can be written in the form of a column matrix B given by

$B = \begin{bmatrix} 5/100 \\ 6/100 \end{bmatrix}$, which is a 2 × 1 matrix.

Here the interest has been calculated per rupee annually.

Now, the interest to be obtained annually is a single number *i.e.,* a matrix of order 1 × 1 and the same can be obtained by the product matrix AB, since this product matrix would be a 1 × 1 matrix.

(Note)

$$\text{Here } AB = [x \quad 50{,}000 -] \times \begin{bmatrix} 5/100 \\ 6/100 \end{bmatrix}$$

$$= \left[x.\frac{5}{100} + (50{,}000 - x).\frac{6}{100} \right]$$

$$= \left[300 - \frac{x}{100} \right]$$

Also we are given that the annual interest = Rs. 2,780.

$\therefore$ We must have $\left[300 - \dfrac{x}{100} \right] = [2780]$ **(Note)**

$$\Rightarrow \quad 3000 - \frac{x}{100} = 2780$$

$$\Rightarrow \quad x = (3000 - 2780) \times 100$$

$$\Rightarrow \quad x = 220 \times 100 = 22{,}000$$

Hence the required amounts are

Rs. 22,000 and Rs. (50,000 – 22,000) *i.e.,*

Rs. 22, 000 and Rs. 28,000. **Ans.**

Example 31(b):

If A and B are both skew-symmetric matrices of same ofer such that AB = Ba, then show that AB is symmetric.

Solution:

If A and B are both skew-symmetric matrices,

then A = – A' and B = – B' ...(1)

Also given that AB = BA

= (–B') (–A'), from (1)

= B' A = (AB)'

⇒ AB = (AB)' i.e., l AB is a symmetric matrix. **Hence proved.**

Example 31(c):

If A is a symmetic matrix, then whow that kA is also symmetric for any scalar k.

Solution:

Here (kA)' = kA', = kA, A' = A, A being symmetic

Hence kA is symmetric, if **A** is so.

Example 31(d):

Find the symmetric and skew-symmetric parts of the matrix

$$A = \begin{bmatrix} 1 & 2 & 4 \\ 6 & 8 & 1 \\ 3 & 5 & 7 \end{bmatrix}$$

Solution:

Here A' = transpose of A

$$= \begin{bmatrix} 1 & 6 & 3 \\ 2 & 8 & 5 \\ 4 & 1 & 7 \end{bmatrix}$$

The symmetric part of **A** = $\frac{1}{2}$ **(A + A')**

$$= \frac{1}{2}\begin{bmatrix} 1 & 2 & 4 \\ 6 & 8 & 1 \\ 3 & 5 & 7 \end{bmatrix} + \begin{bmatrix} 1 & 6 & 3 \\ 2 & 8 & 5 \\ 4 & 1 & 7 \end{bmatrix}$$

$$= \frac{1}{2}\begin{bmatrix} 1+1 & 2+6 & 4+3 \\ 6+2 & 8+8 & 1+5 \\ 3+4 & 5+1 & 7+7 \end{bmatrix} = \frac{1}{2}\begin{bmatrix} 2 & 8 & 7 \\ 8 & 16 & 6 \\ 7 & 6 & 14 \end{bmatrix}$$

$$= \begin{bmatrix} 1 & 4 & \frac{7}{8} \\ 4 & 8 & 3 \\ \frac{7}{8} & 3 & 7 \end{bmatrix}$$ **Ans.**

And the skew-symmetric part of A = $\frac{1}{2}$ (A – A')

$$= \frac{1}{2}\begin{bmatrix} 1 & 2 & 4 \\ 6 & 8 & 1 \\ 3 & 5 & 7 \end{bmatrix} - \begin{bmatrix} 1 & 6 & 3 \\ 2 & 8 & 5 \\ 4 & 1 & 7 \end{bmatrix}$$

$$= \frac{1}{2}\begin{bmatrix} 1-1 & 2-6 & 4-3 \\ 6-2 & 8-8 & 1-5 \\ 3-4 & 5-1 & 7-7 \end{bmatrix} = \frac{1}{2}\begin{bmatrix} 0 & -4 & 1 \\ 4 & 0 & -4 \\ -1 & 4 & 0 \end{bmatrix}$$

$$= \begin{bmatrix} 0 & -2 & \frac{1}{2} \\ 2 & 0 & -2 \\ -\frac{1}{2} & 2 & 0 \end{bmatrix}$$ **Ans.**

Example 32:

If A is any square matrix, show that AA' is a symmetric matrix.

Solution:

(A'A) = transpose of AA'

= (A')' A'

= AA'

i.e., AA = (AA'). Hence AA' is a symmetric matrix by definition.

Example 33:

If A be a square matrix, show that A + A' is symmetric and A – A' is a skew-symmetric matrix.

Solution:

If **A** is a square matrix, then

(A + A') = A' + (A')'

= A' + A

$= A + A'$, by commutative law of addition

Hence by definition $A + A'$ is symmetric.

Again $(A - A')' = A' - (A')'$,

$= A' - A$,

$= -(A - A)'$

Hence by definintion $A - A'$ is skew-symmetric.

Example 34(a):

If A is a skew-symmetric matrix, then show that $AA' = A'A$ and A^2 symmetric.

Solution:

If A is a skew-symmetric matrix, then we know that

$$A' = -A \quad \text{...(1)}$$

Pre-multiplying both sides of (1) by A, we get

$$AA' = -AA = -A^2 \quad \text{...(2)}$$

Post-multiplying both sides of (1) by A, we get

$$A'A = -AA = -A^2 \quad \text{...(3)}$$

From (2) and (3) we conclude that $AA' = A'A$.

Further we can prove that AA' and $A'A$ are symmetric matrices Hence from (2) and (3) we find that $-A^2$ is a symmetic matrix or A^2 is a symmetric matrix, as we know is a symmetric matrix or A^2 is a symmetric matrix, as we know that kA is also symmetric if k is scalar and A is symmetric.

Example 34(b):

If A is a Skew-Hermitian Matrix, then shwo that ia is Hermitian.

Solution:

If A is a Skew-Hermitian Matrix, then

we have $-A = \overline{\mathbf{A}}$

Also $\overline{\mathbf{A}} = \mathbf{A}^{\Theta}$

$\therefore$ $-A = \overline{\mathbf{A}'} = \mathbf{A}^{\Theta}$

Now $(i\mathbf{A})^{\Theta} = -i\mathbf{A}^{\Theta}$ $\quad \because \bar{i} = -i$

$= -i(-A)$, from (i)

$\Rightarrow \qquad (i\mathbf{A})^{\Theta} = iA \qquad ...(2)$

We know that if A is a Hermitian matrix, then $\overline{\mathbf{A}'} = A = \mathbf{A}^{\Theta}$, from (1)

And from (2) we find that (iA) hence i**A** is a Hermitian matrix.

Example 34(c):

If A is any square matrix, show that $\mathbf{AA}^{\Theta}$ *and* $\mathbf{A}^{\Theta}A$ *are Hermitian.*

Solution:

$$(\mathbf{AA}^{\Theta})^{\Theta} = (\mathbf{A}^{\Theta})^{\Theta}\ \mathbf{A}^{\Theta}$$

$$= \mathbf{AA}^{\Theta}$$

$\therefore$ By definition $\mathbf{AA}^{\Theta}$ Hermitian.

Similarly $(A^{\Theta}A)^{\Theta} = A^{\Theta}\ (A^{\Theta})^{\Theta}$

$$= A^{\Theta}\ A$$

$\therefore$ By definition A^{Θ} A is Hermitian.

Example 34(d):

Show that A is Hermitan iff $\overline{A}$ *is Hermitian.*

Solution:

Let **A** be Hermitian, then $\mathbf{A} = \mathbf{A}^{\varnothing}$ ...(1)

Now $(\overline{A}^{\Theta})$ = transposed conjugate of $\overline{A}$

= Transposed matrix of **A**, since $(\overline{\overline{A}}) = \mathbf{A}$

= A = $(A^{\Theta})'$, by (1)

= thanspose of transposed conjugate of A

= conjugate of **A**, $\qquad \because \quad$ **(B')', = B**

i.e., $(\overline{A})^{\Theta} = \overline{A}$

Hence by defintion, $\overline{A}$ is a Hermitian matrix,

Again if $\overline{A}$ is Hermitian, then we have

$$\overline{A} = (\overline{A})^{\Theta}$$

$= \text{transposed conjugate of } \overline{A}$

$= \text{transpose of } \mathbf{A}$

$\Rightarrow \quad \overline{A} = \mathbf{A}'$

Now $\mathbf{A}^{\varnothing} = (\overline{A})'$ by definition

$= (A')'$ by (ii)

$= \mathbf{A}^{\Theta} = \mathbf{A}$

Hence by defintion A is Hermitian. **Hence proved.**

Example 35:

If $A = \begin{bmatrix} 3 & 2-3i & 3+5i \\ 2+3i & 5 & i \\ 3-5i & -i & 7 \end{bmatrix}$, *then prove that* $\overline{A}$ *is Hermitian*

Solution:

$$\overline{A} = \begin{bmatrix} 3 & 2+3i & 3+5i \\ 2-3i & 5 & -i \\ 3+5i & i & 7 \end{bmatrix} = B \text{ (say)}$$

$$\text{Then } B' = \begin{bmatrix} 3 & 2-3i & 3+5i \\ 2+3i & 5 & i \\ 3-5i & -i & 7 \end{bmatrix}$$

$$\therefore \quad \overline{B}' = \begin{bmatrix} 3 & 2-3i & 3-5i \\ 2-3i & 5 & -i \\ 3+5i & -i & 7 \end{bmatrix} = B$$

$\therefore$ B *i.e.,* $\overline{A}$ is Hermitian Matrix.

Example 36(a):

Prove that the matrix $A = \begin{bmatrix} 1 & 1-i & 2 \\ 1+i & 3 & i \\ 2 & -i & 0 \end{bmatrix}$ *is Hermitian.*

Solution:

$$A' = \begin{bmatrix} 1 & 1+i & 2 \\ 1-i & 3 & -i \\ 2 & i & 0 \end{bmatrix}$$

$$\overline{A}' = \begin{bmatrix} 1 & 1-i & 2 \\ 1+i & 3 & i \\ 2 & -i & 0 \end{bmatrix} = A$$

$\therefore$ A is Hermitian Matrix.

Example 36(b):

If A is a Hermitian matrix, then show that iA is Skew-Hermitian Matrix.

Solution:

If A is a Hermitian matrix, then

we heve $\quad A = \overline{\mathbf{A}}'$

Also $\quad \overline{\mathbf{A}}' = \mathbf{A}^{\Theta}$

$\therefore$ Here $\quad A = \overline{\mathbf{A}}' = \mathbf{A}^{\Theta}$ ...(1)

Now $\quad (i\mathbf{A})^{\Theta} = -i\mathbf{A}^{\Theta}, \qquad \because \bar{i} = -i$

$$= -\left(i\mathbf{A}^{\Theta}\right)$$

$\Rightarrow \quad (i\mathbf{A})^{\Theta} = -(iA)$, from (1) ...(2)

We know that if A is a Skew Hermitian Matrix, then $\overline{\mathbf{A}}' = -A = \mathbf{A}^{\Theta}$, from (1).

And from (2), we find that $-(iA) = (i\mathbf{A})^{\Theta}$, hence (i**A**) is a Skew Hermitian Matrix.

Example 36(c):

If A and B are Hermitian, then show that AB is Hermitian if and only if A and B commute.

Solution:

If A and B are Hermitian matrices, then we have

$A = (\overline{\mathbf{A}})' = \mathbf{A}^{\Theta}$ and $B = (\overline{\mathbf{B}})' = \mathbf{B}^{\Theta}$ ...(1)

Then $\quad (\mathbf{AB})^{\Theta} = \mathbf{B}^{\Theta}\ \mathbf{A}^{\Theta}$,

= BA, by (1) above

= AB, if A and B commute

i.e., $(\mathbf{AB})^{\Theta} = \mathbf{AB} \Rightarrow (\overline{\mathbf{AB}})' = \mathbf{AB}$, $\because A^{\Theta} = (\overline{A})'$

Hence, by definition **AB** is Hermitian.

Converse of this can be proved to be true by reversing the above calculations.

Example 37:

If the product of two non-zero square matrices is a zero matrix, then pove that both of them are singular matrices.

Solution:

Let A and B be two non-zero n × n martices.

Given that AB = O, who O is the n × n null matrix

Let us suppose that B is a non-singular matrix then B^{-1} exists.

Then $AB = O \Rightarrow (AB)\, B^{-1}\, OB^{-1}$,

post multiplying both sides by B^{-1}

$\Rightarrow A\,(BB^{-1}) = O$,

by associative law of multiplication. (Note)

$\Rightarrow AI = O, \quad \because \; BB^{-1} = I$

$\Rightarrow A = O$,

which is against hypothesis as A is a non-zero matrix.

Hence B is not a non-singular matrix i.e., B is a singular matrix.

Similarly we can prove that A is also a singular matrix.

Example 38:

Show that

$$\begin{bmatrix} 1 & 2 & 3 \\ 2 & 5 & 7 \\ -2 & -4 & -5 \end{bmatrix} \text{ is the inverse of } \begin{bmatrix} 3 & -2 & -1 \\ -4 & 1 & -1 \\ 2 & 0 & 1 \end{bmatrix}$$

Solution:

$$\begin{bmatrix} 1 & 2 & 3 \\ 2 & 5 & 7 \\ -2 & -4 & -5 \end{bmatrix} \times \begin{bmatrix} 3 & -2 & -1 \\ -4 & 1 & -1 \\ 2 & 0 & 1 \end{bmatrix}$$

$$= \begin{bmatrix} 1.3 + 2(-4) + 3.2 & 1(-2) + 2.1 + 3.0 & 1(-1) + 2(-1) + 3.1 \\ 2.3 + 5(-4) + 7.2 & 2(-2) + 5.1 + 7.0 & 2(-1) + 5(-1) + 7.1 \\ -2.3 - 4(-4) - 5.2 & -2(-2) - 4.1 - 5.0 & -2(-1) - 4.1(-1) - 5.1 \end{bmatrix}$$

$= \begin{bmatrix} 1 & 0 & 0 \\ 0 & 1 & 0 \\ 0 & 0 & 1 \end{bmatrix}$, which is an unit matrix.

Hence $\begin{bmatrix} 1 & 2 & 3 \\ 2 & 5 & 7 \\ -2 & -4 & -5 \end{bmatrix}$ is the iverse of $\begin{bmatrix} 3 & -2 & 1 \\ -4 & 1 & -1 \\ 2 & 0 & 1 \end{bmatrix}$

Example 39(a):

If A is a non-singular matrix, then prove that AB = AC ⇒ B = C, where B and C are square matrices of the same order.

Solution:

Since A is a non-singular matrix, so A^{-1} exists.

Now $AB = AC \Rightarrow A^{-1}(AB) = A^{-1}(AC)$,

premeultiplying both sides by A^{-1}

$\Rightarrow (A^{-1}A)B = (A^{-1}A)C$,

by associative law of multiplication

$\Rightarrow IB = IC, \quad \because \quad A^{-1}A = I$

$\Rightarrow B = C, \quad \because \quad IB = B$ etc. **Hence proved.**

Example 39(b):

Find a if

$[a \quad 4 \quad 1] \times \begin{bmatrix} 2 & 1 & 0 \\ 1 & 0 & 2 \\ 0 & 2 & 4 \end{bmatrix} \times \begin{bmatrix} a \\ 4 \\ 1 \end{bmatrix} = O$, *where O 1 × 1 null matrix.*

Solution:

$[a \quad 4 \quad 1] \times \begin{bmatrix} 2 & 1 & 0 \\ 1 & 0 & 2 \\ 0 & 2 & 4 \end{bmatrix}$

$= [2a + 4 + 0 \quad a + 0 + 2 \quad 0 + 8 + 4]$ **(Note)**

$= [2a + 4 \quad a + 2 \quad 12]$

$$\therefore \quad [a \quad 4 \quad 1] \times \begin{bmatrix} 2 & 1 & 0 \\ 1 & 0 & 2 \\ 0 & 2 & 4 \end{bmatrix} \times \begin{bmatrix} a \\ 4 \\ \_1 \end{bmatrix}$$

$$= [2a + 4 \quad a + 2 \quad 12] \times \begin{bmatrix} a \\ 4 \\ \_1 \end{bmatrix}$$

$= [\{(2a + 4) \times a\} + (a + 2).4 + 12\,(-1)]$ **(Note)**

$= [2a^2 + 4a + 4a + 8 - 12] = [2a^2 + 8a - 4] = [0]$; given

$\therefore \quad 2a^2 + 8a - 4 = 0 \quad \Rightarrow \quad a^2 + 4a - 2 = 0$

$\Rightarrow \qquad a = \frac{1}{2}\,[-4 \pm \sqrt{(4 + 8)}] = -2 \pm \sqrt{6}$ **Ans.**

Example 39(c):

If $A = \begin{bmatrix} 2 & 3 & 4 \\ 1 & 2 & 3 \\ -1 & 1 & 2 \end{bmatrix}$ *and* $B = \begin{bmatrix} 1 & 3 & 0 \\ -1 & 2 & 1 \\ 0 & 0 & 2 \end{bmatrix}$ *find AB and BA.*

Solution:

$$AB = \begin{bmatrix} 2 & 3 & 4 \\ 1 & 2 & 3 \\ -1 & 1 & 2 \end{bmatrix} \times \begin{bmatrix} 1 & 3 & 0 \\ -1 & 2 & 1 \\ 0 & 0 & 2 \end{bmatrix}$$

$$= \begin{bmatrix} 2.1 + 3.(-1) + 4.0 & 2.3 + 3.2 + 4.0 & 2.0 + 3.1 + 4.2 \\ 1.1 + 2.(-1) + 3.0 & 1.3 + 2.2 + 3.0 & 1.0 + 2.1 + 3.2 \\ -1.1 - 1.1 + 2.0 & -1.3 + 1.2 + 2.0 & -1.0 + 1.1 + 2.2 \end{bmatrix}$$

$$= \begin{bmatrix} 2 - 3 + 0 & 6 + 6 + 0 & 0 + 3 + 8 \\ 1 - 2 + 0 & 3 + 4 + 0 & 0 + 2 + 6 \\ -1 - 1 + 0 & -3 + 2 + 0 & 0 + 1 + 4 \end{bmatrix} = \begin{bmatrix} -1 & 12 & 11 \\ -1 & 7 & 8 \\ -2 & - & 5 \end{bmatrix}$$

$$\text{And } BA = \begin{bmatrix} 1 & 3 & 0 \\ -1 & 2 & 1 \\ 0 & 0 & 2 \end{bmatrix} \times \begin{bmatrix} 2 & 3 & 4 \\ 1 & 2 & 3 \\ 1 & 1 & 2 \end{bmatrix}$$

$$= \begin{bmatrix} 1.2 + 3.1 + 0.(-1) & 1.3 + 3.2 + 0.1 & 1.4 + 3.3 + 0.2 \\ (-1).2 + 2.1 + (-1) & (-1).3 + 2.2 + 1.1 & (-1).4 + 2.3 + 1.2 \\ 0.2 + 0.1 + 2.(-1) & 0.3 + 0.2 + 2.1 & 0.4 + 0.3 + 2.2 \end{bmatrix}$$

$$= \begin{bmatrix} 2+3+0 & 3+6+0 & 4+9+0 \\ -2+2-1 & -3+4+1 & -4+6+2 \\ 0+0-2 & 0+0+2 & 0+0+4 \end{bmatrix} = \begin{bmatrix} 5 & 9 & 13 \\ -1 & 2 & 4 \\ -2 & 2 & 4 \end{bmatrix}$$

Example 39(d):

If $A = \begin{bmatrix} i & 0 \\ 0 & i \end{bmatrix}$, *prove that* $A^n = I_2, A, -I_2 - A$ *according as* $n = 4p$, $4p + 1$, $4p + 2$ *and* $4p + 3$ *respectively.*

Solution:

Given $A = \begin{bmatrix} i & 0 \\ 0 & i \end{bmatrix}$ ...(1)

$$\therefore \quad A^2 = A.A = \begin{bmatrix} i & 0 \\ 0 & i \end{bmatrix} \times \begin{bmatrix} i & 0 \\ 0 & i \end{bmatrix}$$

$$= \begin{bmatrix} i.i + 0.0 & i.0 + 0.i \\ 0.i + i.0 & 0.0 + i.i \end{bmatrix} = \begin{bmatrix} i^2 & 0 \\ 0 & i^2 \end{bmatrix} \quad ...(2)$$

$$A^3 = A^2.\ A = \begin{bmatrix} i^2 & 0 \\ 0 & i^2 \end{bmatrix} \times \begin{bmatrix} i & 0 \\ 0 & i \end{bmatrix}$$

$$= \begin{bmatrix} i^2.i + 0.0 & i^2.0 + 0.i \\ 0.i + i^2.0 & 0.0 + i^2.i \end{bmatrix} = \begin{bmatrix} i^3 & 0 \\ 0 & i^3 \end{bmatrix} \quad ...(3)$$

From (2) and (3) we get $A^2 = \begin{bmatrix} i^2 & 0 \\ 0 & i^2 \end{bmatrix}$, $A^3 = \begin{bmatrix} i^3 & 0 \\ 0 & i^3 \end{bmatrix}$

Let us assume that $A^n = \begin{bmatrix} i^n & 0 \\ 0 & i^n \end{bmatrix}$ ...(4)

and also assume that (4) is true when n = k.

i.e $A^k = \begin{bmatrix} i^k & 0 \\ 0 & i^k \end{bmatrix}$ ...(5)

$\therefore \quad A^{k+1} = A^k.A = \begin{bmatrix} i^k & 0 \\ 0 & i^k \end{bmatrix} \times \begin{bmatrix} i & 0 \\ 0 & i \end{bmatrix}$

$$= \begin{bmatrix} i^k.i + 0.0 & i^k.0 + 0.i \\ 0.i + i^k.0 & 0.0 + i^k.i \end{bmatrix} = \begin{bmatrix} i^{k+1} & 0 \\ 0 & i^{k+1} \end{bmatrix}$$

$\therefore$ (4) is true for n = k + 1 provided (5) is true.

Also we have shown in (2) and (3) that (4) is true for n = 2 and 3. So it is true for 3 + 1 *i.e.,* 4 and so on.

Hence (4) is true for all positive integrate values of n.

Also if n = 4p then from (4) we get

$$A^n = \begin{bmatrix} i^{4p} & 0 \\ 0 & i^{4p} \end{bmatrix} = \begin{bmatrix} 1 & 0 \\ 0 & 1 \end{bmatrix}, \text{ Since } i^{4p} = (i^4)^p = (1)^p = 1, \text{ where } i = \sqrt{(-1)}$$

i.e., $A^n = I_2$. **Hence proved.**

If n = 4p + 1, then $i^n = i^{4p+1} = (i)^{4p}.\ i = 1.\ i = i$

$\therefore$ From (4), we get $A^n = \begin{bmatrix} i^n & 0 \\ 0 & i^n \end{bmatrix} = \begin{bmatrix} i & 0 \\ 0 & i \end{bmatrix} = A.$ **Hence proved.**

If n = 4p + 2, then $i^n = i^{4p+2} = i^{4p} = \times\ i^2$

$= (1)\ (-1)$, since $i^{4p} = 1,\ i^2 = -1$

$= -1.$

$\therefore$ From (4), we get

$$A^n = \begin{bmatrix} i^n & 0 \\ 0 & i^n \end{bmatrix} = \begin{bmatrix} -1 & 0 \\ 0 & -1 \end{bmatrix} = -\begin{bmatrix} 1 & 0 \\ 0 & 1 \end{bmatrix}$$

$\Rightarrow$ $A^n = -I_2$. **Hence proved.**

If n = 4p + 3, then $i^n = i^{4p+3} = (i^{4p+2}).\ i = (-1)\ i$, as above

$\therefore$ From (4), we get

$$A^n = \begin{bmatrix} i^n & 0 \\ 0 & i^n \end{bmatrix} = \begin{bmatrix} -i & 0 \\ 0 & -i \end{bmatrix} = -\begin{bmatrix} i & 0 \\ 0 & i \end{bmatrix}$$

$\Rightarrow$ $A^n = -A$, from (1). **Hence proved.**

Example 40:

If $AB = BA$ then prove that $(AB)^n = A^nB^m$.

Solution:

We shall prove this by mathematical induction.

If $n = 1$, then $(AB)^n = A^nB^n \Rightarrow (AB)^1 = AB$, which is true.

If $n = 2$, then

$(AB)^n = (AB)^2 = (AB)(AB)$

$= (ABA)B$, by associative law

$= (AAB)B$, $\because$ $BA = AB$, given

$= A^2B^2$

Hence $(AB)^n = A^nB^n$ is true for $n = 2$.

Now suppose that it is true for $n = m$ *i.e.,* $(AB)^m = A^mB^m$

$\Rightarrow$ $(AB)^m(AB) = (A^mB^m)(AB)$

$\Rightarrow$ $(AB)^{m+1} = A^m(B^mA)B$, by associative law

$\Rightarrow$ $(AB)^{m+1} = A^m(B^{m-1}BA)B$, $\because$ $B^m = B^{m-1}B$

$= A^m(B^{m-1}AB)B$, $\because$ $BA = AB$, given

$= A^m(B^{m-2}BAB)B$, $\because$ $B^{m-1} = B^{m-2}B$

$= A^m(B^{m-2}ABB)B$, $\because$ $BA = AB$, given

$= A^m(B^{m-2}AB^2)B$

$= A^m(B^{m-2}B^2)B = (A^mA)(B^{m-2}B^2B)$

$= A^{m+1}B^{m+1}$

i.e., If $(AB)^n$ A^nB^n is true for $n = m$, it is true for $n = m + 1$.

Also we have proved that it is true for $n = 1$ and 2.

Hence, by mathematical induction it is true for all + ve integral values of n.

Example 41:

If A and B are two matrices such that AB and A + B are both defined, then prove that A and B are square matrices.

Solution:

Let A be an $m \times n$ matrix.

Since A + B is defined *i.e.,* A and B are conformable to addition, so B must also be an m × n matrix.

Again AB is defined *i.e.,* A and B are comfortable to multiplication and hence the number of columns in A must be equal to the number of rows in B *i.e.,* n = m.

Hence A and B are m × m matrices *i.e.,* square matrices.

Example 42:

If $p(x) = \begin{bmatrix} \cos x & \sin x \\ -\sin x & \cos x \end{bmatrix}$, *then show that* $P(x) \cdot P(y) = P(x + y)$ $= P(y)\, P(x)$.

Solution:

$$P(x) \cdot P(y)$$

$$= \begin{bmatrix} \cos x & \sin x \\ -\sin x & \cos x \end{bmatrix} + \begin{bmatrix} \cos y & \sin y \\ -\sin y & \cos y \end{bmatrix}$$

$$= \begin{bmatrix} \cos x \cos y - \cos x \sin y & \cos x \sin y + \sin x \cos y \\ -\sin x \cos y - \cos x \sin y & -\sin x \sin y + \cos x \cos y \end{bmatrix}$$

$$= \begin{bmatrix} \cos(x + y) & \sin(x + y) \\ -\sin(x + y) & \cos(x + y) \end{bmatrix} = P(x + y)$$

Similarly we can prove (to be proved in the exam.) that

P (y) . P(x) = P (y) . P (x)

Hence $P(x) \cdot P(y) = P(x + y) = P(y) \cdot P(x)$ **Hence proved.**

EXERCISES

1. If A and B are idempotent, then A + B will be idempotent if AB = BA = O, where O is the null matrix.
2. Show that the matrix $\begin{bmatrix} ab & b^2 \\ -a^2 & -ab \end{bmatrix}$ is nilpotent.
3. If $A = \begin{bmatrix} 1 & -1 \\ -1 & 1 \end{bmatrix}$ and $B = \begin{bmatrix} 1 & 1 \\ 1 & 1 \end{bmatrix}$, wow that AB is a null matrix.

4. Show that $\begin{bmatrix} 0 & 0 & 1 \\ 0 & 1 & 0 \\ 1 & 0 & 0 \end{bmatrix} \times \begin{bmatrix} 0 & 1 & 0 \\ 0 & 0 & 1 \\ 1 & 0 & 0 \end{bmatrix} = \begin{bmatrix} 1 & 0 & 0 \\ 0 & 0 & 1 \\ 0 & 1 & 0 \end{bmatrix}$

5. If $A = \begin{bmatrix} 1 & 1 & -1 \\ -2 & 3 & -4 \\ 3 & -2 & 3 \end{bmatrix}$, $B = \begin{bmatrix} -1 & -2 & -1 \\ 6 & 12 & 6 \\ 5 & 10 & 5 \end{bmatrix}$ then prove that AB = O but BA ≠ O.

6. Show that $\begin{bmatrix} 1 & 1 & 3 \\ 5 & 2 & 6 \\ -2 & -1 & -3 \end{bmatrix}$ is a nilpotent matrix of order 3.

7. If $A = \begin{bmatrix} 1 & -2 & 3 \\ -4 & 2 & 5 \end{bmatrix}$; $B = \begin{bmatrix} 1 & 2 \\ -1 & 0 \\ 2 & 4 \end{bmatrix}$, then show that (AB)" = B'A'.

8. Show that the matrix $A = \begin{bmatrix} 1 & 2 \\ 3 & 1 \end{bmatrix}$ satisfies the equation $A^2 - 2A - 5I = O$, where O is the 2 × 2 null matrix.

9. Evaluate $A^2 - 3A - 13I$, where I is the 2 × 2 unit matrix and

$A = \begin{bmatrix} 2 & 5 \\ 3 & 1 \end{bmatrix}$

10. Show that matrix $A = \begin{bmatrix} 1 & 0 & 0 \\ 2 & 1 & 0 \\ 3 & 2 & 1 \end{bmatrix}$

satisfies the equation $A^3 - 3A^2 + 3A - I = O$, where I is the unit matrix and O the null matrix of order 3.

11. If $A = \begin{bmatrix} 2 & 3 \\ 4 & -1 \end{bmatrix}$, $B = \begin{bmatrix} 3 & -2 \\ 2 & 1 \end{bmatrix}$, $C = \begin{bmatrix} 1 & 2 \\ 3 & 4 \end{bmatrix}$ verify (1) (AB) C = A (BC); (2) (A + B) C = AC + BC.

12. If $A = \begin{bmatrix} 1 & 2 \\ 3 & 4 \end{bmatrix}$, $B = \begin{bmatrix} 2 & 1 \\ 4 & 2 \end{bmatrix}$, $C = \begin{bmatrix} 5 & 1 \\ 7 & 4 \end{bmatrix}$,

show that A (B + C) = AB + AC.

13. If $A_\alpha = \begin{bmatrix} \cos\alpha & \sin\alpha \\ -\sin\alpha & \cos\alpha \end{bmatrix}$, then show that $A_\alpha{}^n = \begin{bmatrix} \cos n\alpha & \sin n\alpha \\ -\sin n\alpha & \cos n\alpha \end{bmatrix}$,

    where n is any positive integer.

    Also prove that A_α and A_β commute and $A_\alpha . A_\beta = A_{\alpha+\beta}$.

14. A fruit seller has in stock 20 dozen mangoes, 16 dozen apples and 32 dozen bananas. Suppose the selling prices are Rs. 0.35, Rs. 0.75 and Rs. 0.08 per mango, apple and banana respectively. Find the total amount the fruit seller will get by selling his whole stock.

15. Find AB and BA if

    $$A = \begin{bmatrix} 3 & 4 & -2 \\ -2 & -1 & -1 \\ -1 & -3 & -1 \end{bmatrix} \text{ and } B = \begin{bmatrix} -1 & -1 & -1 \\ 2 & 2 & 2 \\ 1 & 1 & 1 \end{bmatrix}$$

16. If A be any square matrix, then show that

    $A + A^\Theta$ is Hermitian.

17. If A and B are symmetric (or skew-symmetric) matrices, then so is A + B.

18. If A and B are symmetric matrices, then prove that AB + BA is symmertic and AB – BA is Skew Symmettic,

19. show that all positive integal powers of s symmetric matrix are symmertic.

20. If A is any matrix, then show that A' A is a symmetric matrix.

21. If A is symmetric matrix, then show that AA' = A'A and A^2 is symmetic.

22. Show that the matrix $A = \frac{1}{\sqrt{2}}\begin{bmatrix} 1 & i \\ -i & -1 \end{bmatrix}$ is unitary.

23. For any two orthogonal matrices A and B, show that BA is an orthogonal matrix.

24. Find AB when $A = \begin{bmatrix} 2 & -1 & 0 \\ 0 & 2 & 1 \\ 1 & 0 & 1 \end{bmatrix}$ and $B = \begin{bmatrix} -2 & 1 & -1 \\ 1 & 2 & -2 \\ 2 & -1 & -4 \end{bmatrix}$

25. If A and B are symmetric and they commute, then $A^{-1} B$ and $A^{-1}B^1$ are symmetic

26. Show that every square matrix can be expressed is one and only one way as P + iQ, where P and Q are Hermitian.

27. If B is any square matrix, show that B' AB is symmetric provided B' AB is defined.

28. Find the matrices A and B, when

$$A + B = \begin{bmatrix} 1 & 0 & 2 \\ 2 & 2 & 2 \\ 1 & 1 & 2 \end{bmatrix} \text{ and } B = \begin{bmatrix} 1 & 4 & 4 \\ 4 & 2 & 0 \\ -1 & -1 & 2 \end{bmatrix}$$

29. If X, Y are two matrices given by the equations $X + Y = \begin{bmatrix} 1 & -2 \\ 3 & 4 \end{bmatrix}$ and $X - Y = \begin{bmatrix} 3 & 2 \\ -1 & 0 \end{bmatrix}$, find X, Y.

30. If $A = \begin{bmatrix} 1 & 2 & 3 \\ 0 & 5 & 7 \\ 6 & 8 & 9 \end{bmatrix}$, $B = \begin{bmatrix} 2 & 0 & 3 \\ 3 & 0 & 5 \\ 5 & 7 & 0 \end{bmatrix}$ evaluate 2A – 3B.

31. Multiply $[4 \quad 5 \quad 6]$ and $\begin{bmatrix} 2 \\ 3 \\ -1 \end{bmatrix}$

3

Basic Concept of Integration

INTRODUCTION

The process inverse to differentiation is defined as integration.

Thus $\frac{d}{dx}F(x) = f(x)$, we say that F(x) is an *integal* or a *primitive* of f(x) and, in symbols, we write

$$\int f(x)dx = F(x).$$

The letter x in dx denotes that the integration is to be performed with respect to the variable x.

The process of determing an integral of a function is called integration and the function to be integrated is called integrand.

SOME INTEGRALS FORMS

1. $\int \frac{f'(x)}{f(x)}dx = \log f(x).$

Put f(x) = t; differentiating we have f '(x) dx = dt.

$$\therefore \int \frac{f'(x)}{f(x)}dx = \int \frac{dt}{t} = \log t = \log f(x).$$

Thus **(Remember)**

the integral of a fraction whose numerator is the exact derivative of its denominator is equal to the logarithm of its denominator.

For example $\int \frac{4x^3}{1+x^4}dx = \log\left(1+x^4\right),$

as in this case numerator is the exact derivative of the denominator.

Similarly $\int \frac{e^x}{1+e^x}dx = \log\left(1+e^x\right).$

Integrals of tan x, cot x, sec x and cosec x

(i) $\int \tan x\, dx = \int \frac{\sin x}{\cos x} dx = -\int \frac{(-\sin x)}{\cos x} dx$,

adjusting the numerator as the exact diff. coeff. of the denominator

$= -\log x \cos = \log(\cos x)^{-1} = \log(\sec x)$.

(ii) Similarly, $\int \cot x\, dx = \int \frac{\cos x\, dx}{\sin x} = \log(\sin x)$.

(iii) $\int \operatorname{cosec} x\, dx = \int \frac{dx}{\sin x} = \int \frac{dx}{2\sin\frac{1}{2}x\cos\frac{1}{2}x} = \int \frac{\sec^2\frac{1}{2}x\, dx}{2\tan\frac{1}{2}x}$,

$\left[\text{dividing Nr. and Dr. by } \cos^2\frac{1}{2}x\right]$

$= \log\left(\tan\frac{1}{2}x\right)\left(\because \frac{1}{2}\sec^2\frac{1}{2}x \text{ is the diff. coeff. of } \tan\frac{1}{2}x\right)$.

(iv) $\int \sec x\, dx = \int \operatorname{cosec}\left(\frac{1}{2}\pi + x\right) dx$.

Now proceeding as in the case of $\int \operatorname{cosec} x\, dx$, we have

$\int \sec x\, dx = \log \tan\left(\frac{x}{2} + \frac{\pi}{4}\right)$.

Alternative Method

$\int \sec x\, dx = \int \frac{\sec x(\sec x + \tan x)}{\sec x + \tan x} dx$,

[We have multiplied the Nr. and Dr. both by (sec x + tanx)]

$= \int \frac{\sec x \tan x + \sec^2 x}{\sec x + \tan x}$, [Here Nr. is the diff. coeff. of Dr.]

$= \log(\sec x + \tan x)$.

2. $\int [f(x)]^n f'(x)\, dx = \frac{[f(x)]^{n+1}}{n+1}$, when $n \neq -1$. **(Power formula)**

Putting f(x) = t, so that f '(x) dx = dt, we get

$\int [f(x)]^n f'(x) dx = \int t^n dt = \frac{t^{n+1}}{n+1}$, (for $n \neq -1$) $= \frac{[f(x)]^{n+1}}{n+1}$.

Thus *Remember: If the integrand consists of the product of a constant power of a function f(x) and the derivative f '(x) of f(x), to obtain the integral we increase the index by unity and then divide by the increased index.* This is known as *Power formula.* The students are advised to have a lot of practice of applying this formula.

3. $\int f'(ax+b)\,dx = \dfrac{f(ax+b)}{a}$.

METHODS OF INTEGRATION

There are various mehtods of integration by which we can reduce the given integral to one of the fundamental or known integrals. Following are the four principal methods of integration:

(i) Integration by substitution,

(ii) Integration by parts,

(iii) Integration by decomposition into sum,

(iv) Integration by successive reduction.

(i) Integration by Substitution : A change in the variable of integration often reduces an integral to one of the fundamental integrals. The method in which we change the variable to some other variable is called the *Method of substitution.*

Let $I = \int f(x)\,dx$; then by differentiation w.r.t. x, we have

$$\frac{dI}{dx} = f(x).$$

Now put $x = \phi(t)$,

so that $\dfrac{dx}{dt} = \phi'(t)$.

Then $\dfrac{dI}{dt} = \dfrac{dI}{dx} \cdot \dfrac{dx}{dt} = f(x).$

$$\phi'(t) = f\{\phi(t)\}\ \phi'(t), \text{ for } x = \phi(t).$$

This gives, $I = \int f\{\phi(t)\}.\phi'(t)\,dt$.

Rule to Remember

To evaluate $\int f\{\phi(x)\}.\phi'(x)dx$,

put $\phi(x) = t$ and $\phi'(x)\,dx = dt$,

where $\phi'(x)$ is the different coefficient of $\phi(x)$ w.r.t. x.

Important : The success of the method of substitution depends on choosing the substitution $x = \phi(t)$ so that the new integrand $f\{\phi(t)\}.\phi'(t)$ is of a form whose integral is known. This is done by guess rather than in according with some rule. However, try to put that expression of x equal to t whose differential coefficient is multiplied with dx.

INTEGRALS OF e^{ax} cos bx AND e^{ax} sin bx

Let $I = \int e^{ax} \sin bx\,dx$.

Integrating by parts taking sin bx as the second function, we get

$$I = -\int \frac{e^{ax}\cos bx}{b} - \int ae^{ax}\left(-\frac{\cos bx}{b}\right)dx$$

$$= -\frac{e^{ax}\cos bx}{b} + \frac{a}{b}\int e^{ax}\cos bx\,dx.$$

Again integrating by parts taking cos bx as the second function, we get

$$I = -\frac{e^{ax}\cos bx}{b} + \frac{a}{b}\left[\frac{e^{ax}\sin bx}{b} - \int ae^{ax}\frac{\sin bx}{b}dx\right]$$

or $I = -\dfrac{e^{ax}\cos bx}{b} + \dfrac{a}{b^2}e^{ax}\sin bx - \dfrac{a^2}{b^2}\int e^{ax}\sin bx\,dx$

or $I = \dfrac{e^{ax}}{b^2}(a\sin bx - b\cos bx) - \dfrac{a^2}{b^2}I.$ $\left[\because \int e^{ax}\sin bx\,dx = I\right]$

Transposing the term $-\dfrac{a^2}{b^2}I$ to the left hand side, we get

$$\left(1 + \frac{a^2}{b^2}\right)I = \frac{e^{ax}}{b^2}\text{ (a sin bx – b cos bx)}$$

or $\dfrac{1}{b^2}\left(a^2 + b^2\right)I = \dfrac{1}{b^2}e^{ax}(a\sin bx - b\cos bx).$

$\therefore\ I = \dfrac{e^{ax}}{a^2 + b^2}$ (a sin bx – b cos bx).

Thus $\int e^{ax}\sin bx\,dx = \dfrac{e^{ax}}{a^2 + b^2}$ (a sin bx – b cos bx).

Similarly $\int e^{ax}\cos bx\,dx = \dfrac{e^{ax}}{a^2 + b^2}$ (a cos bx + b sin bx).

Alternative forms of $\int e^{ax}\sin bx\,dx$ and $\int e^{ax}\cos bx\,dx$

We have $\int e^{ax}\sin bx\,dx = \frac{e^{ax}}{a^2+b^2}$ (a sin bx – b cos bx).

Put a = r cos θ

and b = r sin θ. Then

$r = \sqrt{(a^2+b^2)}$ and $\theta = \tan^{-1}\left(\frac{b}{a}\right)$.

Now we have

$$\int e^{ax}\sin bx\,dx = \frac{e^{ax}}{r^2}\,(r\cos\theta\sin bx - r\sin\theta\cos bx)$$

$$= \frac{e^{ax}}{r}\sin(bx-\theta).$$

Thus $\int e^{ax}\sin bx\,dx = \frac{e^{ax}}{\sqrt{(a^2+b^2)}}\sin\left(bx - \tan^{-1}\frac{b}{a}\right)$.

Similarly $\int e^{ax}\cos bx\,dx = \frac{e^{ax}}{\sqrt{(a^2+b^2)}}\cos\left(bx - \tan^{-1}\frac{b}{a}\right)$.

INTEGRAL OF THE PRODUCT OF TWO FUNCTIONS

Integration by Parts: Let u and v be two functions of x. Then we have from differential calculus

$$\frac{d}{dx}(uv) = u\cdot\frac{dv}{dx} + v\cdot\frac{du}{dx} \qquad ...(1)$$

Integrating both sides of (1) with respect to x, we have

$$uv = \int u\cdot\frac{dv}{dx}dx + \int v\cdot\frac{du}{dx}dx.$$

By transposition, we have

$$\int u\frac{dv}{dx}dx = uv - \int v\frac{du}{dx}dx. \qquad ...(2)$$

Now put $u = f_1(x)$ and $v = \int f_2(x)\,dx$,

so that $\frac{dv}{dx} = f_2(x)$.

Then from (2), we have

$$\int f_1(x) f_2(x) dx = f_1(x) \cdot \int f_2(x) dx - \int \left[\left\{ \frac{d}{dx} f_1(x) \right\} \cdot \int f_2(x) dx \right] dx$$

i.e., the integral of the product of two functions

= first function × integral of second function

– integral of {diff. coeff. of first function × Integral of second function}.

Notes:

1. Care must be taken in choosing the first function and the second function. Obviously we must take that function as the second function whose integral is well known to us. Thus to evaluate $\int x \log x \, dx$ we shall take x as the second function because we so far do not know the integral of log x. But to evaluate $\int x \sin x \, dx$ we must take sin x as the second function and x as the first function. Here if we take x as the second function, then the new integral will become more complicated. Thus to evaluate integrals of the type $\int x^2 e^x dx$ $\int x^3 \cos x \, dx$ etc., the function of the type x^n must be taken as the first function. In certain cases we can take unity (*i.e.*, 1) as the second function. Thus to evaluate $\int \log x \, dx$ we shall take 1 as the second function. To evaluate $\int e^x \sin x \, dx$ we can take either e^x or sin x as the second function.
2. The formula of integration by parts can be applied more than once if necessary.
3. *Integration by parts as applied to the functions of the type $e^x [f(x) + f'(x)]$.*

Let $I = \int e^x [f(x) + f'(x)] dx = \int e^x f(x) dx + \int e^x f'(x) dx$.

Integrating the first integral by parts regarding e^x as the 2nd function, we have

$$I = \left[f(x) e^x - \int f'(x) e^x dx \right] + \int f'(x) e^x dx = e^x f(x).$$

[Note that we have left the other integral unchanged because the last two integrals cancel each other].

SUCCESSIVE INTEGRATION BY PARTS

If u is a function of the type

$a_0 x^n + a_1 x^{n-1} + \ldots + a_{n-1} x + a_n$, where n is a positive integer, the following formula for successive integration by parts can be applied. While writing this formula the successive differential coefficients of u have been denoted by u', u", U"' etc., while the successive integrals of v have been denoted by v_1, v_2, v_3 etc. Thus

$$\int uv\,dx = uv_1 - u'v_2 + u''v_3 - u'''v_4 + \ldots$$

This process of successive integration by parts will be continued till on being differentiated successively the differential coefficient of u becomes zero. The following examples will make the process clear.

SOME MORE STANDARD INTEGRALS

(i) To evaluate $\int \left[\frac{1}{\sqrt{(a^2+x^2)}}\right] dx$.

For complete solution of this problem 5. The result is

$$\int \frac{dx}{\sqrt{(a^2+x^2)}} = \sinh^{-1}\left(\frac{x}{a}\right) = \log\left\{x+\sqrt{(x^2+a^2)}\right\}.$$

(ii) To evaluate $\int \frac{dx}{\sqrt{(a^2-x^2)}}$.

The result is $\int \frac{dx}{\sqrt{(a^2-x^2)}} = \sin^{-1}\left(\frac{x}{a}\right)$.

(iii) To evaluate $\int \frac{dx}{\sqrt{(x^2-a^2)}}$.

Put x = a cosh q

so that dx = a sinh q dq.

Then the given integral $= \int \frac{dx}{\sqrt{(x^2-a^2)}} = \int \frac{a\sinh\theta\, d\theta}{\sqrt{(a^2\cosh^2\theta - a^2)}}$

$$= \int \frac{a\sinh\theta\, d\theta}{a\sqrt{(\cosh^2\theta - 1)}} = \int \frac{\sinh\theta\, d\theta}{\sinh\theta} = \int d\theta = \theta = \cosh^{-1}\left(\frac{x}{a}\right)$$

$$= \log\left[\frac{x}{a} + \sqrt{\left\{\left(\frac{x}{a}\right)^2 - 1\right\}}\right] = \log\left\{\frac{x+\sqrt{(x^2-a^2)}}{a}\right\}$$

$$= \log\left\{x+\sqrt{(x^2-a^2)}\right\} - \log a = \log\left\{x+\sqrt{(x^2-a^2)}\right\},$$

because the constant term – log a may be added to the constant of integration c which we usually do not write.

Thus $\int \frac{dx}{\sqrt{(x^2-a^2)}} = \cosh^{-1}\left(\frac{x}{a}\right) = \log\left\{x+\sqrt{(x^2-a^2)}\right\}.$

(iv) To evaluate $\int \sqrt{(x^2+a^2)}\,dx.$

Put $x = a \sinh \theta$

so that $dx = a \cosh \theta \, d\theta$.

Then the given integral $= \int \sqrt{(a^2 \sinh^2\theta + a^2)}.a\cosh\theta\, d\theta$

$$= \int a^2\cosh^2\theta\, d\theta = \int \frac{1}{2}a^2(1+\cosh 2\theta)\,d\theta,$$

$[\because \cosh 2\theta = 2\cosh^2\theta - 1]$

$$= \frac{1}{2}a^2\int(1+\cosh 2\theta)\,d\theta = \frac{1}{2}a^2\left[\theta + \frac{1}{2}\sinh 2\theta\right]$$

$$= \frac{1}{2}a^2[\theta + \sinh\theta\cosh\theta], \qquad [\because \sinh 2\theta = 2\sinh\theta\cosh\theta]$$

$$= \frac{1}{2}a^2\left[\sinh^{-1}\frac{x}{a} + \frac{x}{a}\sqrt{\left(1+\frac{x^2}{a^2}\right)}\right]$$

$$= \frac{a^2}{2}\sinh^{-1}\left(\frac{x}{a}\right) + \frac{a^2}{2}\cdot\frac{x}{a^2}\sqrt{(a^2+x^2)}.$$

Thus $\int \sqrt{(x^2+a^2)}\,dx = \frac{x}{2}\sqrt{(x^2+a^2)} + \frac{a^2}{2}\sinh^{-1}\left(\frac{x}{a}\right)$

$$= \frac{x}{2}\sqrt{(x^2+a^2)} + \frac{a^2}{2}\log\left\{x+\sqrt{(x^2+a^2)}\right\}.$$

(v) To evaluate $\int \sqrt{(a^2-x^2)}\,dx.$

Put $x = a \sin \theta$

so that $dx = a \cos \theta \, d\theta$.

Then the given integral $= \int a\cos\theta.a\cos\theta\, d\theta$

$$= \frac{1}{2}a^2\left(\theta + \frac{1}{2}\sin 2\theta\right) = \frac{1}{2}a^2(\theta + \sin\theta\cos\theta)$$

$$= \frac{1}{2}a^2\left[\theta + \sin\theta\sqrt{(1-\sin^2\theta)}\right]$$

$$= \frac{1}{2}a^2\sin^{-1}\left(\frac{x}{a}\right) + \frac{1}{2}a^2\cdot\frac{x}{a}\sqrt{\left(1-\frac{x^2}{a^2}\right)}.$$

Thus $\int \sqrt{(a^2 - x^2)}dx = \frac{1}{2}x\sqrt{(a^2 - x^2)} + \frac{1}{2}a^2 \sin^{-1}\left(\frac{x}{a}\right)$.

(vi) To evaluate $\int \sqrt{(x^2 - a^2)}dx$.

Put x = a cosh θ

so that dx = a sinh θ dθ.

Then the given integral $= \int \sqrt{(a^2 \cosh^2 \theta - a^2)} a \sinh\theta \, d\theta$

$= \int a^2 \sinh^2 \theta \, d\theta = \int \frac{1}{2} a^2 (\cosh 2\theta - 1) d\theta,$

$[\because 2 \sinh^2\theta = \cosh 2\theta - 1]$

$= \frac{1}{2}a^2\left[\frac{1}{2}\sinh 2\theta - \theta\right] = \frac{1}{2}a^2[\sinh\theta \cosh\theta - \theta]$

$= \frac{1}{2}a^2\left[\sqrt{(\cosh^2 \theta - 1)}\cosh\theta - \theta\right]$

$= \frac{1}{2}a^2\left[\sqrt{\left(\frac{x^2}{a^2} - 1\right)} \cdot \frac{x}{a} - \cosh^{-1}\frac{x}{a}\right] \cdot \left[\because \cosh\theta = \frac{x}{a}\right]$

Thus $\int \sqrt{(x^2 - a^2)}dx = \frac{1}{2}x\sqrt{(x^2 - a^2)} - \frac{1}{2}a^2 \cosh^{-1}\left(\frac{x}{a}\right)$

$= \frac{1}{2}x\sqrt{(x^2 - a^2)} - \frac{1}{2}a^2 \log\left\{x|\sqrt{(x^2 - a^2)}\right\}.$

HYPERBOLIC FUNCTIONS

Following fundamental properties of hyperbolic functions should be *committed to memory* by the students as we shall make frequent use of them during the study of integral calculus.

$\sinh x = (e^x - e^{-x})/2$, $\cosh x = (e^x + e^{-x})/2$

$\tanh x = (e^x - e^{-x})/(e^x + e^{-x})$,

$\coth x = (e^x + e^{-x})/(e^x - e^{-x})$

$\text{sech } x = 2/(e^x + e^{-x})$, $\text{cosech } x = 2/(e^x - e^{-x})$

$\cosh^2 x - \sinh^2 x = 1$, $\text{sech}^2 x = 1 - \tanh^2 x$

$\text{cosech}^2 x = \coth^2 x - 1$, $\sinh 2x = 2 \sinh x \cosh x$

$\cosh^2 x = \cosh^2 x + \sinh^2 x = 1 + 2 \sinh^2 x = 2 \cosh^2 x - 1.$

Logarithmic values of inverse hyperbolic functions.

(i) $\sinh^{-1} x = \log\left[x + \sqrt{(x^2+1)}\right]$;

$$\sinh^{-1}\left(\frac{x}{a}\right) = \log\left[\left\{x + \sqrt{(x^2+a^2)}\right\}/a\right]$$

(ii) $\cosh^{-1} = \log\left[x + \sqrt{(x^2-1)}\right]$;

$$\cosh^{-1}\left(\frac{x}{a}\right) = \log\left[\left\{x + \sqrt{(x^2-a^2)}\right\}/a\right]$$

(iii) $\tanh^{-1} x = \frac{1}{2}\log\left\{\frac{(1+x)}{(1-x)}\right\}$;

$$\tanh^{-1}\left(\frac{x}{a}\right) = \frac{1}{2}\log\left\{\frac{(a+x)}{(a-x)}\right\}, (x < a)$$

(iv) $\coth^{-1} x = \frac{1}{2}\log\left\{\frac{(x+1)}{(x-1)}\right\}$;

$$\coth^{-1}\left(\frac{x}{a}\right) = \frac{1}{2}\log\left\{\frac{(x+a)}{(x-a)}\right\}, (x > a).$$

CONSTANT OF INTEGRATION

As the differential coefficient of a constant is zero, we have

$$\frac{d}{dx}[F(x)+C] = f(x), \text{ if } \frac{d}{dx}F(x) = f(x);$$

therefore $\int f(x)dx = F(x) + c$.

This constant c is called a *constant of integration* and can take any constant value. Also $\int f(x)dx$ is called the *Indefinite integral* of f(x) w.r.t. 'x'; for by giving different values to the constant of integration the indefinite nature is preserved.

For example, we know that

$\frac{d}{dx}\sin^{-1} x = \frac{1}{\sqrt{(1-x^2)}}$ and ; is follows, when the omit the constant of integration, that $\int \frac{1}{\sqrt{(1-x^2)}}dx$ is equal to $-\cos^{-1}x$.

But it is wrong to conclude from above that $\sin^{-1}x$ is equal to $-\cos^{-1}x$. The correct inference is that the two integrals, given above differ in their constant of integration.

The correct result, as we see from trigonometry, is that

$$\sin^{-1} x = \frac{1}{2}\pi - \cos^{-1} x.$$

The arbitrary constant of integration may be imaginary also. Generally such a constant is added to make the result real.

TWO SIMPLE THEOREMS

Theorem 1:

The integral of a sum of difference of a finite number of functions is equal to the sum or difference of the integrals of the functions. Symbolically

$$\int [f_1(x) \pm f_2(x) \pm \ldots \pm f_n(x)]dx$$

$$= \int f_1(x)dx \pm \int f_2(x)dx \pm \ldots \pm \int f_n(x)dx.$$

Proof:

Let $\int f_1(x)dx = F_1(x)$, $\int f_2(x)dx = F_2(x)$, ..., $\int f_n(x)dx = F_n(x)$.

Clearly $\frac{d}{dx}\{F_1(x) \pm F_2(x)\} = \frac{d}{dx}F_1(x) \pm \frac{d}{dx}F_2(x)$

$= f_1(x) \pm f_2(x\_,$

Hence from the definition of the integral, we have

$$\int \{f_1(x) \pm f_2(x)\}dx = F_1(x) \pm F_2(x)$$

$$\int f_1(x)dx \pm \int f_2(x)dx.$$

Thus repeating the above process to other functions, we have

$$\int \{f_1(x) \pm f_2(x) \pm \ldots \pm f_n(x)\}dx$$

$$\int f_1(x)dx \pm \int f_2(x)dx \pm \ldots \pm \int f_n(x)dx.$$

Theorem 2:

The integral of the product of a constant and a function is equal to the product of the constant and the integral of the function.

Thus if λ is a constant, then $\int \lambda f(x)dx = \lambda \int f(x)dx$.

Proof:

Let $\int f(x)dx = F(x)$. Then $\frac{d}{dx}F(x) = f(x)$.

By differential calculus, $\frac{d}{dx}\{\lambda F(x)\} = \lambda \frac{d}{dx}F(x) = \lambda f(x)$.

$\therefore$ by definition of integral

$\int \lambda f(x)dx = \lambda F(x) = \lambda \int f(x)dx$.

EXTENDED FORMS OF FUNDAMENTAL FORMULAE

Suppose we know that $\int f(x)dx = F(x)$ and we want to find $\int f(ax+b)dx$.

Let $I = \int f(ax+b)dx$.

Put ax + b = t so that a dx = dt.

Then $$I = \int f(t)\frac{dt}{a} = \frac{1}{a}\int f(t)dt = \frac{1}{a}F(t) = \frac{1}{a}F(ax+b)$$

Thus if $\int f(x)dx + F(x)$, then $\int f(ax+b)dx + \frac{1}{a}F(ax+b)$.

From this we conclude the following results:

(i) $\int \sec^2(ax+b)dx = \frac{1}{a}\tan(ax+b)$,

(ii) $\int \operatorname{cosec}^2(ax+b)dx = -\frac{1}{a}\cot(ax+b)$,

(iii) $\int \sec(ax+b)\tan(ax+b)dx = \frac{1}{a}\sec(ax+b)$,

(iv) $\int \operatorname{cosec}(ax+b)\cot(ax+b)dx = -\frac{1}{a}\operatorname{cosec}(ax+b)$,

(v) $\int \frac{1}{\sqrt{(a^2-x^2)}}dx = \sin^{-1}\frac{x}{a}$ and $\int \frac{1}{a^2+x^2}dx = \frac{1}{a}\tan^{-1}\frac{x}{a}$.

(vi) $\int (ax+b)^n dx = \frac{1}{a}\frac{(ax+b)^{n+1}}{(n+1)}$,

(vii) $\int \frac{1}{(ax+b)^n}dx = -\frac{1}{a(n-1)(ax+b)^{n-1}}$,

(viii) $\int \frac{1}{ax+b} dx = \frac{1}{a}\log(ax+b)$,

(ix) $\int e^{ax+b} dx = \frac{1}{a} e^{ax+b}$ and $\int a^{px+q} dx = \frac{1}{p}\frac{a^{px+q}}{\log_e a}$,

(x) $\int \sin(ax+b)\,dx = -\frac{1}{a}\cos(ax+b)$,

To evaluate $\int \frac{1}{\sqrt{(a^2-x^2)}} dx$.

Put $x = a \sin \theta$,

so that $dx = a \cos \theta \, d\theta$.

Also $a^2 - x^2 = a^2 (1 - \sin^2 \theta) = a^2 \cos^2 \theta$.

Thus $\int \frac{1}{\sqrt{(a^2-x^2)}} dx = \int \frac{1}{a\cos\theta} a\cos\theta \, d\theta$

$$= \int 1.d\theta = \theta = \sin^{-1}\left(\frac{x}{a}\right).$$

And to evaluate $\int \frac{1}{\sqrt{(a^2+x^2)}} dx$.

Put $x = a \sinh \theta$;

so that $dx = a \cosh \theta \, d\theta$.

Also we have $a^2 + x^2 = a^2 (1 + \sinh^2 \theta) = a^2 \cosh^2 \theta$.

Thus $\int \frac{1}{\sqrt{(a^2+x^2)}} dx = \int \frac{1}{a\cosh\theta} a \cosh\theta \, d\theta$

$$= \int 1.d\theta = \theta = \sin^{-1}\left(\frac{x}{a}\right) = \log\left\{\frac{x+\sqrt{(x^2+a^2)}}{a}\right\}$$

$$= \log\left\{x+\sqrt{(x^2+a^2)}\right\} - \log a = \log\left\{x+\sqrt{(x^2+a^2)}\right\},$$

omitting the constant term – log a because it may be added to the constant of integration c which we usually do not write.

Similarly,

$$\int \frac{1}{\sqrt{(x^2-a^2)}} dx = \cosh^{-1}\left(\frac{x}{a}\right) = \log\left\{x+\sqrt{(x^2-a^2)}\right\}$$

$$\int\sqrt{(a^2-x^2)}dx=\frac{x\sqrt{(a^2-x^2)}}{2}+\frac{a^2}{2}\sin^{-1}\left(\frac{x}{a}\right)$$

$$\int\sqrt{(a^2+x^2)}dx=\frac{x\sqrt{(a^2+x^2)}}{2}+\frac{a^2}{2}\sinh^{-1}\left(\frac{x}{a}\right)$$

$$\int\sqrt{(x^2-a^2)}dx=\frac{x\sqrt{(x^2-a^2)}}{2}-\frac{a^2}{2}\cosh^{-1}\left(\frac{x}{a}\right).$$

FUNDAMENTAL FORMULAE

We have read in differential calculus that

$$\frac{d}{dx}\left(\frac{x^{n+1}}{n+1}\right)=\frac{(n+1)x^n}{(n+1)}=x^n.$$

Thus $$\int x^n dx+\frac{x^{n+1}}{n+1},\ (n\neq-1).$$

The above formula is very important and shall be frequently used in this book. This may be remembered like this:

"To find the integral of x^n w.r.t. 'x', increase the index (power) of x by one (unity) and divide by the increased index."

Thus $$\int x^3dx=\frac{x^4}{4},\ \int x^{5/2}dx=\frac{x^{(5/2)+1}}{(5/2)+1}=\frac{2}{7}x^{7/2};$$

$$\int\frac{1}{x^5}dx=\int x^{-5}dx=\frac{x^{-5+1}}{-5+1}=-\frac{1}{4}x^{-4}=-\frac{1}{4x^4};$$

$$\int\frac{1}{x^{1/2}}dx=\int x^{-1/2}dx=\frac{x^{(-1/2)}+1}{(-1/2)+1}=2x^{1/2}=2\sqrt{x};$$

and $$\int dx=\int 1.dx=\int x^0dx=\frac{x^{0+1}}{0+1}=x.$$

Thus $\int a\,dx=ax$ *i.e., the integral of a constant is equal to the constant multiplied by the variable.*

However if n = – 1, we have

$$\int x^{-1}dx=\int\frac{1}{x}dx=\log x,\qquad\left[\because\frac{d}{dx}\log x=\frac{1}{x}\right].$$

SOLVED EXAMPLES

Example 1:

Integrate $e^x + 2\sin x - 3\cos x$.

Solution:

Here $I = \int (e^x + 2\sin x - 3\cos x)dx$

$$= \int e^x dx + 2\int \sin x\, dx - 3\int \cos x\, dx$$

$$= e^x - 2\cos x - 3\sin x.$$

Example 2:

Integrate $10^x + 3e^x + x^3$.

Solution:

Here $I = \int (10^x + 3e^x + x^3)dx$

$$= \int 10^x dx + 3\int e^x dx + \int x^3 dx = \left\{\frac{10^x}{(\log_e 10)}\right\} + 3e^x + \frac{1}{4}x^4.$$

Example 3:

Integrate $5\cos x + 2\sec^2 x - 10$.

Solution:

Here $I = \int (5\cos x + 2\sec^2 x - 10)dx$

$$= \int 5\cos x\, dx + 2\int \sec^2 x\, dx - \int 10 dx$$

$$= 5\sin x + 2\tan x - 10x.$$

Example 4(a):

Integrate $\left\{\frac{6}{\sqrt{(1-x^2)}}\right\} + 3\sec^2 x$.

Solution:

Here $I = \int \left[\left\{\frac{6}{\sqrt{(1-x^2)}}\right\} + 3\sec^2 x\right]dx$

$$= \int \left\{ \frac{6}{\sqrt{(1-x^2)}} \right\} dx + 3\int \sec^2 x\, dx = 6\sin^{-1} x + 3\tan x.$$

Example 4(b):

Integrate sec x tan x – 5 cosec²x.

Solution:

Here $\quad I = \int (\sec x \tan x - 5\operatorname{cosec}^2 x)\, dx$

$$= \int \sec x \tan x\, dx - 5\int \operatorname{cosec}^2 x\, dx = \sec x + 5\cot x.$$

Example 4(c):

Integrae $\left\{ \frac{(2\cos x)}{(3\sin^2 x)} \right\} + 1.$

Solution:

Here $I = \int \left(\frac{2\cos x}{3\sin^2 x} \right) dx = \int \frac{2\cos x}{3\sin^2 x} dx + \int 1.dx$

$$= \frac{2}{3}\int \operatorname{cosec} x \cot x\, dx + \int 1.dx = -\frac{2}{3}\operatorname{cosec} x + x.$$

Example 4(d):

Integrate $1 + x + \frac{x^2}{2!} + \frac{x^3}{3!} + \ldots$

Solution:

Here $\quad I = \int \left(1 + x + \frac{x^2}{2!} + \frac{x^3}{3!} + \ldots \right) dx$

$$= \int 1.dx + \int x.dx + \int \frac{1}{2}x^2 dx + \int \frac{1}{6}x^3 dx + \ldots$$

$$= x + \frac{x^2}{2} + \frac{x^3}{6} + \frac{x^4}{24} + \ldots$$

Example 5:

Integrate $\frac{(2x^3 + 3x - 7)}{x^{2/3}}.$

Solution:

Here $I = \int \left(2x^3 + 3x - 7\right)x^{-2/3}dx$

$$= \int \left(2x^{7/3} + 3x^{1/3} - 7x^{-2/3}\right)dx$$

$$= \frac{2x^{(7/3)+1}}{(7/3)+1} + \frac{3x^{(1/3)+1}}{(1/3)+1} - \frac{7x(-2/3)+1}{(-2/3)+1}$$

$$= \frac{3}{5}x^{10/3} + \frac{9}{4}x^{4/3} - 21x^{1/3}.$$

Example 6:

Integrate $\frac{a}{x^2} + \frac{b}{x} + c$.

Solution:

Here $I = \int \left(\frac{a}{x^2} + \frac{b}{x} + c\right)dx = \int \frac{a}{x^2}dx + \int \frac{b}{x}dx + \int cdx$

$$= \int ax^{-2}dx + \int \frac{b}{x}dx + \int c\,dx$$

$$= \frac{ax^{-1}}{-1} + b\log x + cx = -\frac{a}{x} + b\log x + cx.$$

Example 7:

Integrate $(x^2 + 8)^2/x^4$.

Solution:

Here $I = \int \frac{\left(x^2+8\right)^2}{x^4}dx = \int \frac{\left(x^4 + 16x^2 + 64\right)}{x^4}dx$

$$= \int \left(1 + 16x^{-2} + 64x^{-4}\right)dx = x + 16\frac{x^{-2+1}}{-2+1} + 64\frac{x^{-4+1}}{-4+1}$$

$$= x - \left(\frac{16}{x}\right) - \frac{64}{\left(3x^3\right)}.$$

Example 8:

Integrate $\frac{(x+a)^3}{\sqrt{x}}$.

Solution:

Here $I = \int \frac{(x+a)^3}{\sqrt{x}}dx = \int \left\{\frac{\left(x^3 + 3ax^2 + 3a^2x + a^3\right)}{\sqrt{x}}\right\}dx$

$$= \int \left\{x^{5/2} + 3ax^{3/2} + 3a^2x^{1/2} + a^3x^{-1/2}\right\}dx$$

$$= \frac{2}{7}x^{7/2} + 3.\frac{2}{5}ax^{5/2} + 3.\frac{2}{3}a^2x^{3/2} + 2a^3.x^{1/2}.$$

Example 9(a):

Integrate $\frac{1}{x^{3/4}} + \frac{1-x^4}{1-x} + \sec x \tan x.$

Solution:

Here I = $I = \int x^{-3/4}dx + \int (1+x+x^2+x^3)dx + \int \sec x \tan x\, dx,$

$$\left[\because \frac{(1-x^n)}{(1-x)} = 1+x+x^2+x^3+ \ldots + x^{n-1}\right]$$

$$\therefore\ I = 4x^{1/4} + x + \left(\frac{x^2}{2}\right) + \left(\frac{x^3}{3}\right) + \left(\frac{x^4}{4}\right) + \sec x.$$

Example 9(b):

Integrate $\frac{5\cos^3 x + 2\sin^3 x}{2\sin^2 x \cos^2 x} + \sqrt{(1+\sin 2x)} + \frac{1+2\sin x}{\cos^2 x} + \frac{1-\cos 2x}{1+\cos 2x}.$

Solution:

The given expression may be written as

$$\frac{5\cos x}{2\sin^2 x} + \frac{\sin x}{\cos^2 x} + \sqrt{(\cos^2 x + \sin^2 x + 2\sin x \cos x)}$$
$$+ \frac{1}{\cos^2 x} + \frac{2\sin x}{\cos^2 x} + \frac{2\sin^2 x}{2\cos^2 x}$$

$$= \frac{5}{2} \text{ cosec } x \cot x + \sec \tan x + \cos x + \sin x + \sec^2 x$$
$$+ 2 \sec x \tan x + 2 (\sec^2 x - 1)$$

$$= \frac{5}{2} \text{ cosec } x \cot x + 3 \sec x \tan x + \cos x + \sin x + 3 \sec^2 x - 2.$$

Now integrating, we get

$$I = \frac{5}{2}\int \text{cosec}\, x \cot x\, dx + 3\int \sec x \tan x\, dx$$
$$+ \int \cos x\, dx + \int \sin x\, dx + 3\int \sec^2 x\, dx - 2\int dx$$

$$= -\frac{5}{2}\text{cosec } x + 3 \sec x + \sin x - \cos x + 3 \tan x - 2x.$$

Example 9(c):

Integrate (i) $(3x + 4x^2)$, (ii) $(5x + 7)/x$.

Solution:

(i) $$I = \int \left(3x + 4x^2\right) dx.$$

Then $$I = \int 3x\, dx + \int 4x^2\, dx = \frac{3}{2}.x^2 + \frac{4}{3}x^3$$

(ii) Here $$I = \int \left\{ \frac{(5x+7)}{x} \right\} dx$$

$$= \int 5dx + \int \left(\frac{7}{x}\right) dx = 5x + 7\log x.$$

Example 9(d):

Integrate $\sqrt{(2x+3)}$.

Solution:

Here $$I = \int \sqrt{(2x+3)}dx = \frac{1}{2}\frac{(2x+3)^{3/2}}{3/2} = \frac{1}{3}(2x+3)^{3/2}.$$

Example 9(e):

Integrate $\sin\left(2x + \frac{1}{4}\pi\right)$.

Solution:

Here $$I = \int \sin\left(2x + \frac{1}{4}\pi\right) dx = \frac{1}{2}\left[-\cos\left(2x + \frac{1}{4}\pi\right)\right].$$

Example 10:

Integrate $\int \frac{1}{4 + (2-3x)^2} dx$.

Solution:

We know that

$$\int \frac{1}{a^2 + x^2} dx = \frac{1}{a}\tan^{-1}\frac{x}{a}.$$

Here in place of x we have $(2 - 3x)$; therefore after applying the standard result we shall divide in the end by the coefficient of x in $(2 - 3x)$ *i.e.*, by -3. Also here $a^2 = 4$; $\therefore$ $a = 2$. Thus the required integral

$$I = -\frac{1}{3}\cdot\frac{1}{2}\tan^{-1}\left(\frac{2-3x}{2}\right) = -\frac{1}{6}\tan^{-1}\left(\frac{2-3x}{2}\right).$$

Example 11:

Integrate $\dfrac{1}{\sqrt{\left[7-\left(\frac{1}{2}x-3\right)^2\right]}}$.

Solution:

We know that, $\int \frac{1}{\sqrt{(a^2-x^2)}}dx = \sin^{-1}\frac{x}{a}$.

Here in place of x we have $\left(\frac{1}{2}x-3\right)$; therefore after applying the standard result we shall divide in the end by the coefficient of x in $\frac{1}{2}x-3$ *i.e.* by $\frac{1}{2}$. Also here $a^2 = 7$; $\therefore$ $a = \sqrt{7}$.

Thus the required integral $I = \int \frac{1}{\sqrt{\left[7-\left(\frac{1}{2}x-3\right)^2\right]}}dx$

$$= \frac{1}{(1/2)}\sin^{-1}\left\{\frac{(1/2x-3)}{\sqrt{7}}\right\} = 2\sin^{-1}\left\{\frac{(x-6)}{2}\sqrt{7}\right\}.$$

Example 11(a):

Integrate $\dfrac{1}{(1+\sin x)}$.

Solution:

Here $I = \int \frac{1}{1+\sin x}dx = \int \frac{dx}{1-\cos\left(\frac{1}{2}\pi + x\right)}$

$$= \int \frac{dx}{2\sin^2\left(\frac{1}{4}\pi+\frac{1}{2}x\right)} = \frac{1}{2}\int \operatorname{cosec}^2\left(\frac{1}{2}x+\frac{1}{4}\pi\right)dx$$

$$= -\cot\left(\frac{1}{2}x+\frac{1}{4}\pi\right).$$

Example 12(b):

Integrate $\dfrac{(1+\sin x)}{(1-\cos x)}$.

Solution:

Here $$I=\int\frac{1+\sin x}{1-\cos x}dx=\int\frac{1+2\sin\frac{1}{2}x\cos\frac{1}{2}x}{2\sin^2\frac{1}{2}x}dx$$

$$=\frac{1}{2}\int\operatorname{cosec}^2\frac{1}{2}x\,dx+\int\cos\frac{1}{2}x\,dx$$

$$=-\cot\frac{1}{2}x+2\log\left(\sin\frac{1}{2}x\right).$$

$$\left[\because\int\cot x\,dx=\log(\sin x)\right]$$

$$=-\cot\frac{1}{2}x+2\log\left(\sin\frac{1}{2}x\right).$$

Example 12(c):

Integrate $\dfrac{1}{\sqrt{(2x^2+3x+4)}}$.

Solution:

We have $2x^2 + 3x + 4 = 2\left[x^2+\frac{3}{2}x+2\right]$

[Note that we have made the coefficient of x^2 as 1]

$$=2\left[\left(x+\frac{3}{4}\right)^2+2-\frac{9}{16}\right]=2\left[\left(x+\frac{3}{4}\right)^2+\left(\frac{\sqrt{23}}{4}\right)^2\right].$$

Hence the given integral

$$I=\int\frac{1}{\sqrt{(2x^2+3x+4)}}dx=\int\frac{dx}{\sqrt{2}.\sqrt{\left[\left(x+\frac{3}{4}\right)^2+\left(\sqrt{\frac{23}{4}}\right)^2\right]}}$$

$$=\frac{1}{\sqrt{2}}\sinh^{-1}\left\{\frac{x+\left(\frac{3}{4}\right)}{\sqrt{\frac{23}{4}}}\right\}=\frac{1}{\sqrt{2}}\sinh^{-1}\left(\frac{4x+3}{\sqrt{23}}\right).$$

Example 12(d):

Integrate $\dfrac{1}{\sqrt{(-2x^2+3x+4)}}$.

Solution:

Here $-2x^2 + 3x + 4 = -2\left[x^2 - \frac{3}{2}x - 2\right]$

$$= -2\left[\left(x - \frac{3}{4}\right)^2 - \left(\frac{\sqrt{41}}{4}\right)^2\right] = 2\left[\left(\frac{\sqrt{41}}{4}\right)^2 - \left(x - \frac{3}{4}\right)^2\right].$$

$$\therefore \quad I = \int \frac{1.dx}{\sqrt{\left(-2x^2 + 3x + 4\right)}} = \frac{1}{\sqrt{2}} \int \frac{dx}{\sqrt{\left\{\left(\frac{\sqrt{41}}{4}\right)^2 - \left(x - \frac{3}{4}\right)^2\right\}}}$$

$$= \frac{1}{\sqrt{2}} \sin^{-1}\left\{\frac{x - \left(\frac{3}{4}\right)}{\sqrt{\frac{41}{4}}}\right\} = \frac{1}{\sqrt{2}} \sin^{-1}\left\{\frac{4x - 3}{\sqrt{(41)}}\right\}.$$

Example 13:

Evaluate $\int \left\{\frac{1}{\left(c^2 + b^2 y^2\right)}\right\} dy.$

Solution:

Put by = t; $\therefore$ b dy = dt

or $\quad dy = \left(\frac{1}{b}\right) dt$

$$\therefore \quad I = \frac{1}{b} \int \frac{dt}{c^2 + t^2} = \frac{1}{bc} \tan^{-1} \frac{t}{c} = \frac{1}{bc} \tan^{-1}\left(\frac{by}{c}\right).$$

Example 14:

Evaluate $\int \left\{\frac{(\cos ax)}{\sin^2 ax}\right\} dx.$

Solution:

We have $I = \int \frac{\cos ax\, dx}{\sin^2 ax} = \int \frac{\cos ax\, dx}{\sin ax \sin ax}$

$$= \int \cos ax.\text{cosec } ax\, dx.$$

Now put ax = t; $\therefore$ adx = dt

or $\quad dx = \left(\frac{1}{a}\right) dt.$

Then $I = \frac{1}{a} \int \cot t \text{ cosec } t\, dt = -\frac{1}{a} \text{cosec } t = -\frac{1}{a} \text{cosec } ax.$

Example 15:

Evaluate $\int -\dfrac{cosec^2 x}{\sqrt{(cot^2 x - 16)}} dx$.

Solution:

Put cot x = t, so that – cosec² x dx = dt.

Hence $I = \int \dfrac{dt}{\sqrt{(t^2 - 16)}} = \cosh^{-1}\dfrac{t}{4} = \cosh^{-1}\left(\dfrac{\cot x}{4}\right)$.

Example 16:

Evalulate $\int e^x \cos e^x \, dx$.

Solution:

Put $e^x = t$,

so that $e^x\, dx = dt$.

$\therefore \quad I = \int e^x \cos e^x dx = \int \cos t\, dt = \sin t = \sin e^x$.

Example 17:

Evaluate $\int \sin^2 x \cos x \, dx$.

Solution:

Put sin x = t, so that cos x dx = dt.

$\therefore\ I = \int \sin^2 x \cos x\, dx = \int t^2 dt = \dfrac{1}{3}t^3 = \dfrac{1}{3}\sin^3 x$.

Example 18:

Evaluate $\int \left[\dfrac{4x^3}{(1+x^8)}\right] dx$.

Solution:

Put $x^4 = t$ so that $4x^3\, dx = dt$.

$\therefore\ I = \int \left[\dfrac{4x^3}{(1+x^8)}\right] dx = \int \left[\dfrac{1}{(1+t^2)}\right] dt = \tan^{-1} t = \tan^{-1} x^4$.

Example 19:

Evaluate $\int x \sec^2 x \, dx$.

Solution:

Put $x^2 = t$, so that $2xdx = dt$ or $x\,dx = \frac{1}{2}dt$.

$$\therefore \int x\sec^2 x^2 dx = \frac{1}{2}.\int \sec^2 t\,dt = \frac{1}{2}\tan t = \frac{1}{2}\tan x^2$$

Example 20:

Evaluate $-\int \frac{\cos^{-1} x}{\sqrt{(1-x^2)}} dx.$

Solution:

Put $\cos^{-1} x = t$,

so that $\frac{1}{\sqrt{(1-x^2)}} dx = dt$.

$\therefore$ the given integral $= \int t\,dt = \frac{1}{2}t^2 = \frac{1}{2}\left(\cos^{-1} x\right)^2.$

Example 21:

Evalulate $\int \left[\frac{1}{(cx+d)^4}\right] dx.$

Solution:

Put $cx + d = t$;

then $c\,dx = dt$,

or $dx = \frac{dt}{c}$.

Thus the given integral

$$= \int (cx+d)^{-4} dx = \int t^{-4}\frac{dt}{c} = \frac{1}{c}\int t^{-4} dt = \frac{1}{c}\cdot\frac{t^{-3}}{-3}$$

$$= -\frac{1}{3c(cx+d)^3}.$$

Example 22(a):

Evaluate $\int 20^{5x} dx.$

Solution:

Put $5x = t$; then $5dx = dt$ or $dx = \frac{1}{5}dt$.

$$\therefore \qquad I = \int 20^{5x}\,dx = \int \frac{1}{5}\cdot 20^t\,dt$$

$$= \frac{1}{5}\frac{20^t}{\log 20} = \frac{1}{5}\frac{20^{5x}}{\log 20}.$$

Example 22(b):

Evaluate $\int \frac{dx}{\sqrt{\{1-(cx+d)^2\}}}$.

Solution:

Put cx + d = t, so that cdx = dt.

Hence the given integral

$$= \int \frac{1}{c}\frac{dt}{\sqrt{(1-t^2)}} = \frac{1}{c}\sin^{-1} t = \frac{1}{c}\sin^{-1}(cx+d).$$

Example 22(c):

Integrate (i) $\frac{cosec^2 x}{1+\cot x}$, *(ii)* $\frac{1}{(1+x^2)\tan^{-1}x}$.

Solution:

(i) Here $I = \int \left\{\frac{\text{cosec}^2 x}{(1+\cot x)}\right\} dx$.

Putting 1 + cot x = t,

so that $-$ cosec^2 x dx = dt,

we have $I = -\int\left(\frac{1}{t}\right)dt = -\log t = -\log(1+\cot x)$.

(ii) Here $I = \int \frac{dx}{(1+x^2)\tan^{-1} x}$.

Putting $\tan^{-1}$ x = t, so that $\left[\frac{1}{(1+x^2)}\right]dx = dt$, we have

$I = \int\left(\frac{1}{t}\right)dt$ = log t = log ($\tan^{-1}$x).

Example 23:

Integrate (i) $\frac{ax+b}{ax^2+2bx+c}$. *(ii)* $\frac{ax^{n-1}}{x^n+b}$.

Solution:

(i) Let $I=\int\frac{ax+b}{ax^2+2bx+c}dx$.

[Put $ax^2 + 2bx + c = t$, so that $(2ax + 2b)\,dx = dt$].

$$\therefore\ I=\frac{1}{2}\int\frac{dt}{t}=\frac{1}{2}\log t=\frac{1}{2}\log\left(ax^2+2bx+c\right).$$

(ii) Let $I=\int\frac{ax^{n-1}dx}{x^n+b}=\frac{1}{n}\int\frac{dt}{t}$

[Putting $x^n + b = t$, so that $nx^{n-1}\,dx = dt$]

$$=\left(\frac{a}{n}\right).\log(t)=\left(\frac{a}{n}\right).\log\left(x^n+b\right).$$

Example 24(a):

Integrate (i) $\frac{e^x-e^{-x}}{e^x+e^{-x}}$, *(ii)* $\frac{10x^9+10x.\log_e 10}{10^x+x^{10}}$.

Solution:

(i) Let $I=\int\frac{e^x-e^{-x}}{e^x+e^{-x}}dx$.

Now putting $e^x + e^{-x} = t$,

so that $(e^x - e^{-x})\,dx = dt$,

we have $I=\int\left(\frac{1}{t}\right)dt=\log t=\log\left(e^x+e^{-x}\right)$.

(ii) Here $I=\int\frac{10x^9+10^x.\log_e 10}{10^x+x^{10}}dx$.

Now putting $10^x + x^{10} = t$, and $(10^x \log_e 10 + 10x^9)\,dx = dt$,

we have $I=\int\left(\frac{1}{t}\right)dt = \log t = \log(10^x + x^{10})$.

Example 24(b):

Integrate (i) $\frac{\sin x}{a+b\cos x}$, *(ii)* $\frac{\sin x\cos x}{a\cos^2 x+b\sin^2 x}$

Solution:

(i) Here $I=\int\frac{\sin x\,dx}{a+b\cos x}=-\frac{1}{b}\int\frac{-b\sin x}{a+b\cos x}dx$.

Putting $a + b\cos x = t$,

so that – b sin x dx = dt,

we have $I = \left(-\frac{1}{b}\right)\int\left(\frac{1}{t}\right)dt$

$$= -\left(\frac{1}{b}\right).\log(t) = -\left(\frac{1}{b}\right).\log(a + b\cos x).$$

(ii) Let $I = \int\left(\frac{\sin x \cos x}{a\cos^2 x + b\sin^2 x}\right)dx$

Put a $\cos^2$ x + b $\sin^2$ x = t,

so that (– 2a cos x sin x + 2b sin x cos x) dx = dt,

or {2 (b – a) sin x cos x} dx = dt.

We have $I = \frac{1}{2(b-a)}\int\frac{dt}{t} = \frac{1}{2(b-a)}\log t$

$$= \frac{1}{2(b-a)} \log (a \cos^2 x + b \sin^2 x).$$

Example 24(c):

Integrate (i) $\frac{\sin x}{1+\cos^2 x}$, *(ii)* $\frac{\sin(a+b\log x)}{x}$.

Solution:

(i) Here $I = \int\frac{\sin x}{1+\cos^2 x}dx$.

Putting cos x = t, so that – sin x dx = dt, we have

$$I = -\int\frac{dt}{(1+t^2)} = -\tan^{-1} t = -\tan^{-1} (\cos x).$$

(ii) Here $I = \int\left[\frac{\{\sin(a+b\log x)\}}{x}\right]dx$.

Putting a + b log x = t, so that $\left(\frac{b}{x}\right)$ dx = dt, we have

$$I = \frac{1}{b}\int \sin t\, dt = -\frac{1}{b}\cos t = -\frac{1}{b}\cos(a + b\log x).$$

Example 24(d):

Integrate (i) $\frac{1}{x\cos^2(1+\log x)}$ *(ii)* $\frac{1}{x(1+\log x)^m}$.

Solution:

(i) Here $I = \int \frac{dx}{\{x\cos^2(1+\log x)\}}$.

Putting 1 + log x = t, so that $\left(\frac{1}{x}\right)$ dx = dt, we have

$$I = \int \frac{dt}{\cos^2 t} = \int \sec^2 t\, dt = \tan t = \tan(1 + \log x).$$

(ii) Here $I = \int \frac{dx}{\{x(1+\log x)^m\}}$.

Putting 1 + log x = t, so that $\left(\frac{1}{x}\right)$ dx = dt, we have

$$I = \int \frac{dt}{t^m} = \frac{t^{-m+1}}{-m+1} = \frac{(1+\log x)^{-m+1}}{(1-m)}$$

$$= \frac{1}{(1-m)}(1+\log x)^{1-m}.$$

Example 24(e):

Integrate (i) $\frac{e^{\tan^{-1}}x}{1+x^2}$ *(ii)* $\frac{\sin(\tan^{-1}x)}{1+x^2}$.

Solution:

(i) Putting $\tan^{-1}$ x - t, so that $\left[\frac{1}{(1+x^2)}\right]dx = dt$, we have

$$I = \int \frac{e^{\tan^{-1}}x}{1+x^2}dx = \int e^t.dt = e^t = e^{\tan^{-1}x}.$$

(ii) Putting $\tan^{-1}$ x = t so that $\left[\frac{1}{(1+x^2)}\right]dx = dt$, we have

$$I = \int \frac{\sin(\tan^{-1}x)}{1+x^2}dx = \int \sin t\, dt = -\cos t = -\cos(\tan^{-1} x).$$

Example 25:

Integrate (i) x $\cos^3 x^2.\sin x^2$ (ii) $x^3 \tan^4 x^4.\sec^2 x^4$.

Solution:

(i) Here $I = \int x\cos^3 x^2 \sin x^2 dx$.

First, putting $x^2 = t$,
so that $2x\, dx = dt$, we have

$$I = \frac{1}{2}\int \cos^3 t \sin t\, dt.$$

Now putting $\cos t = u$, so that $-\sin t\, dt = du$, we have

$$I = -\frac{1}{2}\int u^3 du = -\frac{1}{2}\frac{u^4}{4} = -\frac{1}{8}u^4$$

$$= -\frac{1}{8}\cos^4 t = -\frac{1}{8}\cos^4 x^2. \qquad [\because t = x^2]$$

Note: The students should also solve this problem by making the single substitution $\cos x^2 = t$.

(ii) Here $I = \int x^3 \tan^4 x^4 \sec^2 x^4 dx$.

Putting $\tan x^4 = t$, so that $(\sec^2 x^4).4x^3\, dx = dt$, we have

$$I = \int \frac{t^4}{4} \cdot dt = \frac{1}{4}\left(\frac{t^5}{5}\right) = \frac{(\tan x^4)^5}{20}.$$

Example 26:

Integrate $\dfrac{x^3 \tan^{-1} x^4}{1 + x^8}$.

Solution:

Putting $\tan^{-1} x^4 = t$ so that $\frac{1}{1+x^8}.4x^3 dx = dt$, we have

$$I = \int \frac{x^3 \tan^{-1} x^4}{1+x^8} dx = \frac{1}{4}\int t\, dt = \frac{1}{8}t^2 = \frac{1}{8}\left(\tan^{-1} x^4\right)^2.$$

Example 27(a):

Evaluate $\displaystyle\int \frac{e^x (1+x)}{\sin^2 (xe^x)}$.

Solution:

Putting $xe^x = t$ so that $(e^x + xe^x)\, dx = dt$
or $e^x(1 + x)\, dx = dt$, we have

$$I = \int \frac{dt}{\sin^2 t} = \int \operatorname{cosec}^2 t\, dt = -\cot t = -\cot(xe^x).$$

Example 27(b):

Integrate (i) $\frac{1}{(e^x+1)}$, (ii) $\frac{1}{(e^x-1)}$.

Solution:

(i) Here $I=\int\frac{dx}{e^x+1}=\int\frac{e^{-x}}{1+e^{-x}}dx$,

[Multiplying Nr. and Dr. both by e^{-x}]

$$=-\int\frac{-e^{-x}}{1+e^{-x}}dx=-\log\left(1+e^{-x}\right),\ [\because \text{ Nr. is diff. coeff. of Dr.}]$$

$$=-\log\left(\frac{1+e^x}{e^x}\right)=-\left[\log\left(1+e^x\right)-\log e^x\right]$$

$$= x \log e - \log (1 + e^x) = x - \log (1 + e^x).$$

(ii) Similarly $\int\frac{dx}{e^x-1}=\log\left(1-e^{-x}\right)$.

Example 27(c):

Integrate (i) $\frac{\cot x}{\log(\sin x)}$, (ii) $\frac{\tan x}{\log(\sec x)}$.

Solution:

(i) Here $\frac{d}{dx}(\log\sin x)=\frac{1}{\sin x}\cos x=\cot x$.

$$\therefore\ I=\int\frac{\cot x\,dx}{\log\sin x}=\log(\log\sin x),\ [\because \text{ Nr. is diff. coeff. of Dr.}]$$

(ii) Similarly, we have

$$\int\frac{\tan x\,dx}{\log\sec x}=\log(\log\sec x).$$

Example 27(d):

Integrate (i $\sqrt{(1+\sin x)}$, (ii) $\frac{1}{\sqrt{(1+\sin x)}}$.

Solution:

(i) We have

$$I=\int\sqrt{(1+\sin x)}\,dx=\int\sqrt{\left\{1-\cos\left(\frac{1}{2}\pi+x\right)\right\}}\,dx$$

$$= \int \sqrt{\left[2\sin^2\left(\frac{1}{4}\pi + \frac{1}{2}x\right)\right]}dx = \sqrt{2}\int \sin\left(\frac{1}{2}x + \frac{1}{4}\pi\right)dx.$$

Now putting $\frac{1}{2}x + \frac{1}{4}\pi = t$,

so that $\frac{1}{2}dx = dt$ or $dx = 2\ dt$, we have

$$I = \sqrt{2}\int 2\sin t\, dt = -2\sqrt{2}\cos t = -2\sqrt{2}\cos\left(\frac{1}{2}x + \frac{1}{4}\pi\right).$$

(ii) Here $I = \int \frac{dx}{\sqrt{1(1+\sin x)}} = \frac{1}{\sqrt{2}}\int \operatorname{cosec}\left(\frac{1}{2}x + \frac{1}{4}\pi\right)dx,$

Now putting $\frac{1}{2}x + \frac{1}{4}\pi = t$ so that $dx = 2dt$, we have

$$I = \frac{2}{\sqrt{2}}\int \operatorname{cosec} t\, dt = \sqrt{2}\log\tan\left(\frac{1}{2}t\right)$$

$$= \sqrt{2}\log\tan\left(\frac{1}{4}x + \frac{1}{8}\pi\right).$$

Example 28:

Integrate (i) $\frac{\sin x}{\sqrt{(1+\sin x)}}$ *(ii)* $\sqrt{(1-\cos x)}$.

Solution:

(i) Here $I = \int \frac{\sin x\, dx}{\sqrt{(1+\sin x)}} = \int \frac{(1+\sin x)-1}{\sqrt{(1+\sin x)}}dx$

$$= \int \sqrt{(1+\sin x)}\,dx - \int \frac{1}{\sqrt{(1+\sin x)}}dx$$

$$= -2\sqrt{2}\cos\left(\frac{1}{2}x + \frac{1}{4}\pi\right) - \sqrt{2}\log\tan\left(\frac{1}{4}x + \frac{1}{8}\pi\right),$$

(ii) Here $I = \int \sqrt{(1-\cos x)}dx = \sqrt{2}\int \sin\frac{1}{2}x\, dx$

$$= -\sqrt{2}\frac{\cos\left(\frac{x}{2}\right)}{\left(\frac{1}{2}\right)} = -2\sqrt{2}\cos\left(\frac{x}{2}\right).$$

Example 29:

Integrate (i) $\frac{\sec x}{a + b\tan x}$ *(ii)* $\frac{\sec x}{\sqrt{3} + \tan x}$.

Solution:

(i) Let $I = \int \frac{\sec x}{a + b\tan x}dx = \int \frac{dx}{a\cos x + b\sin x}$.

Now let a = r sin ϕ

and b = r cos ϕ. This gives

$r = \sqrt{(a^2 + b^2)}$, and $\phi = \tan^{-1}\left(\frac{a}{b}\right)$.

$$\therefore \quad I = \int \frac{dx}{r\sin(x+\phi)} = \frac{1}{\sqrt{(a^2+b^2)}}\int \operatorname{cosec}(x+\phi)\,dx$$

$$= \frac{1}{\sqrt{(a^2+b^2)}}\log\tan\left(\frac{1}{2}x + \frac{1}{2}\tan^{-1}\frac{a}{b}\right).$$

(ii) We have $\int \frac{\sec x\,dx}{\sqrt{3} + \tan x} = \int \frac{\sec x\,dx}{\sqrt{3} + \left(\frac{\sin x}{\cos x}\right)}$

$$= \int \frac{dx}{\sqrt{3}\cos x + \sin x} = \int \frac{dx}{2\left[\left(\frac{\sqrt{3}}{2}\right)\cos x + \frac{1}{2}\sin x\right]}$$

$$= \frac{1}{2}\int \frac{dx}{\sin\left(x + \frac{1}{3}\pi\right)} = \frac{1}{2}\int \operatorname{cosec}\left(x + \frac{1}{3}\pi\right)dx$$

$$= \frac{1}{2}\log\tan\left[\frac{1}{2}x + \left(\frac{\pi}{6}\right)\right].$$

Example 30:

Integrate (i) $\frac{\cos 2x}{\sin x}$, *(ii)* $\frac{\cos 2x}{\cos x}$.

Solution:

(i) $I = \int \frac{\cos 2x}{\sin x}dx = \int \frac{1 - 2\sin^2 x}{\sin x}dx$ [$\because$ cos 2x = 1 – 2 sin²x]

$$\int (\operatorname{cosec} x - 2\sin x)\,dx = \log\tan\left(\frac{1}{2}x\right) + 2\cos x.$$

(ii) Here $I = \int \frac{\cos 2x}{\cos x}dx = \int \frac{2\cos^2 x - 1}{\cos x}dx$,

[$\because$ $\cos 2x = 2\cos^2 x - 1$]

$$\int(2\cos x-\sec x)\,dx=2\sin x-\log\tan\left(\frac{1}{2}x+\frac{1}{4}\pi\right).$$

Example 31(a):

Integrate $\frac{1}{(1+3\sin^2)}$.

Solution:

Dividing Nr. and Dr. by $\cos^2 x$, we have

$$I=\int\frac{dx}{1+3\sin^2 x}=\int\frac{\sec^2 x\,dx}{\sec^2 x+3\tan^2 x}$$

$$=\int\frac{\sec^2 x\,dx}{(1+\tan^2 x)+3\tan^2 x}$$

$$=\int\frac{\sec^2 x\,dx}{1+4\tan^2 x}.$$

Now putting 2 tan x = t

so that $2\sec^2 x\,dx = dt$, we have

$$I=\frac{1}{2}\int\frac{dt}{1+t^2}=\frac{1}{2}\tan^{-1}t=\frac{1}{2}\tan^{-1}(2\tan x).$$

Example 31(b):

Evaluate $\int\left[\frac{(\cos x)}{(a^2+b^2\sin^2 x)}\right]dx$.

Solution:

Putting b sin x = t so that b cos x dx = dt, we have

$$I=\int\frac{\cos x\,dx}{a^2+b^2\sin^2 x}=\frac{1}{b}\int\frac{dt}{a^2+t^2}=\frac{1}{b}\cdot\frac{1}{a}\tan^{-1}\left(\frac{t}{a}\right)$$

$$=\frac{1}{ab}\tan^{-1}\left(\frac{b\sin x}{a}\right).$$

Example 31(c):

Evaluate $\int\frac{\cot(\log x)}{x}dx$.

Solution:

Putting log x = t so that $\left(\frac{1}{x}\right)dx=dt$, we have the given integral

$$I = \int \cot t\, dt = \log \sin t = \log\{\sin(\log x)\}.$$

Example 32:

Evaluate the following integrals:

$$(i)\ \int \frac{d\theta}{a^2 \sin^2\theta + b^2 \cos^2\theta} \quad (ii)\ \int \frac{dx}{4\sin^2 x + 5\cos^2 x}.$$

Solution:

(i) Dividing Nr. and Dr. by $\cos^2 \theta$, we have

$$I = \int \frac{d\theta}{a^2 \sin^2\theta + b^2 \cos^2\theta} = \int \frac{\sec^2\theta\, d\theta}{a^2 \tan^2\theta + b^2} = \frac{1}{a^2}\int \frac{\sec^2\theta\, d\theta}{\tan^2\theta + \left(\frac{b^2}{a^2}\right)}$$

$$= \frac{1}{a^2}\int \frac{dt}{t^2 + \left(\frac{b^2}{a^2}\right)} \quad \text{[Putting } t = \tan\theta, \text{ so that } dt = \sec^2\theta\, d\theta]$$

$$= \frac{1}{a^2}\cdot\frac{a}{b}\tan^{-1}\left(\frac{t}{b/a}\right) = \frac{1}{ab}\tan^{-1}\left(\frac{a}{b}\tan\theta\right).$$

(ii) Let $I = \int \frac{dx}{4\sin^2 x + 5\cos^2 x}$

$$= \int \frac{\sec^2 x\, dx}{4\tan^2 x + 5}, \text{ dividing the Nr. and Dr. by } \cos^2 x$$

$$= \int \frac{dt}{4t^2 + 5}, \text{ putting } \tan x = t \text{ so that } \sec^2 x\, dx = dt$$

$$= \frac{1}{4}\int \frac{dt}{t^2 + \left(\sqrt{\frac{5}{2}}\right)^2} = \frac{1}{4}\frac{1}{\left(\sqrt{\frac{5}{2}}\right)}\tan^{-1}\frac{t}{\sqrt{\frac{5}{2}}}$$

$$= \frac{1}{2\sqrt{5}}\tan^{-1}\frac{2t}{\sqrt{5}} = \frac{1}{2\sqrt{5}}\tan^{-1}\frac{2\tan x}{\sqrt{5}}.$$

Example 33:

Evaluate $\int \frac{e^y\, dy}{\sqrt{(1+e^{2y})}}.$

Solution:

Put $e^y = t$

so that $e^y\,dy = dt$.

$\therefore$ the given integral

$$= \int \left\{ \frac{1}{\sqrt{(1+t^2)}} \right\} dt = \sinh^{-1} t = \sinh^{-1} (e^y).$$

Example 34:

Evaluate $\int \cos x \sqrt{(4 - \sin^2 x)}\, dx$.

Solution:

Put $\sin x = t$

so that $\cos x\, dx = dt$.

Then the given integral $= \int \sqrt{(4 - t^2)}\, dt = \int \sqrt{(2^2 - t^2)}\, dt$

$$= \frac{1}{2} t \sqrt{(4 - t^2)} + \frac{2^2}{2} \sin^{-1}\left(\frac{t}{2}\right)$$

$$= \frac{1}{2} \sin x . \sqrt{(4 - \sin^2 x)} + 2 \sin^{-1}\left(\frac{1}{2} \sin x\right).$$

Example 35:

Evalulate $\int \sec x \tan x \sqrt{(\sec^2 x + 1)}\, dx$.

Solution:

Put $\sec x = t$

so that $\sec x \tan x\, dx = dt$.

Then the given integral $= \int \sqrt{(t^2 + 1)}\, dt$

$$= \frac{1}{2} t \sqrt{(t^2 + 1)} + \frac{1}{2} \sinh^{-1} t$$

$$= \frac{1}{2} \sec x \sqrt{(\sec^2 x + 1)} + \frac{1}{2} \sinh^{-1} (\sec x).$$

Example 36:

Evaluate $\int \frac{x}{\sqrt{(x^4 + 4)}}\, dx$.

Solution:

Put $x^2 = t$; $\therefore$ $2x\, dx = dt$.

$\therefore$ the given integral $=\frac{1}{2}\int \frac{1}{\sqrt{(t^2+4)}}dt$

$$=\frac{1}{2}\sinh^{-1}\left(\frac{t}{2}\right)=\frac{1}{2}\sinh^{-1}\left(\frac{x^2}{2}\right).$$

Example 37(a):

Evaluate $\int\left\{\frac{x^2}{\sqrt{(x^6-9)}}\right\}dx.$

Solution:

Put $x^3 = t$, $\therefore$ $3x^2\,dx = dt$.

$\therefore$ the given integral $=\frac{1}{3}\int \frac{dt}{\sqrt{(t^2-9)}}$

$$=\frac{1}{3}\cosh^{-1}\left(\frac{t}{3}\right)=\frac{1}{3}\cosh^{-1}\left(\frac{x^3}{3}\right).$$

Example 37(b):

Evaluate $\int x\sqrt{(x^4+9)}dx.$

Solution:

Put $x^2 = t$; $\therefore$ $2x\,dx = dt$.

$\therefore$ the given integral $=\frac{1}{2}\int \sqrt{(t^2+3^2)}\,dt$

$$=\frac{1}{2}\left[\frac{1}{2}t\sqrt{(t^2+9)}+\frac{9}{2}\sinh^{-1}\left(\frac{t}{3}\right)\right]$$

$$=\frac{1}{4}x^2\sqrt{(x^4+9)}+\frac{9}{4}\sinh^{-1}\left(\frac{x^2}{3}\right).$$

Example 37(c):

Evaluate $\int x^2\sqrt{(x^6-1)}\,dx.$

Solution:

Put $x^3 = t$; $\therefore$ $3x^2\,dx = dt$.

$\therefore$ the given integral $=\frac{1}{3}\int \sqrt{(t^2-1)}\,dt$

$$=\frac{1}{3}\left[\frac{t}{2}\sqrt{\left(t^2-1\right)}-\frac{1}{2}\cosh^{-1}\left(\frac{t}{1}\right)\right]$$

$$=\frac{1}{3}\left[\frac{x^3}{2}\sqrt{\left(x^6-1\right)}-\frac{1}{2}\cosh^{-1}x^3\right]$$

$$=\frac{1}{6}x^3\sqrt{\left(x^6-1\right)}-\frac{1}{6}\cosh^{-1}x^3.$$

Example 38:

Evaluate $\int x^5 e^x dx$.

Solution:

Here e^x will be successively integrated and x^5 will be successively differentiated. Thus applying successive integration by parts, the given integral

$I = x^5 e^x = (5x^4) e^x + (20x^3) e^x - (60x^2) e^x + (12^x) e^x - 120e^x,$

the process of successive integration by parts terminates because the differential coefficient of 120 is zero.

$\therefore \quad I = x^5e^x - (5x^4 + 20x^3 - 60x^2 + 120x - 120).$

Example 39(a):

Evaluate $\int x^4 \sin x\, dx$.

Solution:

Applying successive integration by parts, the given integral

$I = x^4 - (-\cos x) - (4x^3)\cdot(-\sin x) + (12x^2)\cdot(\text{cox } x)$
$\quad - (24x)\cdot(\sin x) + (24)\cdot(-\cos x).$

$= -x^4\cos x + 4x^3\sin x + 12x^2\cos x - 24\, x \sin x - 24 \cos x.$

Remark:

While applying successive integration by parts the successive differential coefficients and the successive integrals must at the first stage be put within brackets.

Example 39(b):

Evaluate $\int x^3 e^{-x} dx$.

Solution:

Applying successive integration by parts, the given integral

$$I = (x^3)\cdot(-e^{-x}) - (3x^2)\cdot(e^{-x}) + (6x)\cdot(-e^{-x}) - (6)\cdot(e^{-x})$$
$$= -x^3e^{-x} - 3x^2e^{-x} - 6xe^{-x} - 6e^{-x}.$$
$$= -(x^3 + 3x^2 + 6x + 6)\,e^{-x}.$$

Example 39(c):

Integrate (i) x log x, (ii) $\frac{(\log x)}{x^2}$, *(iii)* x^n *log x.*

Solution:

(i) Here x should be taken as the second function because the integral of log x cannot be easily written down.

We have $\int x\log x\,dx = \int(\log x)x\,dx$

$$= \frac{1}{2}x^2\log x - \frac{1}{2}\int x\,dx = \frac{1}{2}x^2\log x - \frac{1}{2}\cdot\left(\frac{x^2}{2}\right)$$

$$= \frac{1}{4}x^2\log x^2 - \frac{1}{4}x^2\log e, \qquad [\because \log e = 1] \textbf{ Note}$$

$$= \frac{1}{4}x^2\log\left(\frac{x^2}{e}\right).$$

(ii) We have $\int\left[\frac{(\log x)}{x^2}\right]dx = \int(\log x)\left(\frac{1}{x^2}\right)dx$

$$= (\log x)\left(-\frac{1}{x}\right) - \int\left(\frac{1}{x}\right)\cdot\left(-\frac{1}{x}\right)dx,$$

[Integrating by parts taking $1/x^2$ as the second function]

$$= -\left(\frac{1}{x}\right)\log x - \left(\frac{1}{x}\right) = -\left(\frac{1}{x}\right)(\log x + \log e) \qquad \textbf{Note}$$

$$= -\left(\frac{1}{x}\right)\log(xe).$$

(iii) We have $\int x^n\log x\,dx = \int(\log x).x^n dx$

$$= (\log x)\cdot\frac{x^{n+1}}{n+1} - \int\frac{1}{x}\cdot\frac{x^{n+1}}{n+1}dx,$$

[Integrating by parts taking x^n as the second function]

$$= (\log x)\cdot\frac{x^{n+1}}{n+1} - \int\frac{x^n}{n+1}dx$$

$$= (\log x)\cdot\frac{x^{n+1}}{n+1} - \frac{x^{n+1}}{(n+1)^2}.$$

Example 39(d):

Integrate (i) $tan^{-1}x$, (ii) $cot^{-1}x$, (iii) $sin^{-1}x$.

Solution:

(i) As there is only one function here, unity should be taken as the 2nd function. We have $\int \tan^{-1} x\,dx = \int (\tan^{-1} x).1\,dx$.

Integrating by parts regarding 1 as the secon function, we have

$$\int (\tan^{-1} x).1\,dx = (\tan^{-1} x).-\int \left\{\frac{1}{(1+x^2)}\right\} x\,dx$$

$$= x\tan^{-1} x - \frac{1}{2}\int \frac{2x}{1+x^2}dx = x\tan^{-1} x - \frac{1}{2}\log(1+x^2),$$

(∵ Nr. 2x is the diff. coeff. of the Dr. $1 + x^2$)

(ii) We have $\int \cot^{-1} x\,dx = \int (\cot^{-1} x).1\,dx$

$$= (\cot^{-1} x).x - \int \left\{-\frac{1}{(1+x^2)}\right\}.x\,dx,$$

[Integrating by parts taking unity as the second function]

$$= x\cot^{-1} x + \frac{1}{2}\int \left\{\frac{2x}{(1+x^2)}\right\}dx = x\cot^{-1} x + \frac{1}{2}\log(1+x^2).$$

(iii) We have $\int \sin^{-1} x\,dx = \int (\sin^{-1} x).1\,dx$

$$= (\sin^{-1} x).x - \int \left\{\frac{1}{\sqrt{(1-x^2)}}\right\} x\,dx$$

$$= x\sin^{-1} x + \frac{1}{2}\int (1-x^2)^{-1/2}(-2x)\,dx$$ **Note**

$$= x\sin^{-1} x + \frac{1}{2}\frac{(1-x^2)^{1/2}}{1/2},$$

$$= x \sin^{-1} x + (1 - x^2)^{1/2}.$$

Example 40:

Integrate (i) x^2e^{mx}, (ii) $x^2 \sin x$, (iii) $x^2 \cos 2x$.

Solution:

(i) Integrating by parts taking e^{mx} as the second function, we have

$$\int x^2 e^{mx} dx = x^2 \cdot \frac{e^{mx}}{n} - \int 2x.\frac{e^{mx}}{m} dx$$

$$= \frac{x^2}{m} e^{mx} - \frac{2}{m}\int x.e^{mx} dx = \frac{x^2}{m} e^{mx} - \frac{2}{m}\left\{x \cdot \frac{e^{mx}}{m} - \int 1.\frac{e^{mx}}{m} dx\right\},$$

[again integrating by parts]

$$= \frac{x^2}{m} e^{mx} - \frac{2}{m^2} x e^{mx} + \frac{2}{m^3} e^{mx}$$

$= e^{mx}\, m^{-3}\, (m^2x^2 - 2mx + 2).$

(ii) Integrating by parts taking sin x as second function, we have

$$\int x^2 \sin x\, dx = -x^2 \cos x - \int 2x(-\cos x) dx$$

$$= -x^2 \cos x + 2\int x \cos x\, dx$$

$$= -x^2 \cos x + 2x \sin x - 2\int \sin x\, dx$$

$= -x^2 \cos x + 2x \sin x + 2 \cos x.$

Similarly $\int x^2 \cos x\, dx = x^2 \sin x + 2x \cos x - 2 \sin x.$

(iii) Integrating by parts taking cos 2x as second function, we have

$$\int x^2 \cos 2x\, dx = x^2 \cdot \left(\frac{1}{2}\sin 2x\right) - \int 2x \cdot \frac{1}{2} \sin 2x\, dx$$

$$= \frac{1}{2} x^2 \sin 2x - \int x \sin 2x\, dx$$

$$= \frac{1}{2} x^2 \sin 2x - \left\{x \cdot \frac{1}{2}(-\cos 2x) + \frac{1}{2}\int \cos 2x\, dx\right\},$$

(Again integrating by parts taking sin 2x as the 2nd function)

$$= \frac{1}{2} x^2 \sin 2x + \frac{1}{2} x \cos 2x - \frac{1}{2}\int \cos 2x\, dx$$

$$= \frac{1}{2} x^2 \sin 2x + \frac{1}{2} x \cos 2x - \frac{1}{4} \sin 2x$$

$$= \frac{1}{2}\left(x^2 - \frac{1}{2}\right) \sin 2x + \frac{1}{2} x \cos 2x.$$

Example 41:

Integrate (i) log x, (ii) $(\log x)^2$, (iii) $x^n (\log x)^2$.

Solution:

(i) As there is only one function here, unity should be taken as the second function. We have $\int \log x\, dx = \int (\log x) \cdot 1\, dx$

$$= (\log x)\cdot x - \int\left(\frac{1}{x}\right)\cdot x\,dx = x\log x - \int 1.dx = x\log x - x$$

$$= x\ (\log x - 1) = x \log\left(\frac{x}{e}\right).$$

(ii) We have $\int(\log x)^2 dx = \int(\log x)^2 .1dx$

$$= (\log x)^2 .x - \int\left\{(2\log x).\left(\frac{1}{x}\right)\right\}.x\,dx,$$

[Integrating by parts taking 1 as the second function]

$$= x(\log x)^2 - 2\int(\log x).1dx$$

$$= x(\log x)^2 - 2\left[(\log x).x - \int\frac{1}{x}\cdot x\,dx\right],$$

[Again integrating by parts taking 1 as the 2nd function]

$$= x\ (\log x)^2 - 2x \log x + 2\int 1.dx$$

$$= x\ (\log x)^2 - 2x \log x + 2x.$$

(iii) We have $\int x^n(\log x)^2 dx = \int(\log x)^2 .x^n dx$

$$= (\log x)^2 \frac{x^{n+1}}{n+1} - \int\left\{(2\log x).\frac{1}{x}\right\}\frac{x^{n+1}}{n+1}dx,$$

[Integrating by parts taking x^n as the second function]

$$= \frac{1}{(n+1)}x^{n+1}(\log x)^2 - \frac{2}{n+1}\int(\log x).x^n dx$$

$$= \frac{1}{(n+1)}x^{n+1}(\log x)^2 - \frac{2}{(n+1)}\left[(\log x).\frac{x^{n+1}}{n+1} - \int\frac{1}{x}\cdot\frac{x^{n+1}}{n+1}\cdot dx\right],$$

[Again integrating by parts taking x^n as the second function]

$$= \frac{1}{(n+1)}x^{n+1}.(\log x)^2 - \frac{2}{(n+1)^2}(\log x).x^{n+1} + \frac{2}{(n+1)^2}\int x^n\,dx$$

$$= \frac{1}{(n+1)}x^{n+1}(\log x)^2 - \frac{2}{(n+1)^2}(\log x).x^{n+1} + \frac{2}{(n+1)^3}x^{n+1}$$

$$= x^{n+1}\left[\frac{(\log x)^2}{(n+1)} - \frac{2\log x}{(n+1)^2} + \frac{1}{(n+1)^3}\right].$$

Example 42(a):

Integrate (i) $e^x \sin x$ (ii) $e^x \cos x$, (iii) $e^{2x} \sin x$, (iv) $e^{3x} \cos 4x$.

Solution:

(i) Integrating by parts taking sin x as the second function, we have

$$\int e^x \sin x\,dx = -e^x \cos x + \int e^x \cos x\,dx.$$

Now taking cos x as the second function we again apply integration by parts to the integral on the right hand side. Thus, we get

$$\int e^x \sin x\,dx = -e^x \cos x + e^x \sin x - \int e^x \sin x\,dx.$$

Transposing the last term on the right hand side to the left and dividing by 2, we get

$$\int e^x \sin x\,dx = \frac{1}{2} e^x (\sin x - \cos x).$$

(ii) Here $\int e^x \cos x\,dx = e^x.\sin x - \int e^x \sin x\,dx$,

[Integrating by parts taking cos x as the second function]

$$= e^x (\sin x + \cos x) - \int e^x \cos x\,dx.$$

Transposing the last term on the right hand side to the left and dividing by 2, we get

$$\int e^x \cos x\,dx = \frac{1}{2} e^x (\sin x + \cos x).$$

(iii) We have

$$\int e^{2x} \sin x\,dx = e^{2x}.(-\cos x) - \int 2e^{2x}.(-\cos x)\,dx,$$

[Integrating by parts taking sin x as the second function]

$$= -e^{2x} \cos x + 2\int e^{2x} \cos x\,dx$$

$$= -e^{2x} \cos x + 2\left[e^{2x}.\sin x - \int 2e^{2x} \sin x\,dx\right]$$

$$= -e^{2x} \cos x + 2e^{2x} \sin x - 4\int e^{2x} \sin x\,dx.$$

Transposing and dividing by 5, we get

$$\int e^{2x} \sin x\,dx = \frac{1}{5} e^{2x} [2\sin x - \cos x].$$

(iv) Integrating by parts taking cos 4x as the second function, we have

$$\int e^{2x}\cos 4x\,dx + e^{3x}.\left(\frac{\sin 4x}{4}\right) - \int (3e^{3x})\cdot\left(\frac{\sin 4x}{4}\right)dx$$

$$= \frac{1}{4}e^{3x}\sin 4x - \frac{3}{4}\int e^{3x}\sin 4x\,dx$$

$$= \frac{1}{4}e^{3x}\sin 4x - \frac{3}{4}\left[e^{3x}.\left(-\frac{\cos 4x}{4}\right) - \int 3e^{3x}.\left(-\frac{\cos 4x}{4}\right)dx\right]$$

$$= \frac{1}{4}e^{3x}\sin 4x + \frac{3}{16}e^{3x}\cos 4x - \frac{9}{16}\int e^{3x}\cos 4x\,dx.$$

Transposing the last term on the right hand side to the left, we have

$$\left(1+\frac{9}{16}\right)\int e^{3x}\cos 4x\,dx = \frac{1}{4}e^{3x}\sin 4x + \frac{3}{16}e^{3x}\cos 4x$$

or $\dfrac{25}{16}\int e^{3x}\cos 4x\,dx = \dfrac{1}{4}e^{3x}\sin 4x + \dfrac{3}{16}e^{3x}\cos 4x$

or $\int e^{3x}\cos 4x\,dx = \dfrac{4}{25}e^{3x}\sin 4x + \dfrac{3}{25}e^{3x}\cos 4x$

$$= \frac{1}{25}e^{3x}(4\sin 4x + 3\cos 4x).$$

Alternative Solution

$\int e^{3x}\cos 4x\,dx$ is of the form

$\int e^{ax}\cos bx\,dx$, where a = 3 and b = 4.

Now $\int e^{ax}\cos bx\,dx = \dfrac{e^{ax}}{\sqrt{(a^2+b^2)}}\cos\left(bx - \tan^{-1}\dfrac{b}{a}\right).$

$$\therefore \int e^{3x}\cos 4x\,dx = \frac{e^{3x}}{\sqrt{(9+16)}}\cos\left(4x - \tan^{-1}\frac{4}{3}\right)$$

$$= \frac{e^{3x}}{5}\cos\left(4x - \tan^{-1}\frac{4}{3}\right).$$

Example 42(b):

$$\int e^x(n\cos x + \sin x)\,dx.$$

Solution:

$$\int e^x(n\cos x + \sin x)\,dx = n\int e^x\cos x\,dx + \int e^x\sin x\,dx.$$

Now using results of Ex. 5 (i), and (ii), we get the required value of the given integral

$$= \frac{1}{2} ne^{x}(\cos x + \sin x) + \frac{1}{2} e^{x}(\sin x - \cos x).$$

Example 42(c):

Evaluate $\int \frac{xe^{x}}{(x+1)^{2}} dx$.

Solution:

We have

$$\int \frac{xe^{x}}{(x+1)^{2}} dx = \int (xe^{x}) \frac{1}{(x+1)^{2}} dx.$$

Integrating by parts taking $\frac{1}{(x+1)^{2}}$ as the second function and xe^{x} as the first function, we have

$$\int \frac{xe^{x}}{(x+1)^{2}} dx = (xe^{x})\left(-\frac{1}{x+1}\right) - \int (e^{x} + xe^{x})\left(-\frac{1}{x+1}\right) dx,$$

$$\left[\text{Note that the integral of } \frac{1}{(x+1)^{2}} \text{ is} -\frac{1}{x+1}\right]$$

$$= -\frac{xe^{x}}{x+1} + \int e^{x}(x+1) \cdot \frac{1}{x+1} dx$$

$$= -\frac{xe^{x}}{x+1} + \int e^{x} dx = -\frac{xe^{x}}{x+1} + e^{x}$$

$$= e^{x}\left[1 - \frac{x}{x+1}\right] = e^{x} \frac{x+1-x}{x+1} = \frac{e^{x}}{x+1}.$$

Alternative Solution

We have $\int \frac{xe^{x}}{(x+1)^{2}} dx = \int e^{x} \frac{(x+1)-1}{(x+1)^{2}} dx$

$$= \int e^{x}\left[\frac{1}{x+1} - \frac{1}{(x+1)^{2}}\right] dx$$

$$= \int e^{x}[f(x) + f'(x)] dx, \text{ where } f(x) = \frac{1}{x+1}$$

$$= \int e^x f(x)dx + \int e^x f'(x)dx$$

$$= e^x f(x) - \int e^x f'(x)dx + \int e^x f'(x)dx,$$

applying integration by parts to the first integral taking e^x as the second function

$$= e^x f(x) = e^x \frac{1}{x+1}.$$

Example 42(d):

Evaluate $\int e^x \frac{(x^2+1)}{(x+1)^2} dx.$

Solution:

We have $\int e^x \frac{(x^2+1)}{(x+1)^2} dx$

$$= \int e^x \frac{(x^2-1)+2}{(x+1)^2} dx = \int e^x \frac{(x+1)(x-1)+2}{(x+1)^2} dx$$

$$= \int e^x \left[\frac{x-1}{x+1} + \frac{2}{(x+1)^2}\right] dx$$

$$= \int e^x \left[\frac{(x+1)-2}{x+1} + \frac{2}{(x+1)^2}\right] dx$$

$$= \int e^x \left[\left\{1 - \frac{2}{x+1}\right\} + \frac{2}{(x+1)^2}\right] dx$$

$= \int e^x [f(x) + f'(x)] dx$, where $f(x) = 1 - \frac{2}{x+1}$

so that $\quad f'(x) = \frac{2}{(x+1)^2}$

$$= e^x f(x) = e^x \left[1 - \frac{2}{x+1}\right] = e^x \frac{x-1}{x+1}.$$

Example 42(e):

Evaluate $\int e^x \frac{(1-x)^2}{(1+x^2)^2} dx.$

Solution:

We have $\int e^x \frac{(1-x)^2}{\left(1+x^2\right)^2} dx = \int e^x \frac{\left(1+x^2\right)-2x}{\left(1+x^2\right)^2} dx$

$$= \int e^x \left[\frac{1}{1+x^2} - \frac{2}{\left(1+x^2\right)^2}\right] dx$$

$$= \int e^x [f(x) + f'(x)] dx,$$

where $\quad f(x) = \frac{1}{1+x^2}$

and $\quad f'(x) = \frac{-2x}{\left(1+x^2\right)^2}$

$$= e^x f(x) = e^x . \frac{1}{1+x^2}.$$

Example 42(f):

Evaluate $\int e^x \frac{x^2+3x+3}{(x+2)^2} dx.$

Solution:

We have $\int e^x \frac{x^2+3x+3}{(x+2)^2} dx$

$$= \int e^x \frac{(x+2)(x+1)+1}{(x+2)^2} dx$$

$$= \int e^x \left[\frac{x+1}{x+2} + \frac{1}{(x+2)^2}\right] dx$$

$= \int e^x [f(x) + f'(x)] dx$, where

$$f(x) = \frac{x+1}{x+2} = \frac{(x+2)-1}{x+2} = 1 - \frac{1}{x+2}$$

so that $f'(x) = \frac{1}{(x+2)^2}$

$$= e^x \frac{x+1}{x+2}.$$

Example 43:

Evaluate $\int e^x \frac{x^3 - x + 2}{\left(x^2 + 1\right)^2} dx$.

Solution:

We have $\int e^x \frac{x^3 - x + 2}{\left(x^2 + 2\right)^2} dx$

$$= \int e^x \frac{\left(x^2 + 1\right)(x + 1) - x^2 - 2x + 1}{\left(x^2 + 1\right)^2} dx$$

$$= \int e^x \left[\frac{x + 1}{x^2 + 1} + \frac{1 - x^2 - 2x}{\left(x^2 + 1\right)^2} \right] dx$$

$= \int e^x [f(x) + f'(x)] dx,$ where $f(x) = \frac{x + 1}{x^2 + 1}$ so that

$$f'(x) = \frac{1.\left(x^2 + 1\right) - 2x(x + 1)}{\left(x^2 + 1\right)^2} = \frac{1 - x^2 - 2x}{\left(x^2 + 1\right)^2}$$

$$= e^x f(x) = e^x \frac{x + 1}{x^2 + 1}.$$

Example 44(a):

Evaluate $\int \frac{\log x}{(1 + \log x)^2} dx$.

Solution:

Put log x = t. ∴ x = e^t and dx = e^t dt.

Then the given integral

$$I = \int \frac{te^t}{(1 + t)^2} = \int e^t \frac{(1 + t) - 1}{(1 + t)^2} dt$$

$$= \int e^t \left[\frac{1}{1 + t} - \frac{1}{(1 + t)^2} \right] dt$$

$= \int e^t [f(t) + f'(t)] dt$, where $f(t) = \frac{1}{1 + t}$

$$= e^t f(t) = e^t \frac{1}{1 + t} = \frac{x}{1 + \log x}.$$

Example 44(b):

Evaluate $\int \frac{e^x(1+\sin x)}{1+\cos x}dx$.

Solution:

The given integral $I = \int \frac{e^x\,dx}{1+\cos x} + \int \frac{e^x \sin x\,dx}{(1+\cos x)}$

$$= \int \frac{e^x\,dx}{2\cos^2(x/2)} + \int \frac{e^x 2\sin(x/2)\cos(x/2)}{2\cos^2(x/2)}$$

$$= \frac{1}{2}\int e^x \sec^2\left(\frac{x}{2}\right)dx + \int e^x \tan\left(\frac{x}{2}\right)dx$$

$$= \int e^x\left[\tan\frac{x}{2} + \frac{1}{2}\sec^2\frac{x}{2}\right]dx.$$

Since $\frac{d}{dx}\left(\tan\frac{1}{2}x\right) = \frac{1}{2}\sec^2\frac{x}{2}$, therefore this *integral is of the type* $\int e^x[f(x)+f'(x)]dx$.

To evaluate this integral, integrating $e^x \tan\left(\frac{x}{2}\right)$ by parts regarding e^x as the second function, we get

$$I = e^x \tan\frac{x}{2} - \int \frac{1}{2}e^x.\sec^2\frac{x}{2}dx + \int \frac{1}{2}e^x \sec^2\frac{x}{2}dx = e^x \tan\frac{x}{2},$$

because the last two integrals cancel each other.

Example 44(c):

Evaluate $\int e^x \frac{1-\sin x}{1-\cos x}dx$.

Solution:

We have

$$\int e^x \frac{1-\sin x}{1-\cos x}dx = \int e^x\left[\frac{1}{1-\cos x} - \frac{\sin x}{1-\cos x}\right]dx$$

$$= \int e^x\left[\frac{1}{2\sin^2\frac{1}{2}x} - \frac{2\sin\frac{1}{2}x\cos\frac{1}{2}x}{2\sin^2\frac{1}{2}x}\right]dx$$

$$= \int e^x\left[\frac{1}{2}\text{cosec}^2\frac{1}{2}x - \cot\frac{1}{2}x\right]dx$$

$$= \int e^x\left[\left(-\cot\frac{1}{2}x\right) + \frac{1}{2}\text{cosec}^2\frac{1}{2}x\right]dx$$

$= \int e^x [f(x) + f'(x)]dx$ where

$f(x) = -\cot\frac{1}{2}x$

so that $f'(x) = \frac{1}{2}\text{cosec}^2\frac{1}{2}x$

$= e^x f(x) = e^x\left(-\cot\frac{1}{2}x\right) = -e^x \cot\frac{1}{2}x.$

Example 44(d):

Evaluate $\int e^x \frac{2+\sin 2x}{1+\cos 2x}dx.$

Solution:

We have $\int e^x \frac{2+\sin 2x}{1+\cos 2x}dx$

$= \int e^x \left[\frac{2}{1+\cos 2x} + \frac{\sin 2x}{1+\cos 2x}\right]dx$

$= \int e^x \left[\frac{2}{2\cos^2 x} + \frac{2\sin x \cos x}{2\cos^2 x}\right]dx$

$= \int e^x [\sec^2 x + \tan x]dx$

$= \int e^x [\tan x + \sec^2 x]dx$

$= \int e^x [f(x) + f'(x)]dx$, where $f(x) = \tan x$

$= e^x f(x) = e^x \tan x.$

Example 45:

Evaluate $\int e^x (\cot x + \log \sin x)dx.$

Solution:

We have $\int e^x (\cot x + \log \sin x)dx$

$= \int e^x (\log \sin x + \cot x)dx$

$= \int e^x [f(x) + f'(x)]dx$, where

$f(x) = \log \sin x$

so that $f'(x) = \left(\frac{1}{\sin x}\right) \cos x = \cot x$

$= e^x f(x) = e^x \log \sin x.$

Example 46(a):

$$\int e^x[\log(\sec x + \tan x) + \sec x]dx.$$

Solution:

The given integral

$$I = \int e^x \sec x\, dx + \int e^x \log(\sec x + \tan x)dx$$

$$= e^x \log(\sec x + \tan x) - \int e^x \log(\sec x + \tan x)dx$$

$$+ \int e^x \log(\sec x + \tan x)dx,$$

(applying integration by parts to the first integral taking e^x as the second function and keeping the second integral as it is)

$= e^x \log (\sec x + \tan x).$

Example 46(b):

$$\text{Evaluate } \int e^x\left[\frac{1+\sqrt{(1-x^2)}\sin^{-1}x}{\sqrt{(1-x^2)}}\right]dx.$$

Solution:

The given integral

$$I = \int e^x\left[\frac{1}{\sqrt{(1-x^2)}} + \sin^{-1} x\right]dx$$

$$= \int e^x\left[\sin^{-1} x + \frac{1}{\sqrt{(1-x^2)}}\right]dx$$

$= \int e^x[f(x) + f'(x)]dx$, where $f(x) = \sin^{-1}$ so that $f'(x) = \frac{1}{\sqrt{(1-x^2)}}$

$$= e^x f(x) = e^x \cdot \frac{1}{\sqrt{(1-x^2)}}.$$

Example 46(c):

Evaluate $\int \frac{x \sin^{-1} x}{\sqrt{(1-x^2)}} dx$.

Solution:

Put $\sin^{-1} x = t$ (or $x = \sin t$); $\therefore \frac{1}{\sqrt{(1-x^2)}} dx = dt$.

$$\therefore \int \frac{x \sin^{-1} x}{\sqrt{(1-x^2)}} dx = \int (\sin t).t\,dt = \int t \sin dt$$

$$= t.(-\cos t) - \int 1.(-\cos t).dt = -t\cos t + \sin t$$

$$= -\sqrt{(1-\sin^2 t)} + \sin t$$

$$= -\sin^{-1} x.\sqrt{(1-x^2)} + x.$$

Example 47:

Evaluate $\int \frac{x^2 \tan^{-1} x}{1+x^2} dx$.

Solution:

Put $\tan^{-1} x = t$ (or $x = \tan t$); $\therefore \frac{1}{1+x^2} dx = dt$.

$$\therefore \int \frac{x^2 \tan^{-1} x}{1+x^2} dx = \int t \tan^2 t\,dt = \int t(\sec^2 t - 1)dt$$

$$= \int t \sec^2 t\,dt - \int t\,dt$$

$$= t \tan t - \log \sec t - \frac{1}{2}t^2 = t\tan t - \log\sqrt{(1+\tan^2 t)} - \frac{1}{2}t^2$$

$$= x\tan^{-1} x - \log\sqrt{(1+x^2)} - \frac{1}{2}(\tan^{-1} x)^2, \qquad [\because x = \tan t].$$

Example 48:

Evaluate $\int \frac{\sin^{-1} x\, dx}{(1-x^2)^{3/2}}$.

Solution:

Put $\sin^{-1} x = t$,

i.e., $x = \sin t$

so that dx = cos t dt.

$$\therefore \int \frac{\sin^{-1} x\,dx}{\left(1-x^2\right)^{3/2}} = \int \frac{t}{\cos^3 t}\cdot \cos t\,dt = \int t.\sec^2 t\,dt$$

$$= t\tan t - \int 1.\tan t\,dt$$

$$= t\tan t + \log\cos t, \qquad \left[\because \int \tan t\,dt = -\log\cos t\right]$$

$$= t\frac{\sin t}{\cos t} + \log\cos t = t\frac{\sin t}{\sqrt{\left(1-\sin^2 t\right)}} + \log\left\{\sqrt{\left(1-\sin^2 t\right)}\right\}$$

$$= \left(\sin^{-1} x\right)\cdot\frac{x}{\sqrt{\left(1-x^2\right)}} + \log\sqrt{\left(1-x^2\right)}, \qquad [\because x = \sin t]$$

$$= \frac{x\sin^{-1} x}{\sqrt{\left(1-x^2\right)}} + \frac{1}{2}\log\left(1-x^2\right).$$

Example 49:

Find $\int \frac{2x\sin^{-1}\left(x^2\right)}{\sqrt{\left(1-x^4\right)}}dx$.

Solution:

Put $\sin^{-1} x^2 = t$,

so that $\frac{2x\,dx}{\sqrt{\left(1-x^4\right)}} = dt$.

$\therefore$ the given integral $= \int t\,dt = \frac{1}{2}t^2 = \frac{1}{2}\left(\sin^{-1} x^2\right)^2$.

Example 50:

Evaluate $\int \frac{x\tan^{-1} x}{\left(1+x^2\right)^{3/2}}dx$.

Solution:

Put $\tan^{-1} x =- t$,

so that x = tan t.

Also $dx = \sec^2 t\,dt$.

$$\therefore \text{ the given integral } = \int \frac{(\tan t)\, t \sec^2 t\, dt}{\left(1+\tan^2 t\right)^{3/2}} = \int \frac{t \tan t \sec^2 t\, dt}{\sec^3 t}$$

$$= \int \frac{t \tan t}{\sec t} dt = \int t \sin t\, dt.$$

Now integrating by parts regarding sin t as the second function, the given integral = – t cos t + $\int \cos t\, dt$ = – t cos t + sin t

$$= -\frac{\tan^{-1} x}{\sqrt{(1+x^2)}} + \frac{x}{\sqrt{(1+x^2)}} = \frac{x - \tan^{-1} x}{\sqrt{(1+x^2)}}.$$

Example 51(a):

Evaluate $\int x^3 e^{x^2} dx$.

Solution:

The given integral $I = \int x^2 . e^{x^2} . x\, dx$.

Put $x^2 = t$ so that $2x\, dx = dt$ or $x\, dx = \frac{1}{2} dt$.

$\therefore$ the given integral $I = \frac{1}{2} \int t . e^t dt$.

Now integrating by parts regarding e^t as the 2nd function, we have

$$I = \frac{1}{2} t.e^t - \frac{1}{2}\int e^t dt = \frac{1}{2} t.e^t - \frac{1}{2} e^t$$

$$= \frac{1}{2} e^t (t-1) = \frac{1}{2} e^{x^2}\left(x^2 - 1\right).$$

Example 51(b):

Evaluate $\int \frac{x^2 dx}{(x \sin x + \cos x)^2}$.

Solution:

Let $I = \int \frac{x^2}{(x \sin x + \cos x)^2} dx$

$$= \int x^2 (x \sin x + \cos x)^{-2} dx.$$

Here $\frac{d}{dx}$(x sin x + cos x) = sin x + x cos x – sin x = x cos x.

So we adjust the given integral in the form

$$I = \int \frac{x^2}{x\cos x}\{(x \sin x + \cos x)^{-2} (x \cos x)\} dx$$

$$= \int \left(\frac{x}{\cos x}\right) \{(x \sin x + \cos x)^{-2} (x \cos x)\} dx.$$

Now by power formula, the integral of

$(x \sin x + \cos x)^{-2} (x \cos x)$ is $\{(x \sin x + \cos x)^{-1}\}/(-1)$

i.e., $-\dfrac{1}{(x\sin x+\cos x)}$. So applying to I integration by parts taking $(x \sin x + \cos x)^{-2} (x \cos x)$ as the second function, we get

$$I = \left(\frac{x}{\cos x}\right)\left(-\frac{1}{x\sin x+\cos x}\right)$$

$$-\int\left[\left\{\frac{d}{dx}\left(\frac{x}{\cos x}\right)\right\}\cdot\left\{-\frac{1}{x\sin x+\cos x}\right\}\right]dx$$

$$= -\frac{x}{\cos x(x\sin x+\cos x)} + \int\left(\frac{\cos x + x\sin x}{\cos^2 x}\right)\cdot\left(\frac{1}{x\sin x+\cos x}\right)dx$$

$$= -\frac{x}{\cos x(x\sin x+\cos x)} + \int \sec^2 x\, dx$$

$$= -\frac{x}{\cos x(x\sin x+\cos x)} + \tan x = \frac{\sin x}{\cos x} - \frac{x}{\cos x(x\sin x+\cos x)}$$

$$= \frac{\sin x(x\sin x+\cos x) - x}{(x\sin x+\cos x)\cos x} = \frac{\sin x\cos x - x\left(1-\sin^2 x\right)}{(x\sin x+\cos x)\cos x}$$

$$= \frac{\sin x\cos x - x\cos^2 x}{(x\sin x+\cos x)\cos x} = \frac{\cos x(\sin x - x\cos x)}{\cos x(x\sin x+\cos x)}$$

$$= \frac{\sin x - x\cos x}{x\sin x+\cos x}.$$

Example 51(c):

Evaluate $\int sin^{-1}\left[\dfrac{2x}{\left(1+x^2\right)}\right]dx$.

Solution:

Put $x = \tan\theta$,

so that $dx = \sec^2\theta\, d\theta$.

Then the given integral

$$= \int \sin^{-1}\left(\frac{2}{1+x^2}\right)dx = \int\left[\sin^{-1}\left(\frac{2\tan\theta}{1+\tan^2\theta}\right)\right]\sin^2\theta\, d\theta$$

$$= \int \sin^{-1}(\sin 2\theta)\sec^2\theta\, d\theta = \int 2\theta\sec^2\theta\, d\theta = 2\int\theta\sec^2\theta\, d\theta$$

$$= 2\left[\theta\tan\theta - \int 1.\tan\theta\, d\theta\right] = 2[\theta\tan\theta - \log\sec\theta]$$

$$= 2\left[\left(\tan^{-1}x\right).x - \log\sqrt{\left(1+x^2\right)}\right].$$

Example 51(d):

Evaluate (i) $\int \cos^{-1}\frac{1-x^2}{1+x^2}dx$,

(ii) $\int \tan^{-1}\frac{2x}{1-x^2}dx$.

Solution:

Put $x = \tan\theta$,

so that $dx = \sec^2\theta\, d\theta$.

Then each of the two given integrals becomes $= \int\theta\sec^2\theta\, d\theta$

$$= 2\left[x\tan^{-1}x - \log\sqrt{\left(1+x^2\right)}\right].$$

Example 51(e):

Evaluate $\int \sin^{-1}\sqrt{\left(\frac{x}{a+x}\right)}dx$.

Solution:

Put $x = a\tan^2\theta$,

so that $dx = 2a\tan\theta\sec^2\theta\, d\theta$.

Then the given integral

$$= \int\left\{\sin^{-1}\sqrt{\left(\frac{a\tan^2\theta}{a\sec^2\theta}\right)}\right\}2a\tan\theta\sec^2\theta\, d\theta$$

$$= 2a\int\left\{\sin^{-1}(\sin\theta)\right\}.\tan\theta\sec^2\theta\, d\theta = 2a\int\theta\tan\theta\sec^2\theta\, d\theta$$

$= 2a\left[\theta\cdot\frac{\tan^2\theta}{2} - \int 1\cdot\frac{\tan^2\theta}{2}\right]$, [Integrating by parts taking $\tan\theta$ $\sec^2\theta$ as the second function. Note that by power formula the integral of $\tan\theta\sec^2\theta$ is $1/2\tan^2\theta$]

$$= a\left[\theta\tan^2\theta - \int\left(\sec^2\theta - 1\right)d\theta\right] = a\,[\theta\tan^2\theta - \tan\theta + \theta]$$

$$=\left[\left\{\tan^{-1}\sqrt{\left(\frac{x}{a}\right)}\right\}\cdot\frac{x}{a}-\sqrt{\left(\frac{x}{a}\right)}+\tan^{-1}\sqrt{\left(\frac{x}{a}\right)}\right], \qquad \left[\because \tan^2\theta=\frac{x}{a}\right]$$

$$=\left\{\tan^{-1}\sqrt{\left(\frac{x}{a}\right)}\right\}\left\{a\left(\frac{x}{a}\right)+a\right\}-a\sqrt{\left(\frac{x}{a}\right)}$$

$$=(a+x)\tan^{-1}\sqrt{\left(\frac{x}{a}\right)}-\sqrt{(ax)}.$$

Example 51(f):

Evaluate $\int \frac{\log\left(\sec^{-1}x\right)dx}{x\sqrt{\left(x^2-1\right)}}.$

Solution:

Put $\sec^{-1} x = t$

so that $\frac{1}{x\sqrt{\left(x^2-1\right)}}dx = dt.$

Then the given integral

$$=\int \log t\,dt=\int(\log t).1dt=(\log t).t-\int\frac{1}{t}t\,dt$$

$$=t\log t-\int dt=t\log t-t=t(\log t-1)=t.(\log t-\log e)$$

$$=t\log\left(\frac{1}{e}\right)=\left(\sec^{-1}x\right)\log\left(\frac{\sec^{-1}x}{e}\right).$$

Example 51(g):

Evaluate $\int \frac{e^{\sec^{-1}}dx}{x\sqrt{\left(x^2-1\right)}}.$

Solution:

Put $\sec^{-1} x = t,$

so that $\frac{1}{x\sqrt{\left(x^2-1\right)}}dx = dt.$

Then the given integral $=\int e^t\,dt=e^t=e^{\sec^{-1}}x.$

Example 52:

Evaluate $\int \frac{\tan x}{\log \sec x}dx.$

Solution:

Put log sec x = t,

so that tan x dx = dt. $\left[\text{Note that } \frac{d}{dx}(\log\sec x) = \tan x\right]$

Then the given integral $= \int \frac{1}{t} dt = \log t = \log(\log\sec x)$.

Example 53:

Evaluate $\int \frac{\log(\log x)}{x} dx$.

Solution:

Put log x = t, so that $\left(\frac{1}{x}\right)dx = dt$.

Then the given integral $= \int \log t \, dt = \int (\log y).1 \, dt$

$= (\log t).t - \int \frac{1}{t} \cdot t \, dt = t \log\left(\frac{t}{e}\right) = (\log x).\log\left(\frac{\log x}{e}\right)$.

Example 54:

Evaluate $\int \frac{\tan x}{\sqrt{(a + b\tan^2 x)}} dx$, *(b > 0).*

Solution:

The given integral $I = \int \frac{\sin x \, dx}{\sqrt{(a\cos^2 x + b\sin^2 x)}}$

$= \int \frac{\sin x \, dx}{\sqrt{\{b - (b-a)\cos^2 x\}}}$.

Now pub $\sqrt{(b-a)}\cos x = t$ so that $-\sqrt{(b-a)}\sin x \, dx = dt$.

Then $I = -\frac{1}{\sqrt{(b-a)}} \int \frac{dt}{\sqrt{(b-t^2)}} = \frac{1}{\sqrt{(b-a)}} \cos^{-1}\left(\frac{t}{\sqrt{b}}\right)$

$= \frac{1}{\sqrt{(b-a)}} \cos^{-1}\left\{\frac{\sqrt{(b-a)}\cos x}{\sqrt{b}}\right\}$

$= \frac{1}{\sqrt{(b-a)}} \cos^{-1}\left\{\sqrt{\left(\frac{b-a}{b}\right)}\cos x\right\}$.

Example 55(a):

Evaluate $\int \cosh 2x \sin 2x \, dx$.

Solution:

Let $I = \int \cosh 2x \sin 2x \, dx$. Integrating by parts taking sin 2x as the second function, we get

$$I = (\cosh 2x)\left(-\frac{1}{2}\cos 2x\right) - \int (2\sinh 2x)\left(-\frac{1}{2}\cos 2x\right)dx$$

$$= -\frac{1}{2}\cosh 2x \cos 2x + \int \sinh 2x \cos 2x \, dx.$$

Again integrating by parts taking cos 2x as the second function, we get

$$I = -\frac{1}{2}\cosh 2x \cos 2x + \left[(\sinh 2x)\left(\frac{1}{2}\sin 2x\right) - \int (2\cosh 2x)\left(\frac{1}{2}\sin 2x\right)dx\right]$$

$$= -\frac{1}{2}\cosh 2x \cos 2x + \frac{1}{2}\sinh 2x \sin 2x - \int \cosh 2x \sin 2x \, dx$$

$$= -\frac{1}{2}\cosh 2x \cos 2x + \frac{1}{2}\sinh 2x \sin 2x - I.$$

$$\therefore \; 2I = -\frac{1}{2}\cosh 2x \cos 2x + \frac{1}{2}\sinh 2x \sin 2x$$

$$\text{or } I = -\frac{1}{4}\cosh 2x \cos 2x + \frac{1}{4}\sinh 2x \sin 2x.$$

Example 55(b):

Evaluate $\int \sinh 2x \sin 2x \, dx$.

Solution:

Let $I = \int \sinh 2x \sin 2x \, dx$. Integrating by parts taking sin 2x as the second function, we get

$$I = (\sinh 2x).\left(-\frac{1}{2}\cos 2x\right) - \int (2\cosh 2x)\left(-\frac{1}{2}\cos 2x\right)dx$$

$$= -\frac{1}{2}\sinh 2x \cos 2x + \int \cosh 2x \cos 2x \, dx.$$

Again integrating by parts taking cos 2x as the second function, we get

$$I = -\frac{1}{2}\sinh 2x\cos 2x + \left[(\cosh 2x)\left(\frac{1}{2}\sin 2x\right) - \int (2\sinh 2x)\left(\frac{1}{2}\sin 2x\right)dx\right]$$

$$= -\frac{1}{2}\sinh 2x\cos 2x + \frac{1}{2}\cosh 2x\sin 2x - \int \sinh 2x\sin 2x\,dx$$

$$= -\frac{1}{2}\sinh 2x\cos 2x + \frac{1}{2}\cosh 2x\sin 2x - I. \qquad \textbf{(Note)}$$

$$\therefore\ 2I = -\frac{1}{2}\sinh 2x\cos 2x + \frac{1}{2}\cosh 2x\sin 2x$$

or $I = \frac{1}{4}\cosh 2x\sin 2x - \frac{1}{4}\sinh 2x\cos 2x$.

Example 55(c):

Evaluate $\int e^x \sin x\cos x\cos 2x\,dx$

Solution:

The given integral $= \frac{1}{2}\int e^x \sin 2x\cos 2x\,dx$

$$= \frac{1}{4}\int e^x \sin 4x\,dx = \frac{1}{4}\cdot\frac{e^x}{1^2+4^2}\ [1.\sin 4x - 4\cos 4x],$$

$$\left[\because \int e^{ax}\sin bx\,dx = \frac{e^{ax}}{a^2+b^2}(a\sin bx - b\cos bx)\right]$$

$$= \frac{1}{4\times 17}e^x[\sin 4x - 4\cos 4x].$$

Example 55(d):

If u and v are functions of x, prove that

$$\int u\frac{d^2v}{dx^2}dx = u\frac{dv}{dx} - v\frac{du}{dx} + \int v\frac{d^2u}{dx^2}dx.$$

Solution:

Integrating by parts regarding u as first function and $\left(\frac{d^2v}{dx^2}\right)$ as second function, we have

$$\int u\frac{d^2v}{dx^2}dx = u\frac{dv}{dx} - \int \frac{du}{dx}\cdot\frac{dv}{dx}dx$$

$$= u\frac{dv}{dx} - \left[\frac{du}{dx}v - \int \frac{d^2u}{dx^2}\cdot v\,dx\right],$$

[Again integrating by parts regarding (dv/dx) as the second function]

$$= u\frac{dv}{dx} - v\frac{du}{dx} + \int v\frac{d^2u}{dx^2}dx$$

Example 56:

Evaluate $\int x \sin^{-1}\left\{\frac{1}{2}\sqrt{\left(\frac{2a-x}{a}\right)}\right\}dx.$

Solution:

Put $\frac{1}{2}\sqrt{\left(\frac{2a-x}{a}\right)} = \sin\theta$

i.e., $2a - x = 4a\sin^2\theta$

i.e., $x = 2a(1 - 2\sin^2\theta)$

i.e., $x = 2a\cos 2\theta$, so that

$dx = -4a\sin 2\theta\,d\theta.$

$\therefore$ the given integral

$$= \int (2a\cos 2\theta)\sin^{-1}(\sin\theta)\times(-4a\sin 2\theta)d\theta$$

$$= -4a^2\int \theta\, 2\sin 2\theta\cos 2\theta\,d\theta = -4a^2\int \theta\sin 4\theta\,d\theta$$

$$= -4a^2\left[\theta.\left(-\frac{1}{4}\cos 4\theta\right) - \int 1.\left(-\frac{1}{4}\cos 4\theta\right)d\theta\right]$$

$$= a^2\left[\theta\cos 4\theta - \int \cos 4\theta\,d\theta\right] = a^2\left[\theta\cos 4\theta - \frac{1}{4}\sin 4\theta\right]$$

$$= a^2\left[\theta\left(2\cos^2 2\theta - 1\right) - \frac{1}{4}\cdot 2\cos 2\theta\sin 2\theta\right]$$

$$= a^2\left[\theta\left(2\cos^2 2\theta - 1\right) - \frac{1}{2}\cos 2\theta\sqrt{\left(1-\cos^2 2\theta\right)}\right]$$

$$= a^2\left[\left(\frac{1}{2}\cos^{-1}\frac{x}{2a}\right)\left(2\cdot\frac{x^2}{4a^2} - 1\right) - \frac{1}{2}\cdot\frac{x}{2a}\sqrt{\left(1-\frac{x^2}{4a^2}\right)}\right]$$

$$= a^2\left[\frac{1}{2}\left(\cos^{-1}\frac{x}{2a}\right)\left(\frac{x^2-2a^2}{2a^2}\right) - \frac{x}{8a^2}\sqrt{\left(4a^2-x^2\right)}\right]$$

$$= \frac{1}{8}\left[2\left(x^2-2a^2\right)\cos^{-1}\left(\frac{x}{2a}\right) - x\sqrt{\left(4a^2-x^2\right)}\right].$$

Example 57(a):

Evaluate $\int tan^{-1}\sqrt{\left(\frac{1-x}{1+x}\right)}\,dx$.

Solution:

Put x = cos θ

so that dx = – sin θ dθ.

$\therefore$ the given integral $= \int \left\{\tan^{-1}\sqrt{\left(\frac{1-\cos\theta}{1+\cos\theta}\right)}\right\}(-\sin\theta)d\theta$

$= -\int\left\{\tan^{-1}\left(\tan\frac{1}{2}\theta\right)\right\}\sin\theta\,d\theta$

$= -\int\frac{1}{2}\theta\,\sin\theta\,d\theta = -\frac{1}{2}\int\theta\,\sin\theta\,d\theta$

$= -\frac{1}{2}\left[\theta.(-\cos\theta) - \int(-\cos\theta)d\theta\right] = \frac{1}{2}\left[\theta\,\cos\theta - \int\cos\theta\,d\theta\right].$

$= \frac{1}{2}[\theta\,\cos\theta - \sin\theta] = \frac{1}{2}\left[x\cos^{-1}x - \sqrt{(1-x^2)}\right].$

Example 57(b):

Evaluate $\frac{x+\sin x}{1+\cos x}dx$.

Solution:

We have

$$\int\frac{x+\sin x}{1+\cos x}dx = \int\frac{x+2\sin(x/2)\cos(x/2)}{1+2\cos^2(x/2)-1}$$

$$= \frac{1}{2}\int x\sec^2\left(\frac{x}{2}\right)dx + \int\tan\left(\frac{x}{2}\right)dx$$

$$= \frac{1}{2}\left[x\frac{\tan(x/2)}{(1/2)} - \int 1\cdot\frac{\tan(x/2)}{(1/2)}dx\right] + \int\tan\left(\frac{x}{2}\right)dx$$

$$= x\tan\left(\frac{x}{2}\right) - \int\tan\left(\frac{x}{2}\right)dx + \int\tan\left(\frac{x}{2}\right)dx = x\tan\left(\frac{x}{2}\right).$$

Example 57(c):

Evaluate (i) $\int x^2 tan^{-1}x\,dx$

(ii) $\int x^3 tan^{-1}x\,dx$.

Solution:

(i) $= \frac{x^3}{3}\tan^{-1}x - \int \frac{x^3}{3}\cdot\frac{1}{1+x^2}dx,$

integrating by parts taking x^2 as the second function

$$= \frac{x^3}{3}\tan^{-1}x - \frac{1}{3}\int \frac{x(x^2+1)-x}{1+x^2}dx, \ [\because x^3 = x(x^2+1) - x]$$

$$= \frac{x^3}{3}\tan^{-1}x - \frac{1}{3}\int x\,dx + \frac{1}{3}\cdot\frac{1}{2}\int \frac{2x}{1+x^2}dx$$

$$= \frac{x^3}{3}\tan^{-1}x - \frac{1}{6}x^2 + \frac{1}{6}\log(1+x^2).$$

(ii) We have $\int x^3 \tan^{-1}x\,dx$

$$= \frac{x^4}{4}\tan^{-1}x - \int \frac{x^4}{4}\cdot\frac{1}{1+x^2}dx,$$

integrating by parts taking x^3 as the second function

$$= \frac{x^4}{4}\tan^{-1}x - \frac{1}{4}\int\left[x^2 - 1 + \frac{1}{1+x^2}\right]dx,$$

dividing x^4 by $x^2 + 1$, we get $x^2 - 1$ as the quotient and 1 as theremainder

$$= \frac{1}{4}\left[x^4\tan^{-1}x - \frac{x^3}{3} + x - \tan^{-1}x\right]$$

$$= \frac{1}{4}\left[(x^4-1)\tan^{-1}x - \frac{x^3}{3} + x\right].$$

Example 57(d):

Evaluate: (i) $\int \frac{log(1+x)}{\sqrt{(1+x)}}dx$

(ii) $\int log(x+2)^{x+2}\,dx.$

Solution:

(i) We have

$$\int \frac{\log(1+x)}{\sqrt{(1+x)}}dx = \int[\log(1+x)]\,(1+x)^{-1/2}dx$$

$$= [\log(1+x)]\cdot\frac{(1+x)^{1/2}}{1/2} - \int \frac{1}{1+x}\cdot\frac{(1+x)^{1/2}}{1/2}dx,$$

integrating by parts taking $(1+x)^{-1/2}$ as the second function

$$= 2\sqrt{(1+x)}\log(1+x) - 2\int(1+x)^{-1/2}\,dx$$

$$= 2\sqrt{(1+x)}\log(1+x) - 2\cdot\frac{(1+x)^{1/2}}{1/2}$$

$$= 2\sqrt{(1+x)}\left[\log(1+x) - 2\right].$$

(ii) We have $\int \log(x+2)^{x+2}\,dx = \int (x+2)\log(x+2)\,dx$

$$= \frac{(x+2)^2}{2}\log(x+2) - \int \frac{(x+2)^2}{2}\cdot\frac{1}{x+2}\,dx,$$

integrating by parts taking $(x + 2)^1$ as the second function

$$= \frac{(x+2)^2}{2}\log(x+2) - \frac{1}{2}\int (x+2)\,dx$$

$$= \frac{(x+2)^2}{2}\log(x+2) - \frac{1}{2}\frac{(x+2)^2}{2}$$

$$= \frac{(x+2)^2}{4}\left[2\log(x+2) - 1\right].$$

Example 58:

Evaluate: (i) $\int x\sin^{-1}x\,dx$

(ii) $\int x^2\sin^{-1}x\,dx.$

Solution:

(i) We have $\int x\sin^{-1}x\,dx$

$$= \frac{x^2}{2}\sin^{-1}x - \int \frac{x^2}{2}\cdot\frac{1}{\sqrt{(1-x^2)}}\,dx,$$

integrating by parts taking x as the second function

$$= \frac{x^2}{2}\sin^{-1}x + \frac{1}{2}\int \frac{(1-x^2)-1}{\sqrt{(1-x^2)}}\,dx$$

$$= \frac{x^2}{2}\sin^{-1}x + \frac{1}{2}\int \sqrt{(1-x^2)}\,dx - \frac{1}{2}\int \frac{dx}{\sqrt{(1-x^2)}}$$

$$= \frac{x^2}{2}\sin^{-1}x + \frac{1}{2}\int \frac{x}{2}\sqrt{(1-x^2)} + \frac{1}{2}\sin^{-1}x - \frac{1}{2}\sin^{-1}x$$

$$= \frac{1}{2}x^2\sin^{-1}x + \frac{1}{4}x\sqrt{(1-x^2)} - \frac{1}{4}\sin^{-1}x.$$

(ii) We have $\int x^2 \sin^{-1} x\,dx$

$$= \frac{x^3}{3}\sin^{-1} x - \int \frac{x^3}{3}\cdot\frac{1}{\sqrt{(1-x^2)}}dx,$$

integrating by parts taking x^2 as the second function

$$= \frac{x^3}{3}\sin^{-1} x + \frac{1}{3}\int \frac{x(1-x^2)-x}{\sqrt{(1-x^2)}}dx$$

$$= \frac{x^3}{3}\sin^{-1} x + \frac{1}{3}\int x\sqrt{(1-x^2)}\,dx - \frac{1}{3}\int \frac{x}{\sqrt{(1-x^2)}}dx$$

$$= \frac{x^3}{3}\sin^{-1} x - \frac{1}{6}\int (1-x^2)^{1/2}(-2x)dx + \frac{1}{6}\int (1-x^2)^{-1/2}(-2x)dx$$

$$= \frac{x^3}{3}\sin^{-1} x - \frac{1}{6}\cdot\frac{(1-x^2)^{3/2}}{3/2} + \frac{1}{6}\cdot\frac{(1-x^2)^{1/2}}{1/2},$$

integrating by using power formula

$$= \frac{x^3}{3}\sin^{-1} x - \frac{1}{9}(1-x^2)^{3/2} + \frac{1}{3}(1-x^2)^{1/2}$$

$$= \frac{x^3}{3}\sin^{-1} x + \frac{1}{9}(1-x^2)^{1/2}\left[3-(1-x^2)\right]$$

$$= \frac{x^3}{3}\sin^{-1} x + \frac{2+x^2}{9}\sqrt{(1-x^2)}.$$

Example 59(a):

Evaluate: (i) $\int \frac{x^3 \sin^{-1} x}{(1-x^2)^{3/2}}dx$

(ii) $\int \cos 2x \log\frac{\cos x + \sin x}{\cos x - \sin x}dx.$

Solution:

(i) Let $I = \int \frac{x^3 \sin^{-1} x}{(1-x^2)^{3/2}}dx.$

Put $x = \sin\theta$

or $\sin^{-1} x = \theta.$

Then $x = \cos\theta\, d\theta.$

$$\therefore\ I=\int\frac{\sin^3\theta}{\cos^3\theta}\theta\cos\theta\,d\theta=\int\frac{\theta\sin\theta\left(1-\cos^2\theta\right)}{\cos^2\theta}d\theta$$

$$=\int\theta[\sec\theta\tan\theta-\sin\theta]d\theta$$

$$=\theta(\sec\theta+\cos\theta)-\int 1.(\sec\theta+\cos\theta)d\theta,$$

integrating by parts taking sec θ tan θ – sin θ as the second function

$$=\theta\ (\sec\theta+\cos\theta)-\log\ (\sec\theta+\tan\theta)-\sin\theta$$

$$=\theta\left(\frac{1}{\cos\theta}+\cos\theta\right)-\log\left[\frac{1}{\cos\theta}+\frac{\sin\theta}{\cos\theta}\right]-\sin\theta$$

$$=\theta\frac{\left(1+\cos^2\theta\right)}{\cos\theta}-\log\frac{1+\sin\theta}{\sqrt{\left(1-\sin^2\theta\right)}}-\sin\theta$$

$$=\theta\frac{\left(1+1-\sin^2\theta\right)}{\sqrt{\left(1-\sin^2\theta\right)}}-\log\left(\frac{1+\sin\theta}{1-\sin\theta}\right)^{1/2}-\sin\theta$$

$$=\theta\frac{2-\sin^2\theta}{\sqrt{\left(1-\sin^2\theta\right)}}-\frac{1}{2}\log\frac{1+\sin\theta}{1-\sin\theta}-\sin\theta$$

$$=\frac{2-x^2}{\sqrt{\left(1-x^2\right)}}\sin^{-1}x-x-\frac{1}{2}\log\frac{1+x}{1-x}.$$

(ii) We have $\int\cos 2x\log\frac{\cos x+\sin x}{\cos x-\sin x}dx$

$$=\int\cos 2x\log\frac{1+\tan x}{1-\tan x}dx=\int\cos 2x\ \log\tan\left(\frac{1}{4}\pi+x\right)dx$$

$$=\frac{1}{2}\sin 2x\ \log\ \tan\left(\frac{1}{4}\pi+x\right)$$

$$-\int\frac{1}{2}\sin 2x\cdot\frac{1}{\tan\left(\frac{1}{4}\pi+x\right)}\sec^2\left(\frac{1}{4}\pi+x\right)dx$$

integrating by parts taking cos 2x as the second function

$$=\frac{1}{2}\sin 2x\log\ \tan\left(\frac{1}{4}\pi+x\right)-\int\frac{\sin 2x}{2\sin\left(\frac{1}{4}\pi+x\right)\cos\left(\frac{1}{4}\pi+x\right)}dx$$

$$=\frac{1}{2}\sin 2x\ \log\ \tan\left(\frac{1}{4}\pi+x\right)-\int\frac{\sin 2x}{\sin\left(\frac{1}{2}\pi+2x\right)}dx$$

$$=\frac{1}{2}\sin 2x \log \tan\left(\frac{1}{4}\pi + x\right) - \int \frac{\sin 2x}{\cos 2x} dx$$

$$=\frac{1}{2}\sin 2x \log \tan\left(\frac{1}{4}\pi + x\right) - \int \tan 2x \, dx$$

$$=\frac{1}{2}\sin 2x \log \tan\left(\frac{1}{4}\pi + x\right) - \frac{1}{2}\log \sec 2x.$$

Example 59(b):

Evaluate:

(i) $\int \left[\frac{1}{\log x} - \frac{1}{(\log x)^2}\right] dx.$

(ii) $\int \frac{\cosh x + \sinh x . \sin x}{1 + \cos x} dx.$

(iii) $\int \frac{x + \sqrt{(1 - x^2)} \sin^{-1} x}{\sqrt{(1 - x^2)}} dx.$

Solution:

(i) We have $\int \left[\frac{1}{\log x} - \frac{1}{(\log x)^2}\right] dx$

$$= \int \left(\frac{1}{\log x}\right) \cdot 1 \, dx - \int \frac{1}{(\log x)^2} dx$$

$$= x \cdot \frac{1}{\log x} - \int x \cdot \frac{-1}{(\log x)^2 x} dx - \int \frac{1}{(\log x)^2} dx,$$

(applying integration by parts to the first integral taking 1 as the second function and keeping the second integral as it is)

$$= \frac{x}{\log x} + \int \frac{1}{(\log x)^2} dx - \int \frac{1}{(\log x)^2} dx = \frac{x}{\log x}.$$

(ii) We have $\int \frac{\cosh x + \sinh x . \sin x}{1 + \cos x} dx$

$$= \int \frac{\cosh x}{1 + \cos x} dx + \int \sin x \cdot \frac{\sin x}{1 + \cos x} dx$$

$$= \int \frac{\cosh x}{2\cos^2 \frac{1}{2}x} dx + \int \sinh x \cdot \frac{2\sin\frac{1}{2}x \cos\frac{1}{2}x}{2\cos^2\frac{1}{2}x} dx$$

$$= \int (\cosh x).\frac{1}{2}\sec^2\frac{1}{2}x\,dx \int \sinh x.\tan\frac{1}{2}x.\tan\frac{1}{2}x\,dx$$

$$= \int (\cosh x)\tan\frac{1}{2}x - \int \sinh x.\tan\frac{1}{2}x\,dx + \int \sinh x.\tan\frac{1}{2}x\,dx$$

(applying integration by parts to the first integral taking $\frac{1}{2}\sec^2\frac{1}{2}x$ as the second function and keeping the second integral as it is)

$$= \cosh.\tan\frac{1}{2}x.$$

(iii) We have $\int \frac{x+\sqrt{(1-x^2)}\sin^{-1}x}{\sqrt{(1-x^2)}}dx$

$$= \int \frac{x}{\sqrt{(1-x^2)}}dx + \int \sin^{-1}x\,dx$$

$$= x\sin^{-1}x - \int 1.\sin^{-1}x\,dx + \int \sin^{-1}x\,dx,$$

applying integration by parts to the first integral taking $1/\sqrt{(1-x^2)}$ as the second function and keeping the second integral as it is

$= x \sin^{-1} x$.

Example 59(c):

Evaluate: (i) $\int \sin x \log(\sec x + \tan x)dx$

(ii) $\int x \log\left[x+\sqrt{(x^2+a^2)}\right]dx.$

Solution:

(i) We have $\int \sin x \log(\sec x + \tan x)dx$

$= -\cos x \log(\sec x + \tan x) - \int(-\cos x).\sec x\,dx,$

integrating by parts taking sin x as the second function and observing that $(d/dx) \log(\sec x + \tan x) = \sec x$

$$= -\cos x \log(\sec x + \tan x) + \int dx$$

$$= x - \cos x \log(\sec x + \tan x).$$

(ii) We have $\int x \log\left[x+\sqrt{(x^2+a^2)}\right]dx$

$$= \frac{x^2}{2}\log\left[x+\sqrt{(x^2+a^2)}\right] - \int \frac{x^2}{2}\cdot\frac{1}{\sqrt{(x^2+a^2)}}dx,$$

integrating by parts taking x as the second function and observing that $\frac{d}{dx}\log\left[x+\sqrt{(x^2+a^2)}\right]=\frac{1}{\sqrt{(x^2+a^2)}}$

$$=\frac{x^2}{2}\log\left[x+\sqrt{(x^2+a^2)}\right]-\frac{1}{2}\int\frac{(x^2+a^2)-a^2}{\sqrt{(x^2+a^2)}}dx$$

$$=\frac{x^2}{2}\log\left[x+\sqrt{(x^2+a^2)}\right]-\frac{1}{2}\int\sqrt{(x^2+a^2)}\,dx+\frac{a^2}{2}\int\frac{dx}{\sqrt{(x^2+a^2)}}$$

$$=\frac{x^2}{2}\log\left[x+\sqrt{(x^2+a^2)}\right]-\frac{1}{2}\left[\frac{x}{2}\sqrt{(x^2+a^2)}+\frac{a^2}{2}\sinh^{-1}\frac{x}{a}\right]$$

$$=\frac{x^2}{2}\log\left[x+\sqrt{(x^2+a^2)}\right]-\frac{x}{4}\sqrt{(x^2+a^2)}+\frac{a^2}{2}\sinh^{-1}\frac{x}{a}$$

$$+\frac{a^2}{4}\sinh^{-1}\frac{x}{a}.$$

Example 60:

Evaluate $\int\frac{x}{x^2+x-6}dx$.

Solution:

Let $\frac{x}{x^2+x-6}=\frac{x}{(x+3)(x-2)}=\frac{A}{(x+3)}+\frac{B}{(x-2)}$.

$\therefore$ $x \equiv A(x-2)+B(x+3)$

or $x \equiv x(A+B)+3B-2A$. ...(1.1)

Comparing the coefficients of x and the constant terms on both sides of (1), we get $A+B=1$ and $3B-2A=0$., Solving these we get $A=\frac{3}{5}$ and $B=\frac{2}{5}$.

Thus $\int\frac{x\,dx}{x^2+x-6}=\frac{3}{5}\int\frac{dx}{x+3}+\frac{2}{5}\int\frac{dx}{x-2}$

$$=\frac{3}{5}\log(x+3)+\frac{2}{5}\log(x-2)=\frac{1}{5}\log\left\{(x+3)^3(x-2)^2\right\}.$$

Example 61:

Evaluate $\int\frac{(x-1)\,dx}{(x-3)(x-2)}$.

Solution:

Let $\frac{x-1}{(x-3)(x-2)} = \frac{A}{x-3} + \frac{B}{x-2} = \frac{A(x-2)+B(x-3)}{(x-3)(x-2)}$.

Clearly $x - 1 \equiv A(x - 2) + B(x - 3)$ ...(1)

To find A, put $x = 3$ on both sides of (1) and we have

$3 - 1 = A(3 - 2)$; $\therefore A = 2$.

To find B, put $x = 2$ on both sides of (1) and we have

$2 - 1 = B(2 - 3)$; $\therefore B = -1$.

$\therefore \frac{x-1}{(x-3)(x-2)} = \frac{2}{x-3} - \frac{1}{x-2}$.

Thus $\int \frac{(x-1)dx}{(x-3)(x-2)} = \int \frac{2dx}{x-3} - \int \frac{1.dx}{x-2}$

$= 2 \log(x - 3) - \log(x - 2) = \log[(x - 3)^2/(x - 2)]$.

Example 62:

Evaluate $\int \frac{x+1}{(x-1)(x-4)} dx$.

Solution:

Let $\frac{x+1}{(x-1)(x-4)} = \frac{A}{x-1} + \frac{B}{x-4}$

$\Rightarrow x + 1 \equiv A(x - 4) + B(x - 1)$.

$\therefore A = -\frac{2}{3}$ and $B = \frac{5}{3}$.

Now $\int \frac{(x+1)dx}{(x-1)(x-4)} = -\frac{2}{3}\int \frac{dx}{x-1} + \frac{5}{3}\int \frac{dx}{x-4}$

$= -\frac{2}{3}\log(x-1) + \frac{5}{3}\log(x-4)$.

Example 63:

Evaluate $\int \frac{2x\,dx}{(x-1)(x+3)}$.

Solution:

Let $\frac{2x}{(x-1)(x+3)} = \frac{A}{(x-1)} + \frac{B}{(x+3)}$.

Clearly $2x \equiv A(x + 3) + B(x - 1)$. ...(1.1)

Comparing the coefficients of x and the constant terms on both sides of (1), we get

$$2 = A + B \quad ...(1.2)$$

and $3A - B = 0$ ...(1.3)

Solving (2) and (3), we get $A = \frac{1}{2}$ and $B = \frac{3}{2}$.

Thus $\int \frac{2x\,dx}{(x-1)(x+3)} = \int \frac{1}{2(x-1)}dx + \int \frac{3}{2(x+3)}dx$

$= \frac{1}{2}\log(x-1) + \left(\frac{3}{2}\right)\log(x+3) = \frac{1}{2}\log(x-1) + \frac{1}{2}\log(x+3)^3$

$= \frac{1}{2}\log\left\{(x-1)(x+3)^3\right\}$.

Note: An easy way to find the constants A and B etc., corresponding to linear non-repeated factors is like this:

The factor below A is $x - 1$. The equation $x - 1 = 0$ gives $x = 1$. Now suppress $x - 1$ in the given fraction $\frac{2x}{(x-1)(x+3)}$ and put $x = 1$ in the remaining fraction $\frac{2}{(x+3)}$ to get A. Thus $A = \frac{2 \times 1}{(1+3)} = \frac{1}{2}$. Similarly $B = \frac{2 \times (-3)}{(-3-1)} = \frac{-6}{-4} = \frac{3}{2}$.

Example 64:

Evaluate $\int \frac{1}{\left(e^x - 1\right)^2} dx$.

Solution:

We have $\int \frac{1}{\left(e^x - 1\right)^2} dx = \int \frac{e^x}{e^x\left(e^x - 1\right)^2} dx$,

[multiplying the Nr. and Dr. by e^x]

$= \int \frac{dt}{t(t-1)^2}$, putting $e^x = t$ so that $e^x\,dx = dt$.

Now $\frac{1}{t(t-1)^2} \equiv \frac{A}{t} + \frac{B}{t-1} + \frac{C}{(t-1)^2}$,

[on resolving into partial fractions]

$\therefore\ 1 \equiv A(t-1)^2 + Bt(t-1) + Ct.$ ...(1)

To find A, putting t = 0 on both sides of (1), we get A = 1.

To find C, put t = 1 and we get C = 1.

Thus $1 \equiv (t-1)^2 + Bt(t-1) + t$.

Comparing the coefficients of t^2 on both sides, we get

$$0 = 1 + B \text{ or } B = -1.$$

$$\therefore \qquad \frac{1}{t(t-1)^2} = \frac{1}{t} - \frac{1}{t-1} + \frac{1}{(t-1)^2}.$$

Hence $\displaystyle\int \frac{dt}{t(t-1)^2} = \int \frac{1}{t}dt - \int \frac{dt}{t-1} + \int \frac{dt}{(t-1)^2}$

$$= \log t - \log(t-1) - \left\{\frac{1}{(t-1)}\right\}$$

$$= \log e^x - \log(e^x - 1) - \left\{\frac{1}{(e^x-1)}\right\}$$

$$= x - \log(e^x - 1) - \left\{\frac{1}{(e^x-1)}\right\}.$$

Example 65:

Evaluate $\displaystyle\int \frac{x}{(x^2-a^2)(x^2-b^2)}dx.$

Solution:

Put $x^2 = t$ so that $2x\,dx = dt$.

Then $\displaystyle\int \frac{x\,dx}{(x^2-a^2)(x^2-b^2)} = \frac{1}{2}\int \frac{dt}{(t-a^2)(t-b^2)}.$

Now let $\displaystyle\frac{1}{(t-a^2)(t-b^2)} = \frac{A}{(t-a^2)} + \frac{B}{(t-b^2)}.$

$\therefore\ 1 \equiv A(t-b^2) + B(t-a^2)$

or $1 \equiv t(A+B) - (a^2B + b^2A)$,

giving $A + B = 0$ and $Ab^2 + Ba^2 = -1$.

Solving these we get $A = \dfrac{1}{(a^2-b^2)}$

and $$B = -\frac{1}{(a^2-b^2)}.$$

Thus the given integral

$$=\frac{1}{2}\left[\int\frac{dt}{\left(a^2-b^2\right)\left(t-a^2\right)}-\int\frac{dt}{\left(a^2-b^2\right)\left(t-b^2\right)}\right]$$

$$=\frac{1}{2\left(a^2-b^2\right)}\,[\log\,(t-a^2)-\log\,(t-b^2)]$$

$$=\frac{1}{2\left(a^2-b^2\right)}\log\left[\frac{t-a^2}{t-b^2}\right]=\frac{1}{2\left(a^2-b^2\right)}\log\left[\frac{x^2-a^2}{x^2-b^2}\right].$$

Example 66(a):

Evaluate $\int_1^{\infty}\frac{dx}{\left(1+x^2\right)}$.

Solution:

We have $\int_1^{\infty}\frac{dx}{\left(1+x^2\right)}=\left[\tan^{-1}x\right]_1^{\infty}=\left[\tan^{-1}\infty-\tan^{-1}1\right]$

$$=\frac{1}{2}\pi-\frac{1}{4}\pi=\frac{1}{4}\pi.$$

Example 66(b):

Evaluate $\int_0^1\frac{\left(\tan^{-1}x\right)^3}{1+x^2}dx$.

Solution:

Put $\tan^{-1}x=t$

so that $\left\{\frac{1}{\left(1+x^2\right)}\right\}dx=dt$.

$\therefore$ the given integral

$$=\int_0^{\pi/4}t^3dt=\left[\frac{t^4}{4}\right]_0^{\pi/4}=\frac{1}{4}\left[\left(\frac{\pi}{4}\right)^4-0\right]=\frac{1}{4}\left(\frac{\pi}{4}\right)^4.$$

Note: The lower limit of t is zero for when $x=0$, $t=\tan^{-1}0=0$. Similarly the upper limit of t is $\pi/4$ for when $x=1$, $t=\tan^{-1}1=\frac{\pi}{4}$.

Example 66(c):

Evaluate $\int_0^a x^2\sin x^3dx$.

Solution:

Put $x^3 = t$

so that $3x^2dx = dt$.

Also when $x = 0$, $t = 0$

and when $x = a$, $t = a^3$.

Hence $\int_0^a x^2 \sin x^3 dx = \frac{1}{3}\int_0^{a^3} \sin t\, dt = \frac{1}{3}[-\cos t]_0^{a^3}$

$= \frac{1}{3}[-\cos a^3 - (-\cos 0)] = \frac{1}{3}[-\cos a^3 + 1] = \frac{1}{3}[1 - \cos a^3]$.

Example 66(d):

Evaluate $\int_2^3 2x^4 dx$.

Solution:

We have $\int_2^3 2x^4 dx = 2\left[\frac{x^5}{5}\right]_2^3 = \frac{2}{5}[x^5]_2^3$

$= \frac{2}{5}[3^5 - 2^5] = \frac{2}{5}\cdot 243 - 32 = \frac{2}{5}\cdot 211 = \frac{422}{5}$.

Example 67:

Evaluate $\int_0^{\pi/2} \cos^2 x\, dx$.

Solution:

We have $\int_0^{\pi/2} \cos^2 x\, dx = \int_0^{\pi/2} \frac{1}{2}(1 + \cos 2x)dx$

$= \frac{1}{2}\left[x + \frac{1}{2}\sin 2x\right]_0^{\pi/2} = \frac{1}{2}\left[\left(\frac{1}{2}\pi + \frac{1}{2}\sin\pi\right) - \left(0 + \frac{1}{2}\sin 0\right)\right]$

$= \frac{1}{2}\left[\left(\frac{1}{2}\pi + 0\right) - 0\right] = \frac{1}{4}\pi$.

Example 68(a):

Evaluate $\int_0^1 \frac{5x^3 dx}{\sqrt{(1 - x^8)}}$.

Solution:

Put $x^4 = t$

so that $4x^3\, dx = dt$.

Also when x = 0, t = 0

and when x = 1, t = 1.

Hence $\int_0^1 \frac{5x^3 dx}{\sqrt{(1-x^8)}} = \frac{5}{4}\int_0^1 \frac{dt}{\sqrt{(1-t^2)}} = \frac{5}{4}\left[\sin^{-1} t\right]_0^1$

$= \frac{5}{4}\left[\sin^{-1} 1 - \sin^{-1} 0\right] = \frac{5}{4}\left[\frac{1}{2}\pi - 0\right] = \frac{5}{8}\pi.$

Example 68(b):

Evaluate $\int_1^2 \frac{(1+\log x)^4}{x} dx.$

Solution:

Put 1 + log x = t

so that $\left(\frac{1}{x}\right)dx = dt.$

Also when x = 1, t = 1 + log 1 = 1

and when x = 2, t = (1 + log 2).

Hence $\int_1^2 \frac{(1+\log x)^4}{x} dx = \int_1^{(1+\log 2)} t^4 dt = \left[\frac{t^5}{5}\right]_1^{(1+\log 2)}$

$= \frac{1}{5}\left[(1+\log 2)^5 - 1^5\right] = (1+\log 2)^5 - \frac{1}{5}.$

Example 68(c):

Evaluate $\int_0^{\pi/3} \frac{\cos x\, dx}{3+4\sin x}.$

Solution:

Put sin x = t so that cos x dx = dt.

Also when x = 0, t = sin 0 = 0 a

nd when $x = \frac{\pi}{3},\ t = \sin\frac{1}{3}\pi = \sqrt{\frac{3}{2}}.$

Hence $\int_0^{\pi/3} \frac{\cos x\, dx}{3+4\sin x} = \int_0^{\sqrt{3/2}} \frac{dt}{3+4t} = \frac{1}{4}\left[\log(3+4t)\right]_0^{\sqrt{3/2}}$

$= \frac{1}{4}\left[\log\left\{3+4\left(\sqrt{\frac{3}{2}}\right)\right\} - \log(3+4\times 0)\right]$

$= \frac{1}{4}\left[\log(2+2\sqrt{3}) - \log 3\right]$

$$=\frac{1}{4}\left[\log\left(\frac{3+2\sqrt{3}}{3}\right)\right]=\frac{1}{4}\left[\log\left(1+\frac{2}{\sqrt{3}}\right)\right].$$

Example 68(d):

Evaluate $\int_1^3 \frac{\cos(\log x)}{x}\,dx$.

Solution:

Put log x = t

so that $\left(\frac{1}{x}\right)dx = dt$.

Also when x = 1, t = log 1 = 0

and when x = 3, t = log 3.

Hence $\int_1^3 \frac{\cos(\log x)\,dx}{x} = \int_0^{\log 3} \cos t\,dt = [\sin t]_0^{\log 3}$

= [sin (log 3) – sin 0] = sin (log 3).

Example 69:

Evaluate $\int_0^2 x^2 e^{2x}\,dx$.

Solution:

Integrating by parts regarding e^{2x} as the second function, we have

$$\int_0^2 x^2 e^{2x}dx = \left[\frac{x^2e^{2x}}{2}\right]_0^2 - \int_0^2 2x\cdot\frac{e^{2x}}{2}dx$$

$$= 2e^4 - \int_0^2 xe^{2x}dx = 2e^4 - \left\{\left(x\frac{e^{2x}}{2}\right)_0^2 - \int_0^2 \frac{1}{2}e^{2x}dx\right\},$$

[again integrating by parts]

$$= 2e^4 - \left[\frac{xe^{2x}}{2}\right]_0^2 + \int_0^2 \frac{e^{2x}}{2}dx$$

$$= 2e^4 - e^4 + \left[\frac{e^{2x}}{4}\right]_0^2 = e^4 + \frac{e^4}{4} - \frac{1}{4}, \qquad [\because e^0 = 1]$$

$$= \frac{1}{4}\left[4e^4 + e^4 - 1\right] = \frac{1}{4}\left[5e^4 - 1\right].$$

Example 70:

Evaluate $\int_0^{\pi/2} \sin^3 x\,dx$.

Solution:

The given integral

$$I=\int_0^{\pi/2}\sin^2 x\sin x\,dx=\int_0^{\pi/2}\left(1-\cos^2 x\right)\sin x\,dx. \qquad \textbf{(Note)}$$

Now put cos x = t so that – sin x dx = dt.

Also when x = 0, t = 1

and when $x = \frac{\pi}{2}$, t = 0.

Hence $I=-\int_1^0\left(1-t^2\right)dt=\int_0^1\left(1-t^2\right)dt,$

$$\left[\text{Note that }\int_a^b f(x)dx--\int_b^a f(x)dx\right]$$

$$=\left[t-\frac{1}{3}t^3\right]_0^1=1-\frac{1}{3}=\frac{2}{3}.$$

Example 71:

Evaluate $\int_0^{\pi}\cos^3 x\,dx.$

Solution:

We know that cos 3x = 4 $\cos^3$ x – 3 cos x.

Therefore $\cos^3$ x = $\frac{1}{4}$(cos 3x + 3 cos x).

Now $\int_0^{\pi}\cos^3 x\,dx=\int_0^{\pi}\frac{1}{4}(\cos 3x+3\cos x)dx$

$$=\frac{1}{4}\left[\frac{1}{3}\sin 3x+3\sin x\right]_0^{\pi}$$

$$=\frac{1}{4}\left[\left(\frac{1}{3}\sin 3\pi+3\sin\pi\right)-\left(\frac{1}{3}\sin 0+3\sin 0\right)\right]=\frac{1}{4}[0-0]=0.$$

Example 72(a):

Evaluate $\int_0^{\pi/2}\cos^4 x\,dx.$

Solution:

The given integral $I=\frac{1}{4}\int_0^{\pi/2}\left(2\cos^2 x\right)^2 dx$

$$=\frac{1}{4}\int_0^{\pi/2}(1+2\cos 2x)^2\,dx$$

$$= \frac{1}{4}\int_0^{\pi/2}\left(1 + 2\cos 2x + \cos^2 2x\right)dx.$$

$$= \frac{1}{4}\int_0^{\pi/2}\left[1 + 2\cos 2x + \frac{1}{2}\left(2\cos^2 2x\right)\right]dx$$

$$= \frac{1}{4}\int_0^{\pi/2}\left\{1 + 2\cos 2x + \frac{1}{2}(1 + \cos 4x)\right\}dx$$

$$= \frac{1}{4}\int_0^{\pi/2}\left[\frac{3}{2} + 2\cos 2x + \frac{1}{2}\cos 4x\right]dx$$

$$= \frac{1}{4}\left[\frac{3}{2}x + \frac{2\sin 2x}{2} + \frac{1}{2}\frac{\sin 4x}{4}\right]_0^{\pi/2}$$

$$= \frac{1}{4}\left[\left\{\frac{3}{2} \times \frac{\pi}{2} + \sin\left(2 \times \frac{\pi}{2}\right) + \frac{1}{8}\sin\left(4 \times \frac{\pi}{2}\right)\right\} - 0\right]$$

$$= \frac{1}{4}\left[\frac{3}{4}\pi + \sin\pi + \left(\frac{1}{8}\right)\sin 2\pi\right] = \frac{3}{16}\pi.$$

Example 72(b):

Evaluate $\int_0^1 \frac{1-x}{1+x}dx.$

Solution:

We have $\int_0^1 \frac{1-x}{1+x}dx = \int_0^1 \frac{-(1+x)+2}{1+x}dx$

$$= \int_0^1\left[-1 + \frac{2}{1+x}\right]dx = \left[-x + 2\log(1+x)\right]_0^1$$

$= \{-1 + 2\log 2\} - \{-0 + 2\log 1\}$

$= -1 + 2\log 2.$

Example 72(c):

Evaluate $\int_1^2 \frac{dx}{x\left(1+x^4\right)}.$

Solution:

We have $\int_1^2 \frac{dx}{x\left(1+x^4\right)} = \int_1^2 \frac{x\,dx}{x^2\left(1+x^4\right)}.$

Now put $x^2 = t$

so that 2x dx = dt.

Also when x = 1, t = 1

and when x = 2, t = 4.

Thus $\int_1^2 \frac{dx}{x\left(1+x^4\right)} = \frac{1}{2}\int_1^4 \frac{dt}{t\left(1+t^2\right)}$.

Now let $\frac{1}{t\left(1+t^2\right)} = \frac{A}{t} + \frac{Bt+C}{1+t^2}$.

Clearly $1 \equiv A + At^2 + Bt^2 + Ct$. Comparing the coefficients of like powers of t on both sides, we have

A = 1, C = 0 and A + B = 0. Therefore B = – 1.

$\therefore \frac{1}{t\left(1+t^2\right)} = \frac{1}{t} - \frac{t}{1+t^2}$.

Hence the given integral $= \frac{1}{2}\int_1^4 \left[\frac{1}{t} - \frac{t}{1+t^2}\right]dt$

$= \frac{1}{2}\left[(\log t)_1^4 - \frac{1}{2}\left\{\log\left(t^2+1\right)\right\}_1^4\right]$

$= \frac{1}{2}\left[\{\log 4 - \log 1\} - \frac{1}{2}(\log 17 - \log 2)\right]$

$= \frac{1}{2}\left[2\log 2 - \frac{1}{2}\log 17 + \frac{1}{2}\log 2\right]$, [$\because \log 1 = 0$]

$= \log 2 - \frac{1}{4}\log 17 + \frac{1}{4}\log 2 = \frac{5}{4}\log 2 - \frac{1}{4}\log 17$

$= \frac{1}{4}\log 2^5 - \frac{1}{4}\log 17 = \frac{1}{4}\log 32 - \frac{1}{4}\log 17 = \frac{1}{4}\log\left(\frac{32}{17}\right)$.

Example 72(d):

Evaluate $\int_0^{\pi/2} e^x(\sin x + \cos x)\,dx$.

Solution:

Given integral $I = \int_0^{\pi/2} e^x \sin x\,dx + \int_0^{\pi/2} e^x \cos x\,dx$.

Integrating the second integral by parts taking cos x as the second function, we get

$I = \int_0^{\pi/2} e^x \sin x\,dx + \left[\left(e^x \sin x\right)_0^{\pi/2} - \int_0^{\pi/2} e^x \sin x\,dx\right]$

$$= \int_0^{\pi/2} e^x \sin x\, dx + \left(e^x \sin x\right)_0^{\pi/2} - \int_0^{\pi/2} e^x \sin x\, dx$$

$$= \left(e^x \sin x\right)_0^{\pi/2} = e^{\pi/2} \sin\left(\frac{\pi}{2}\right) - e^0 \sin 0$$

$$= (e^{\pi/2}.1) - (1 \times 0) = e^{\pi/2}.$$

Example 72(e):

Integrate and evaluate

$$(i)\ \int_0^{\pi/2} \sin^2 x\, dx,\ (ii)\ \int_0^{\pi/4} \tan^2 x\, dx\ \ (iii)\ \int_0^1 xe^x dx.$$

Solution:

(i) We have $\int_0^{\pi/2} \sin^2 x\, dx = \frac{1}{2}\int_0^{\pi/2} 2\sin^2 x\, dx$

$$= \frac{1}{2}\int_0^{\pi/2} (1 - \cos 2x)\, dx = \frac{1}{2}\left[x - \frac{\sin 2x}{2}\right]_0^{\pi/2}$$

$$= \frac{1}{2}\left[\left(\frac{1}{2}\pi - \frac{1}{2}\sin\pi\right) - 0\right] = \frac{1}{2}\left(\frac{1}{2}\pi - 0\right) = \frac{1}{4}\pi.$$

(ii) We have $\int_0^{\pi/4} \tan^2 x\, dx = \int_0^{\pi/4} \left(\sec^2 x - 1\right) dx$

$$= [\tan x]_0^{\pi/4} - [x]_0^{\pi/4} = \tan\frac{\pi}{4} - \frac{\pi}{4} = 1 - \frac{\pi}{4}.$$

(iii) We have $\int_0^1 xe^x dx = \left[x.e^x\right]_0^1 - \int_0^1 1.e^x dx$

$$= (1.e^1) - (0.e^0) - \int_0^1 e^x dx$$

$$= e - \left[e^x\right]_0^1 = e - [e - e^0] = 0 + 1 = 1, \qquad [\because e^0 = 1].$$

Example 73:

Show that $\int_a^b \frac{\log x}{x} dx = \frac{1}{2}\log\left(\frac{b}{a}\right)\log(ab).$

Solution:

Put log x = t so that $\left(\frac{1}{x}\right)dx = dt$.

Also when x = a, t = log a

and when x = b, t = log b.

$$\therefore \int_a^b \frac{\log x}{x} dx = \int_{\log a}^{\log b} t\,dt = \left[\frac{t^2}{2}\right]_{\log a}^{\log b}$$

$$= \frac{1}{2}\left[(\log b)^2 - (\log a)^2\right] = \frac{1}{2}\left[(\log b - \log a)\cdot(\log b + \log a)\right]$$

$$= \frac{1}{2}\left[\log\left(\frac{b}{a}\right).\log(ab)\right].$$

Example 74(a):

Evaluate $\int_a^{\infty} \frac{dx}{x^4}$.

Solution:

Given integral $= \lim_{b\to\infty} \int_a^b \frac{dx}{x^4} = \lim_{b\to\infty}\left[\frac{-1}{3x^3}\right]_a^b$

$= \lim_{b\to\infty}\left[\frac{-1}{3b^3} + \frac{1}{3a^3}\right] = \frac{1}{3a^3}$. $\left[\because \lim_{b\to\infty} \frac{-1}{3b^3} = 0\right]$

Example 74(b):

Evaluate $\int_a^{\infty} \frac{dx}{\sqrt{x}}$.

Solution:

Given integral $= \lim_{b\to\infty} \int_a^b \frac{dx}{\sqrt{x}} = \lim_{b\to\infty}\left[2\sqrt{x}\right]_a^b$

$= \lim_{b\to\infty} 2\left(\sqrt{b} - \sqrt{a}\right)$.

Since $\lim_{b\to\infty} 2\sqrt{b}$ is not finite *i.e.*, is not a definite number, therefore the integral under consideration is meaningless.

Example 74(c):

Prove that $\int_0^{1/\sqrt{2}} \frac{\sin^{-1}x\,dx}{\left(1-x^2\right)^{3/2}} = \frac{\pi}{4} - \frac{1}{2}\log 2$.

Solution:

Put x = sin θ

so that dx = cos θ dθ.

Also when x = 0,

sin θ = 0 or θ = 0;

and when $x=\frac{1}{\sqrt{2}}$, $\sin\theta=\frac{1}{\sqrt{2}}$ or $\theta=\frac{1}{4}\pi$.

Thus the given integral $=\int_0^{\pi/4}\theta\sec^2\theta\,d\theta$

$$=[\theta\tan\theta]_0^{\pi/4}-\int_0^{\pi/4}1.\tan\theta\,d\theta, \qquad \text{(Integrating by parts)}$$

$$=\left(\frac{\pi}{4}\tan\frac{\pi}{4}\right)=(0\tan 0)-[\log\sec\theta]_0^{\pi/4}$$

$$=\frac{\pi}{4}\cdot 1-\left[\log\frac{1}{4}\pi-\log\sec 0\right]=\frac{1}{4}\pi-\left(\log\sqrt{2}-\log 1\right)$$

$$=\frac{1}{4}\pi-\left(\log\sqrt{2}-0\right)=\frac{1}{4}\pi-\frac{1}{2}\log 2.$$

Example 74(d):

Prove that $\int_0^1 x\left(\tan^{-1}x\right)^2dx=\frac{\pi}{4}\left(\frac{\pi}{4}-1\right)+\frac{1}{2}\log 2.$

Solution:

We have $\int_0^1 x\left(\tan^{-1}x\right)^2dx$

$$=\left[\frac{x^2}{2}\left(\tan^{-1}x\right)^2\right]_0^1-\int_0^1\frac{x^2}{2}\frac{2\tan^{-1}x}{1+x^2}dx,$$

integrating by parts taking x as the second function

$$=\frac{1}{2}\left(\tan^{-1}1\right)^2-\int_0^1\frac{\left(x^2-1\right)-1}{1+x^2}\tan^{-1}x\,dx$$

$$=\frac{\pi^2}{32}-\int_0^1\tan^{-1}x\,dx+\int_0^1\left(\tan^{-1}x\right)\cdot\frac{1}{1+x^2}dx$$

$$=\frac{\pi^2}{32}-\left[x\tan^{-1}x\right]_0^1+\int_0^1 x\cdot\frac{1}{1+x^2}dx+\left[\frac{\left(\tan^{-1}x\right)^2}{2}\right]_0^1,$$

applying integration by parts to the first integral taking 1 as the second function and power formula to the second integral

$$=\frac{\pi^2}{32}-\left[1.\tan^{-1}1\right]+\frac{1}{2}\left[\log\left(1+x^2\right)\right]_0^1+\frac{\pi^2}{32}$$

$$=\frac{\pi^2}{16}-\frac{\pi}{4}+\frac{1}{2}[\log 2-\log 1]=\frac{\pi}{4}\left(\frac{\pi}{4}-1\right)+\frac{1}{2}\log 2.$$

Example 75:

Evaluate (i) $\int_2^4 \frac{x^2+x}{2x+1}dx$

(ii) $\int_0^1 \frac{ab}{[(a-b)x+b]^2}dx.$

Solution:

(i) We have $I = \int_2^4 \frac{x^2+x}{2x+1}dx$

$$= \int_2^4 \left[\frac{1}{2}x + \frac{1}{4} - \frac{1/4}{2x+1}\right]dx,$$

dividing the Nr. by the Dr. we get $\frac{1}{2}x+\frac{1}{4}$ as the quotient and $-\frac{1}{4}$ as the remainder

$$= \left[\frac{x^2}{4} + \frac{x}{4} - \frac{1}{8}\log(2x+1)\right]_2^4$$

$$= 4 + 1 - \frac{1}{8}\log 9 - 1 - \frac{1}{2} + \frac{1}{8}\log 5 = \frac{7}{2} - \frac{1}{8}\log\left(\frac{9}{5}\right).$$

(ii) Let $I = \int_0^1 \frac{ab}{[(a-b)x+b]^2}dx.$

Put (a – b) x + b = t. Then (a – b) dx = dt.

When x = 0, t = b and when x = 1, t = a.

$$\therefore\ I = \int_b^a \left[\frac{ab}{t^2}\cdot\frac{1}{a-b}\right]dt = \frac{ab}{a-b}\int_b^a \frac{1}{t^2}dt$$

$$= \frac{ab}{a-b}\left[-\frac{1}{t}\right]_b^a = \frac{ab}{a-b}\left[-\frac{1}{a}+\frac{1}{b}\right] = \frac{ab}{a-b}\left[\frac{1}{b}-\frac{1}{a}\right]$$

$$= \frac{ab}{a-b}\cdot\frac{a-b}{ab} = 1.$$

Example 76(a):

Prove that $\int_0^1 \frac{x^3\sin^{-1}x}{\sqrt{(1-x^2)}}dx = \frac{7}{9}.$

Solution:

Let $I = \int_0^1 \frac{x^3\sin^{-1}x}{\sqrt{(1-x^2)}}dx.$

Put $x = \sin\theta$ or $\sin^{-1}x = \theta$.

Then $dx = \cos\theta\, d\theta$.

When $x = 0$,

we have $\sin\theta = 0$ giving $\theta = 0$, and when $x = 1$,

we have $\sin\theta = 1$ giving $\theta = \frac{\pi}{2}$.

$$\therefore\ I = \int_0^{\pi/2} \frac{\theta \sin^3\theta}{\cos\theta}\cdot\cos\theta\, d\theta = \int_0^{\pi/2} \theta\sin^3\theta\, d\theta$$

$$= \int_0^{\pi/2} \theta\cdot\frac{3\sin\theta - \sin 3\theta}{4} d\theta, \qquad [\because \sin 3\theta = 3\sin\theta - 4\sin^3\theta]$$

$$= \frac{1}{4}\int_0^{\pi/2} \theta\cdot(3\sin\theta - \sin 3\theta)\, d\theta$$

$$= \frac{1}{4}\left[\theta.\left(-3\cos\theta + \frac{1}{3}\cos 3\theta\right)\right]_0^{\pi/2} - \frac{1}{4}\int_0^{\pi/2} 1.\left(-3\cos\theta + \frac{1}{3}\cos 3\theta\right) d\theta,$$

integrating by parts

$$= \frac{1}{4}\cdot 0 - \frac{1}{4}\left[-3\sin\theta + \frac{1}{9}\sin 3\theta\right]_0^{\pi/2}$$

$$= -\frac{1}{4}\left[-3\sin\frac{1}{2}\pi + \frac{1}{9}\sin\left(\frac{3\pi}{2}\right)\right] = -\frac{1}{4}\left[-3 - \frac{1}{9}\right] = \frac{7}{9}.$$

Example 76(b):

Evaluate $\dfrac{x + \sin x}{1 + \cos x} dx$.

Solution:

We have

$$\int \frac{x + \sin x}{1 + \cos x} dx = \int \frac{x + 2\sin(x/2)\cos(x/2)}{1 + 2\cos^2(x/2) - 1}$$

$$= \frac{1}{2}\int x\sec^2\left(\frac{x}{2}\right)dx + \int \tan\left(\frac{x}{2}\right)dx$$

$$= \frac{1}{2}\left[x\frac{\tan(x/2)}{(1/2)} - \int 1\cdot\frac{\tan(x/2)}{(1/2)}dx\right] + \int \tan\left(\frac{x}{2}\right)dx$$

$$= x\tan\left(\frac{x}{2}\right) - \int \tan\left(\frac{x}{2}\right)dx + \int \tan\left(\frac{x}{2}\right)dx = x\tan\left(\frac{x}{2}\right).$$

Example 76(c):

Evaluate (i) $\int x^2 \tan^{-1}x\, dx$

(ii) $\int x^3 \tan^{-1} x\, dx$.

Solution:

(i) $= \frac{x^3}{3}\tan^{-1} x - \int \frac{x^3}{3}\cdot\frac{1}{1+x^2}dx,$

integrating by parts taking x^2 as the second function

$$= \frac{x^3}{3}\tan^{-1} x - \frac{1}{3}\int \frac{x(x^2+1)-x}{1+x^2}dx, \quad [\because x^3 = x(x^2+1)-x]$$

$$= \frac{x^3}{3}\tan^{-1} x - \frac{1}{3}\int x\,dx + \frac{1}{3}\cdot\frac{1}{2}\int \frac{2x}{1+x^2}dx$$

$$= \frac{x^3}{3}\tan^{-1} x - \frac{1}{6}x^2 + \frac{1}{6}\log(1+x^2).$$

(ii) We have $\int x^3 \tan^{-1} x\,dx$

$$= \frac{x^4}{4}\tan^{-1} x - \int \frac{x^4}{4}\cdot\frac{1}{1+x^2}dx,$$

integrating by parts taking x^3 as the second function

$$= \frac{x^4}{4}\tan^{-1} x - \frac{1}{4}\int\left[x^2 - 1 + \frac{1}{1+x^2}\right]dx,$$

dividing x^4 by $x^2 + 1$, we get $x^2 - 1$ as the quotient and 1 as the remainder

$$= \frac{1}{4}\left[x^4\tan^{-1} x - \frac{x^3}{3} + x - \tan^{-1} x\right]$$

$$= \frac{1}{4}\left[(x^4-1)\tan^{-1} x - \frac{x^3}{3} + x\right].$$

Example 77:

Evaluate: (i) $\int \frac{\log(1+x)}{\sqrt{(1+x)}}dx$

(ii) $\int \log(x+2)^{x+2}\,dx$.

Solution:

(i) We have

$$\int \frac{\log(1+x)}{\sqrt{(1+x)}}dx = \int[\log(1+x)]\,(1+x)^{-1/2}dx$$

$$= [\log(1+x)]\cdot\frac{(1+x)^{1/2}}{1/2} - \int\frac{1}{1+x}\cdot\frac{(1+x)^{1/2}}{1/2}dx,$$

integrating by parts taking $(1 + x)^{-1/2}$ as the second function

$$= 2\sqrt{(1+x)}\log(1+x) - 2\int(1+x)^{-1/2}dx$$

$$= 2\sqrt{(1+x)}\log(1+x) - 2\cdot\frac{(1+x)^{1/2}}{1/2}$$

$$= 2\sqrt{(1+x)}\left[\log(1+x) - 2\right].$$

(ii) We have $\int \log(x+2)^{x+2}dx = \int(x+2)\log(x+2)dx$

$$= \frac{(x+2)^2}{2}\log(x+2) - \int\frac{(x+2)^2}{2}\cdot\frac{1}{x+2}dx,$$

integrating by parts taking $(x + 2)^1$ as the second function

$$= \frac{(x+2)^2}{2}\log(x+2) - \frac{1}{2}\int(x+2)dx$$

$$= \frac{(x+2)^2}{2}\log(x+2) - \frac{1}{2}\frac{(x+2)^2}{2}$$

$$= \frac{(x+2)^2}{4}\left[2\log(x+2) - 1\right].$$

4

Analysis of Determinants

INTRODUCTION

Consider the following two equations:

$$a_1x + b_1y = 0$$

$$a_2x + b_2y = 0$$

$$\therefore \qquad -\frac{a_1}{b_1} = \frac{y}{x} = -\frac{a_2}{b_2}$$

Eliminating x and y we get $-\frac{a_1}{b_1} = -\frac{a_2}{b_2}$

or $a_1b_2 - a_2b_1 = 0$

We shall experts the above eliminate in the form of

$$\begin{vmatrix} a_1 & b_1 \\ a_2 & b_2 \end{vmatrix} = 0 \qquad \text{...(1)}$$

(1) is called a determinant of second order and its value is $a_1b_2 - a_2b_1 = 0$

Let us non eliminate x, y, z from the following equations:

$$a_1x + b_1y + c_1 = 0$$

$$a_2x + b_2y + c_2 = 0$$

$$a_3x + b_3y + c_3 = 0$$

Solving eqn. (2) and (3) by cross multiplication, we have

$$\frac{x}{b_2c_3 - b_3c_2} = \frac{y}{(a_2c_3 - a_3c_2)} = \frac{y}{a_2c_3 - a_3b_2}$$

Substituting the value of x_1 and z in (1), we get

$K[a_1 (b_2c_3 - b_3c_2) - b_1 (a_2c_3 - c_3a_2) + c_1 (a_2b_3 - a_3b_2)] = 0$

$\therefore$ The eliminant is $a_1(b_2c_3 - b_3c_2) - (a_2c_3 - c_3a_2) + c_1(a_2b_3 - a_3b_2) = 0$

$$\begin{vmatrix} a_1 & b_1 & c_1 \\ a_2 & b_2 & c_2 \\ a_3 & b_3 & c_3 \end{vmatrix} = 0 \qquad ...(2)$$

eqn. (2) is called the determinant of order (3).

DETERMINANT OF A SQUARE MATRIX

Let us consider a square matrix A of order n × n given by

$$A = \begin{bmatrix} a_{11} & a_{12} & a_{13} & \cdots & \cdots & a_{1n} \\ a_{21} & a_{22} & a_{23} & \cdots & \cdots & a_{2n} \\ a_{31} & a_{32} & a_{33} & \cdots & \cdots & a_{3n} \\ \cdots & \cdots & \cdots & \cdots & \cdots & \cdots \\ a_{n1} & a_{n2} & a_{n3} & \cdots & \cdots & a_{nn} \end{bmatrix}$$

The product of the elements in the principal diagonal is

$$a_{11}\, a_{22}\, a_{33} \, . \, . \, . \, a_{nn}.$$

This is also called the *trace of the matrix.*

Now obtain n! terms of the above type by operating on the row-subscripts of the elements of the above expression by n! permutations $p = \begin{pmatrix} 1 & 2 & 3 & \ldots & n \\ i_1 & i_2 & i_3 & \ldots & i_n \end{pmatrix}$, where $i_1, i_2, i_3, \ldots i_n$ are one of the n! permutations of the integers 1, 2, 3, . . ., n.

The sum of n! signed terms thus obtained is defined as the *determinants of the matrix* A and is denoted by | A | or $|a_{ij}|$.

Therefore the determinant of the square matrix $A = [a_{ij}]$ of order n × n is given by

$$|a_{ij}| = \Sigma \pm a_{\alpha_1}\, a_{\beta_2}\, a_{\gamma_3}\, a_{\delta_4} \ldots a_{k_n}$$

where + or – sign is taken where α, β, γ, δ, . . ., k is an even or odd permutation of 1, 2, 3, . . ., n, and the summation extends over n! permutations of the row subscripts 1, 2, 3, ...

Note: The determinant of a square matrix of order n is known as a determinant of order n.

DETERMINANT OF ORDER TWO

Let us consider a square matrix $A = \begin{bmatrix} a_{11} & a_{12} \\ a_{21} & a_{22} \end{bmatrix}$ of order 2 × 2.

Then 2! permutations on two symbols 1 and 2 are

$$I = \begin{pmatrix} 1 & 2 \\ 1 & 2 \end{pmatrix},\ p = \begin{pmatrix} 1 & 2 \\ 2 & 1 \end{pmatrix}.$$

The product of the elements of the principal diagonal are $a_{11}\ a_{22}$.

Operating on the row subscripts of $a_{11}\ a_{22}$ by the permutation I we get $+\ a_{11}\ a_{22}$, prefixing + sign as the permutation I is even and by the permutation p we get $-a_{21}\ a_{12}$, prefixing – sign as the permutation is odd. **(Note)**

Hence $\begin{vmatrix} a_{11} & a_{12} \\ a_{21} & a_{22} \end{vmatrix} = a_{11}\ a_{22} - a_{21}\ a_{12}$

DETERMINANT OF ORDER THREE

Let us consider a square matrix $A = \begin{bmatrix} a_{11} & a_{12} & a_{13} \\ a_{21} & a_{22} & a_{23} \\ a_{31} & a_{32} & a_{33} \end{bmatrix}$, of order 3 × 3.

Then 3! permutations on three symbols 1, 2, and 3 are

$$I = \begin{pmatrix} 1 & 2 & 3 \\ 1 & 2 & 3 \end{pmatrix};\ p_1 = \begin{pmatrix} 1 & 2 & 3 \\ 1 & 3 & 2 \end{pmatrix};\ p_2 = \begin{pmatrix} 1 & 2 & 3 \\ 2 & 3 & 1 \end{pmatrix};$$

$$p_3 = \begin{pmatrix} 1 & 2 & 3 \\ 2 & 1 & 3 \end{pmatrix};\ p_4 = \begin{pmatrix} 1 & 2 & 3 \\ 3 & 2 & 1 \end{pmatrix};\ \text{and}\ p_5 = \begin{pmatrix} 1 & 2 & 3 \\ 3 & 2 & 1 \end{pmatrix};$$

Operating on the row subscripts of $a_{11}\ a_{22}\ a_{33}$ by the permutation I, p_1, p_2, . . ., p_5 we have successively

(a) $+a_{11}\ a_{22}\ a_{33}$, prefixing + as permutation I is even,

(b) $-a_{11}\ a_{32}\ a_{23}$, prefixing – as permutation p_1 is odd,

(c) $+a_{21}\ a_{32}\ a_{13}$, prefixing + sign as permutation p_2 is even,

(d) $-a_{21}\ a_{12}\ a_{33}$, prefixing + sign as permutation p_3 is odd,

(e) $+a_{31}\ a_{12}\ a_{23}$, prefixing + sign as permutation p_4 is even,

(f) $-a_{31}\ a_{22}\ a_{13}$, prefixing – sign as permutation p_5 is odd.

Hence we have

$$\begin{bmatrix} a_{11} & a_{12} & a_{13} \\ a_{21} & a_{22} & a_{23} \\ a_{31} & a_{32} & a_{33} \end{bmatrix} = a_{11}\ a_{23}\ a_{33} - a_{11}\ a_{32}\ a_{23} + a_{21}\ a_{32}\ a_{13} - a_{21}\ a_{12}\ a_{33}$$

$$+ a_{31}\ a_{12}\ a_{23} - a_{31}\ a_{22}\ a_{13}$$

$$= a_{11}(a_{22}\,a_{33} - a_{23}\,a_{32}) - a_{13}(a_{21}\,a_{33} - a_{23}\,a_{21}) + a_{13}(a_{21}\,a_{22} - a_{22}\,a_{31})$$

$$= a_{11}\begin{vmatrix} a_{22} & a_{23} \\ a_{32} & a_{33} \end{vmatrix} - a_{12}\begin{vmatrix} a_{21} & a_{23} \\ a_{31} & a_{33} \end{vmatrix} + a_{13}\begin{vmatrix} a_{21} & a_{22} \\ a_{31} & a_{32} \end{vmatrix}.$$

Example 1:

Expand the determinant $\begin{vmatrix} a & h & g \\ h & b & f \\ g & f & c \end{vmatrix}$ *by the elements of 1st row.*

Solution:

Elements of firs row are a, h, g.

Let A, H, and G denote the cofactors of a, h, g.

Then $A = \begin{vmatrix} b & f \\ f & c \end{vmatrix}$; $H = -\begin{vmatrix} h & f \\ g & c \end{vmatrix}$ and $G = \begin{vmatrix} h & b \\ g & f \end{vmatrix}$

Hence $\begin{vmatrix} a & h & g \\ h & b & f \\ g & f & c \end{vmatrix} = aA + hH + gG$

$$= a\begin{vmatrix} b & f \\ f & c \end{vmatrix} - h\begin{vmatrix} h & f \\ g & c \end{vmatrix} + g\begin{vmatrix} h & b \\ g & f \end{vmatrix}$$

$$= a(bc - f^2) - h(ch - fg) + g(hf - bg)$$

$$= abc + 2fgh - af^2 - bg^2 - ch^2.$$ **Ans.**

Example 2:

Find the value of $\begin{vmatrix} a & h & g \\ h & b & f \\ g & f & c \end{vmatrix}$

Solution:

$$\begin{vmatrix} a & h & g \\ h & b & f \\ g & f & c \end{vmatrix} = a\{b'c - f'f\} - h\{hc - f'g\} + g\{hf - bg\}$$

$$= abc - af^2 - ch^2 + fgh + fgh - bg^2$$

$$= abc + 2fgh - af^2 - bg^2 - ch^2.$$ **Ans.**

Example 3:

Evaluate $\begin{vmatrix} 1 & 2 & 3 \\ 4 & 5 & 6 \\ 7 & 8 & 9 \end{vmatrix}$

Solution:

$$\begin{vmatrix} 1 & 2 & 3 \\ 4 & 5 & 6 \\ 7 & 8 & 9 \end{vmatrix} = 1\ \{5.9 - 6.8\} - 2\ \{4.9 - 6.7\} + 3\ \{4.8 - 5.7\},$$

$$= 1\ \{45 - 48\} - 2\{36 - 42\} + 3\ \{32 - 35\}$$

$$= -3 + 12 - 9 = 0.$$ **Ans.**

COFACTOR OF AN ELEMENT

Definition: If in the expansion of a determinant $| a_{ij} |$, all the terms containing a_{ij} as a factor are collected and their sum be denoted by $a_{ij}\ C_{ij}$ then the factor C_{ij} is defined as the cofactor of the element a_{ij}.

From the above definition we find that if $[a_{ij}]$ be the n × n matrix whose determinant is $| a_{ij} |$ then if from $[a_{ij}]$ the element of its ith row and jth column are removed, the terms of C_{ij} are then composed of elements from the remaining (n – 1) × (n – 1) sub-matrix M_{ij} of $[a_{ij}]$.

Hence, the determinant $| a_{ij} |$ can be expressed as a function of the elements of ith row, by collecting all the terms containing $a_{i1}, a_{i2}, a_{i3}, \ldots, a_{in}$ and finding their sum.

i.e., $| a_{ij} | = a_{i1}\ C_{i1} + a_{i2}\ C_{i2} + \ldots + a_{in}\ C_{in}$

$$= \sum_{j-1}^{n} a_{ij}\ C_{ij},$$

which is the expansion of the determinant $| a_{ij} |$ by the elements of the ith row and their cofactors.

In a similar manner we can expand the determinant $| a_{ij} |$ by the elements of the kth column and their cofactor and write as $a_{ij} |= \sum_{i-k}^{n} a_{ik}\ C_{ik}$

Or

The determinant that is left by cancelling the row and columns intersecting at a particular constituent, when the particular constituent is brought in the top left hand corner of a determinant is called its cofactor and is denoted by corresponding capital letter.

PROPERTIES OF DETERMINANTS

Prop. I:

If $A = [a_{ij}]$ is an $n \times n$ matrix then $|A'| = |A|$, where A' is the transpose of the matrix A.

Or

The value of a determinant is not altered by changing the rows into column and column into rows.

Proof:

If $A = [a_{ij}]$, then $A' = [a'_{ij}]$, where $a'_{ij} = a_{ij}$.

Now the product of the elements of the principal diagonal of

$$A' = a'_{11}\ a'_{22}\ a'_{33}\ \ldots\ a'_{nn}.$$

Operating on the row subscripts of the elements of this product by the permutation $p = \begin{pmatrix} 1 & 2 & 3 & n \\ i_1 & i_2 & i_3 \ldots i_n \end{pmatrix}$, where $i_1, t_2, i_3, \ldots$ are $1, 2, 3, \ldots$ in some order, we have $\pm\ a'_{i11}\ a'_{i22}\ a'_{i33}\ \ldots\ a'_{inn}$ as a term of $|A'|$ plus or minus sign to be taken according as p is even or odd.

Now as $a'_{ij} = a_{ji}$, so we have

$a'_{i11}\ a'_{i22}\ a_{i33}\ \ldots\ a'_{inn} = a_{1/1}\ a_{2/2}\ a_{3/3}\ \ldots\ a_{n/n}$

The term $a_{1/1}\ a_{2/2} \ldots a_{n/n}$ can be obtained from the term $a_{i1/1}\ a_{i2/2}\ a_{i3/3}\ \ldots\ a_{in/n}$ by operating on its row subscripts the permutation $p' = \begin{pmatrix} i_1 & i_2 & i_3 \ldots i_n \\ 1 & 2 & 3 & n \end{pmatrix} = p^{-1}$.

Hence p' is even or odd according as p is even or odd. Therefore the term $\pm a'_{i11}\ a'_{i22}\ a'_{i33}\ \ldots\ a'_{inn}$ of $|A'|$ is also a term of $|A|$.

We can thus prove that every one of the n! terms of $|A'|$ is a term of $|A|$. Hence the property.

For example, If $| A | = \begin{vmatrix} a_1 & a_2 & a_3 \\ b_1 & b_2 & b_3 \\ c_1 & c_2 & c_3 \end{vmatrix}$

$$= a_1 \begin{vmatrix} b_2 & b_3 \\ c_2 & c_3 \end{vmatrix} - a_2 \begin{vmatrix} b_1 & b_3 \\ c_1 & c_3 \end{vmatrix} + a_3 \begin{vmatrix} b_1 & b_2 \\ c_1 & c_2 \end{vmatrix}$$

$$= a_1b_2c_3 - a_1b_3c_2 - a_2b_1c_3 + a_2b_3c_1 + a_3b_1c_2 - a_3b_2c_1.$$

Then $| A' | = \begin{vmatrix} a_1 & b_1 & c_1 \\ a_2 & b_2 & c_2 \\ a_3 & b_3 & c_3 \end{vmatrix}$

$$= a_1 \begin{vmatrix} b_2 & c_2 \\ b_3 & c_3 \end{vmatrix} - b_1 \begin{vmatrix} a_2 & c_2 \\ a_3 & c_3 \end{vmatrix} + c_1 \begin{vmatrix} a_2 & b_2 \\ a_3 & b_3 \end{vmatrix}$$

$$= a_1b_2c_3 - a_1b_3c_2 - a_2b_1c_3 + a_3b_1c_2 + a_2b_3c_1 - a_3b_2c_1.$$

$\therefore \quad | A' | = | A |.$

Prop. 2:

If B is obtained from A by interchanging two raws (or columns) then $| B | = - | A |$.

Or

It two adjacent rows or columns of a determinant are interchanged the sign of the determinant is changed whereas its numerical value remarks the same.

Proof:

Let us suppose that sth and rth columns of the determinant A are interchanged where s < 1.

Let $A = \begin{bmatrix} a_{11} & a_{12} & \dots & a_{1s} & \dots & a_{1t} & \dots & a_{1n} \\ a_{21} & a_{22} & \dots & a_{2s} & \dots & a_{2t} & \dots & a_{2n} \\ \dots & \dots & \dots & \dots & \dots & \dots & \dots & \dots \\ a_{i1} & a_{i2} & \dots & a_{is} & \dots & a_{it} & \dots & a_{in} \\ \dots & \dots & \dots & \dots & \dots & \dots & \dots & \dots \\ a_{n1} & a_{n2} & \dots & a_{ns} & \dots & a_{nt} & \dots & a_{nn} \end{bmatrix}$

Then the trace of A *i.e.,* the product of the elements of the principal diagonal of A = $a_{11}\ a_{22} \dots a_{ss} \dots a_{tt} \quad . \ a_{nn}$. If the sth and rth columns are interchanged the product of the elements of the principal diagonal of B

$= a_{11}\ a_{22}\ .\ .\ .\ a_{ss}\ .\ .\ .\ a_{tt}\ .\ .\ .\ a_{nn}$ **(Note)**

In order to have a term of | A |, let us operate on the row subscripts of the trace of A by the permutation

$$p = \begin{pmatrix} 1 & 2 & \ldots & s & \ldots & t & \ldots & n \\ i_1 & i_2 & \ldots & i_s & \ldots & i_t & \ldots & i_n \end{pmatrix}$$

Then we have $a_{i11}\ a_{i22}\ .\ .\ .\ a_{iss}\ .\ .\ .\ a_{itt}\ .\ .\ .\ a_{inn}$.

This term can also be obtained from the trace of B by operating on its row subscripts by the permutation

$$p' = \begin{pmatrix} 1 & 2 & \ldots & s & \ldots & t & \ldots & n \\ i_1 & i_2 & \ldots & i_t & \ldots & i_s & \ldots & i_n \end{pmatrix}$$

Here we observe that $p' = p\ (i_s\ i_t)$, since $(i_s\ i_t)$ is a transposition. Therefore p' is odd or even according as p is even or odd. Hence the term $\pm a_{i11}\ a_{i22}\ .\ .\ .\ a_{iss}\ .\ .\ .\ a_{itt}\ .\ .\ .\ a_{inn}$ is also a term of | B | but with its sign changed. Thus every one of the n! terms of | A | is a term of | B | but with sign changed.

Here | B | = – | A |.

Example:

$$\text{Let } |\ A\ | = \begin{vmatrix} a_1 & a_2 & a_3 \\ b_1 & b_2 & b_3 \\ c_1 & c_2 & c_3 \end{vmatrix}$$

Let the determinant | B | be formed by interchanging second and third columns.

$$\text{Then } |\ B\ | = \begin{vmatrix} a_1 & a_2 & a_3 \\ b_1 & b_2 & b_3 \\ c_1 & c_2 & c_3 \end{vmatrix}$$

Expanding | B | by the elements of its first row, we get

$$|\ B\ | = a_1 \begin{vmatrix} b_3 & b_2 \\ c_3 & c_2 \end{vmatrix} - a_3 \begin{vmatrix} b_1 & b_2 \\ c_1 & c_2 \end{vmatrix} + a_2 \begin{vmatrix} b_1 & b_3 \\ c_1 & c_3 \end{vmatrix}$$

$$= a_1(b_3c_2 - b_2c_3) - a_3(b_1c_2 - b_2c_1) + a_2(b_1c_3 - b_3c_1)$$

$$= -\ [a_1(b_2c_3 - b_3c_2) - a_2(b_1c_3 - b_3c_1) + a_3(b_1c_2 - b_2c_1)$$

$$= -\left[a_1 \begin{vmatrix} b_2 & b_3 \\ c_2 & c_3 \end{vmatrix} - a_2 \begin{vmatrix} b_1 & b_3 \\ c_1 & c_3 \end{vmatrix} + a_3 \begin{vmatrix} b_1 & b_2 \\ c_1 & c_2 \end{vmatrix}\right]$$

$$= -\begin{vmatrix} a_1 & a_2 & a_3 \\ b_1 & b_2 & b_3 \\ c_1 & c_2 & c_3 \end{vmatrix} = -|A|.$$

Prop. 3:

If the elements of ith row (or ith column) of the determinant $| a_{ij} |$ are multiplied by a scalar c then the resulting determinant is c $| a_{ij} |$.

Proof:

We can write $| a_{ij} | = \sum_{j=1}^{n} a_{ij} \ C_{ij} \ . \ . \ .$

In this case, the resulting determinant (when the elements of ith row are multiplied by c)

$$= \sum_{j=1}^{n} ca_{ij} \, C_{ij} = c \sum_{j=1}^{n} a_{ij} \, C_{ij} = c \, | a_{ij} |.$$

Similarly we can prove that statement when elements of kth column are multiplied by c.

Prop. 4:

If two rows or two columns of a determinant $| A |$ are identical then $| A | = 0$.

Proof:

In Prop. 2 above we have proved that if any two rows or columns of a determinant are interchanged then the value of the determinant changes in sign only.

Thus, if the two *identical* columns (or rows) of a determinant are interchanged, then the determinant does not change but its sign only changes.

Hence $| A | = - | A |$ *i.e.,* $| A | + | A | = 0$ *i.e.,* $| A | = 0$.

Example:

Evaluate $\begin{vmatrix} a_1 & a_2 & a_1 \\ b_1 & b_2 & b_1 \\ c_1 & c_2 & c_1 \end{vmatrix}$

Expanding the determinant with respect to the first row, we have the given determinant

$$= a_1 \begin{vmatrix} b_2 & b_1 \\ c_2 & c_1 \end{vmatrix} - a_2 \begin{vmatrix} b_1 & b_1 \\ c_1 & c_1 \end{vmatrix} + a_1 \begin{vmatrix} b_1 & b_2 \\ c_1 & c_2 \end{vmatrix}$$

$$= a_1(b_2c_1 - b_1c_2) - a_2(b_1c_1 - b_1c_1) + a_1(b_1c_2 - b_2c_1)$$

$$= a_1b_2c_1 - a_1b_1c_2 - a_2(0) + a_1b_1c_2 - a_1b_2c_1 = 0.$$

MINOR OF AN ELEMENT

Definition:

If M_{ij} be the $(n - 1) \times (n - 1)$ sub-matrix of the matrix $A = [a_{ij}]$ obtained by removing the ith row and jth column, then the determinant $| M_{ij} |$ is defined as the minor of the element a_{ij} in the determinant $| a_{ij} |$ of order n.

Theorem 1:

The cofactor C_{ij} of the element a_{ij} in the determinant $| a_{ij} |$ is given by $C_{ij} = (-1)^{i+j} | M_{ij} |$.

Proof:

Let us first of all prove the case $C_{11} = (-1)^2 | M_{11} |$ *i.e.*, $C_{11} = | M_{11} |$.

The terms in C_{11} are composed of elements taken from the $(n - 1) \times (n - 1)$ sub-matrix M_{11} of A. The general term of $a_{11}\ C_{11} = \pm\ a_{11}\ a_{i22}\ a_{i33}\ .\ .\ .\ a_{inn}$, where $i_2, i_3, \ldots i_n$ are 2, 3, . . ., n is some order.

This term can also be obtained from the trace of matrix A *i.e.*, the product of the elements of the diagonal of the matrix A *i.e.*, $a_{11}\ a_{22}\ a_{33} \ldots a_{nn}$, by operating on its row subscripts by the permutation $p = \begin{pmatrix} 1 & 2 & 3 & \ldots & n \\ i_1 & i_2 & i_3 & \ldots & i_n \end{pmatrix}$, where $i_2, i_3, \ldots, i_n$ are defined as above.

Thus the permutation p may be regarded as a permutation on the symbols 2, 3, . . ., n. Hence all the terms of $a_{11}\ C_{11}$ can be obtained by running p through the $(n - 1)!$ permutations on the symbols 2, 3, . . ., n keeping 1 fixed.

Thus, the terms of C_{11} can be obtained by operating on the row subscripts of the elements of the product $a_{22}\ a_{33} \ldots a_{nn}$, which is the product of the elements of the diagonal of M_{11} *i.e.*, the trace of M_{11}.

Hence $C_{11} = | M_{11} |$, Now let us prove $C_{ij} = (-1)^{i+j} | M_{11} |$.

Move the jth column of the matrix A to the first column by performing $(j - 1)$ successive interchanges of adjacent columns and move the ith row of the matrix A to the first row by performing $(i - 1)$ successive interchanges of adjacent rows. Then the element a_{ij} is in the first row and first column of the resulting matrix B, say. The sub-matrix of B obtained by removing the first row and first column is the sub-matrix M_{ij} of the matrix A. Hence a_{ij} $|M_{ij}|$ is the term of $|B|$ containing a_{ij}.

Also we know that, if two rows or two columns of a determinant $|A|$ are interchanged, the new determinant $= -|A|$.

$\therefore$ We have $|B| = (-1)^{i-1+j-i}|A|$ **(Note)**

$= (-1)^{i+j}(-1)^{-2}|A|$

$= (-1)^{i+j}|A|$, $\because \quad (-1)^{-2} = 1$

$\Rightarrow \quad |A| = (-1)^{i+j}|B|$ **(Note)**

Equating the coefficient of a_{ij} from both sides, we have

$C_{ij} = (-1)^{i+j}|M_{ij}|$.

Theorem 2:

If C_{ij} is the cofactor of a_{ij} in the determinant $|A| = |a_{ij}|$ of order n, then (i) the sum of the products of the elements of the ith row with the cofactors of the corresponding elements of the kth row is zero provided $i \neq k$.

(ii) Also the sum of the products of the elements of the jth column with the cofactors of the corresponding elements of the kth column is zero provided $j \neq k$,

i.e., (i) $\displaystyle\sum_{j=1}^{n} c_{ij} C_{kj} = 0, \text{ if } i \neq k$

and (ii) $\displaystyle\sum_{j=1}^{n} c_{ij} C_{ik} = 0, \text{ if } j \neq k$

Proof:

(i) The given determinant $|A| = \displaystyle\sum_{j=1}^{n} a_{kj}C_{kj}$.

Now replace the kth row by ith row, then we have the new determinant $= \displaystyle\sum_{j=1}^{n} a_{ij} C_{kj}$.

But the kth and ith rows of the new determinant are identical, hence its values is zero.

$$\sum_{j=1}^{n} a_{ij} C_{kj} = 0$$

(ii) The given determinant $|A| = \displaystyle\sum_{i=1}^{n} a_{ik} C_{ik}$

Now replace the kth column by jth column, then we have the new determinant $= \sum_{i=1}^{n} a_{ij} C_{ik}$.

But the kth and jth columns of the new determinant are identical, hence its value is zero.

$$\therefore \sum_{i=1}^{n} a_{ij} C_{ik} = 0.$$

Example:

In the determinant $|A| = \begin{vmatrix} a_1 & b_1 & c_1 \\ a_2 & b_2 & c_2 \\ a_3 & b_3 & c_3 \end{vmatrix}$ *prove that*

$a_1A_2 + b_1B_2 + c_1C_2 = 0,\ b_1C_1 + b_2C_2 + b_3C_3 = 0$ *and*

$c_1B_1 + c_2B_2 + c_3B_3 = 0$, *where capital letters denote the cofactors of the corresponding small letters. Also prove that,*

$$a_1A_1 + b_1B_1 + c_1C_1 = |A| = a_2A_2 + b_2B_2 + c_2C_2 = a_3A_3 + b_3B_3 + c_3C_3$$

Solution:

In the det $|A| = \begin{vmatrix} a_1 & b_1 & c_1 \\ a_2 & b_2 & c_2 \\ a_3 & b_3 & c_3 \end{vmatrix}$, we have

$$A_1 = \begin{vmatrix} b_2 & c_2 \\ b_3 & c_3 \end{vmatrix};\ A_2 = -\begin{vmatrix} b_1 & c_1 \\ b_3 & c_3 \end{vmatrix};\ B_1 = -\begin{vmatrix} a_2 & c_2 \\ a_3 & c_3 \end{vmatrix};\ B_2 = \begin{vmatrix} a_1 & c_1 \\ a_3 & c_3 \end{vmatrix};$$

$$B_3 = -\begin{vmatrix} a_1 & c_1 \\ a_2 & c_2 \end{vmatrix};\ C_1 = \begin{vmatrix} a_2 & b_2 \\ a_3 & b_3 \end{vmatrix};\ C_2 = -\begin{vmatrix} a_1 & b_1 \\ a_3 & b_3 \end{vmatrix};\ A_3 = \begin{vmatrix} b_1 & c_1 \\ b_2 & c_2 \end{vmatrix};$$

$$C_3 = \begin{vmatrix} a_1 & b_1 \\ a_2 & b_2 \end{vmatrix} \qquad \therefore a_1A_2 + b_1B_2 + c_1C_2$$

$$= -a_1 \begin{vmatrix} b_1 & c_1 \\ b_3 & c_3 \end{vmatrix} + b_1 \begin{vmatrix} a_1 & c_1 \\ a_3 & c_3 \end{vmatrix} - c_1 \begin{vmatrix} a_1 & b_1 \\ a_3 & b_3 \end{vmatrix}$$

$$= -a_1(b_1c_3 - b_3c_1) + b_1(a_1c_3 - a_3c_1) - c_1(a_1b_3 - a_3b_1)$$

$= 0$, on simplifying

$$b_1C_1 + b_2C_2 + b_3C_3 = b_1 \begin{vmatrix} a_2 & b_2 \\ a_3 & b_3 \end{vmatrix} - b_2 \begin{vmatrix} a_1 & b_1 \\ a_3 & b_3 \end{vmatrix} + b_3 \begin{vmatrix} a_1 & b_1 \\ a_2 & b_2 \end{vmatrix}$$

$= b_1(a_2b_3 - a_3b_2) - b_2(a_1b_3 - a_3b_1) + b_3(a_1b_2 - a_2b_1)$

$= 0$, on simplifying.

In a similar way the remaining part can also be proved.

Also $|A| = \begin{vmatrix} a_1 & b_1 & c_1 \\ a_2 & b_2 & c_2 \\ a_3 & b_3 & c_3 \end{vmatrix}$

$$= a_1 \begin{vmatrix} b_2 & c_2 \\ b_3 & c_3 \end{vmatrix} - b_2 \begin{vmatrix} a_2 & c_2 \\ a_3 & c_3 \end{vmatrix} + c_1 \begin{vmatrix} a_2 & b_2 \\ a_3 & b_3 \end{vmatrix} \quad \text{...(i)}$$

expanding with respect to first row.

Again $a_1A_1 + b_1B_1 + c_iC_1$

$$= a_1 \begin{vmatrix} b_2 & c_2 \\ b_3 & c_3 \end{vmatrix} + b_2 \left\{ - \begin{vmatrix} a_2 & c_2 \\ a_3 & c_3 \end{vmatrix} \right\} + c_1 \begin{vmatrix} a_2 & b_2 \\ a_3 & b_3 \end{vmatrix}$$

$$= a_1 \begin{vmatrix} b_2 & c_2 \\ b_3 & c_3 \end{vmatrix} - b_2 \begin{vmatrix} a_2 & c_2 \\ a_3 & c_3 \end{vmatrix} + c_1 \begin{vmatrix} a_2 & b_2 \\ a_3 & b_3 \end{vmatrix}$$

$= |A|$, from (i)

Similarly, $a_2A_2 + b_2B_2 + c_2C_2$

$$= a_2 \left\{ - \begin{vmatrix} b_1 & c_1 \\ b_3 & c_3 \end{vmatrix} \right\} + b_2 \begin{vmatrix} a_1 & c_1 \\ a_3 & c_3 \end{vmatrix} + c_2 \left\{ - \begin{vmatrix} a_1 & b_1 \\ a_3 & b_3 \end{vmatrix} \right\}$$

$$= -a_2 \begin{vmatrix} b_1 & c_1 \\ b_3 & c_3 \end{vmatrix} + b_2 \begin{vmatrix} a_1 & c_1 \\ a_3 & c_3 \end{vmatrix} - c_2 \begin{vmatrix} a_1 & b_1 \\ a_3 & b_3 \end{vmatrix} \quad \text{...(ii)}$$

Also $|A| = \begin{vmatrix} a_1 & b_1 & c_1 \\ a_2 & b_2 & c_2 \\ a_3 & b_3 & c_3 \end{vmatrix}$

$$= -a_2 \begin{vmatrix} b_1 & c_1 \\ b_3 & c_3 \end{vmatrix} + b_2 \begin{vmatrix} a_1 & c_1 \\ a_3 & c_3 \end{vmatrix} - c_2 \begin{vmatrix} a_1 & b_1 \\ a_3 & b_3 \end{vmatrix}, \quad \text{...(ii)}$$

expanding with respect to second row.

$\therefore$ From (ii) and (iii), we get

$$a_2A_2 + b_2B_2 + c_2C_2 = |A|.$$

In a similar way we can prove that $a_2A_2 + b_2B_2 + c_2C_2 = |A|$, by expanding $|A|$ with respect to third row.

AN IMPORTANT PROPERTY OF THE DETERMINANT

(a) If A_i, the ith row of a determinant $| A | = | a_{ij} |$ of order n be replaced by $A_j + cA_k$, where c is a scalar and A_k denotes the kth row of the determinant $| A |$, then the value of the determinant remains unaltered.

Proof:

The determinant $| A | = \sum_{j=1}^{n} a_{ij} C_{ij}$.

Replacing A_i by $A_i + c A_k$ we get the new determinant $| B |$, say

$$\text{Then} \qquad | B | = \sum_{j=1}^{n} (a_{ij} + c\, a_{kj})\, C_{ij}$$

$$= \sum_{j=1}^{n} a_{ij} C_{ij} + c \sum_{j=1}^{n} a_{kj} C_{ij}$$

$$= | A | + c.0$$

i.e., $\qquad | B | = | A |$.

(b) If C_i, the ith column of determinant $| A | = | a_{ij} |$ of order n be replaced by $C_i + \lambda C_k$, where λ is a scalar and C_k denotes the kth column of $| A |$, then the value of the determinant remains unaltered.

Proof is similar to part (a) above.

In the following examples R_1. R_2, R_3, . . . stand for first, second third, . . rows and C_1, C_2,C_3, . . . stand for first, second, third . . . columns.

SOLVED EXAMPLES

Example 1:

Evaluate $\begin{vmatrix} 1 & 1 & 1 & 1 \\ 1 & 1+x & 1 & 1 \\ 1 & 1 & 1+y & 1 \\ 1 & 1 & 1 & 1+z \end{vmatrix}$

Solution:

We have, $= \begin{vmatrix} 1 & 1 & 1 & 1 \\ 0 & x & 0 & 0 \\ 0 & 0 & y & 0 \\ 0 & 0 & 0 & z \end{vmatrix}$, replacing R_2, R_3 and R_4 by $R_4 - R_1$, $R_3 - R_1$ and $R_4 - R_1$ respectively.

$$= \begin{vmatrix} x & 0 & 0 \\ 0 & y & 0 \\ 0 & 0 & z \end{vmatrix}, \text{ expanding with respect to } C_1$$

$$= x \begin{vmatrix} y & 0 \\ 0 & z \end{vmatrix}, \text{ expanding with respect to } R_1$$

$= xyz.$ **Ans.**

Example 2:

Evaluate $\begin{vmatrix} 13 & 16 & 19 \\ 14 & 17 & 20 \\ 15 & 18 & 21 \end{vmatrix}$

Solution:

We have

$$= \begin{vmatrix} 13 & 16 & 3 \\ 14 & 17 & 3 \\ 15 & 18 & 3 \end{vmatrix}, \text{ replacing } C_2 \text{ by } C_3 - C_2$$

$$= \begin{vmatrix} 13 & 3 & 3 \\ 14 & 3 & 3 \\ 15 & 3 & 3 \end{vmatrix}, \text{ replacing } C_2 \text{ by } C_3 - C_1$$

$= 0$, since two columns are identical. **Ans.**

Example 3:

Evaluate $\begin{vmatrix} 1 & 2 & 5 \\ 2 & 3 & 1 \\ -1 & 1 & 1 \end{vmatrix}$

Solution:

The given determinant

$$= \begin{vmatrix} 1 & 3 & 6 \\ 2 & 5 & 3 \\ -1 & 0 & 0 \end{vmatrix}, \text{ replacing } C_2 \text{ and } C_3 \text{ by } C_2 + C_1 \text{ and } C_3 + C_1 \text{ respctively.}$$

$$= \begin{vmatrix} 0 & 3 & 6 \\ 0 & 5 & 3 \\ -1 & 0 & 0 \end{vmatrix}, \text{ replacing } R_1 \text{ by } R_1 + R_3 \text{ and } R_2 \text{ by } R_2 + 2R_3$$

$$= -1 \times \begin{vmatrix} 3 & 6 \\ 5 & 3 \end{vmatrix}, \text{ expanding with respect to first column}$$

$$= -[3\,(3) - 5\,(6)] = 21.$$ **Ans.**

Example 4(a):

Evaluate $\begin{vmatrix} 1 & a & b+c \\ 1 & b & c+a \\ 1 & c & a+b \end{vmatrix}$

Solution:

We have

$$= \begin{vmatrix} 1 & a & a+b+c \\ 1 & b & b+c+a \\ 1 & c & c+a+b \end{vmatrix}, \text{ replacing } C_3 \text{ by } C_3 + C_2$$

$$= (a+b+c) \begin{vmatrix} 1 & a & 1 \\ 1 & b & 1 \\ 1 & c & 1 \end{vmatrix}, \text{ taking out } (a+b+c) \text{ common from } C_3$$

= 0, since two columns are identical. **Ans.**

Example 4(b):

Show that $\begin{vmatrix} (b+c)^2 & a^2 & a^2 \\ b^2 & (c+a)^2 & b^2 \\ c^2 & c^2 & (a+b)^2 \end{vmatrix} = 2abc\,(a+b+c)^2$

Solution:

We have

$$= \begin{vmatrix} (b+c)^2 - a^2 & 0 & a^2 \\ 0 & (c+a)^2 - b^2 & b^2 \\ c^2 - (a+b)^2 & c^2 - (a+b)^2 & (a+b)^2 \end{vmatrix},$$

replacing C_1 and C_2 by $C_1 - C_2$ and $C_2 - C_3$ respectively.

$$= \begin{vmatrix} (b+c+a)(b+c-a) & 0 & a^2 \\ 0 & (c+a+b)(c+a-b) & b^2 \\ (c+a+b)(c-a-b) & (c+a+b)(c-a-b) & (a+b)^2 \end{vmatrix},$$

$$= (a+b+c)^2 \begin{vmatrix} b+c-a & 0 & a^2 \\ 0 & c+a-b & b^2 \\ c-a-b & c-a-b & (a+b)^2 \end{vmatrix},$$ taking out the common factors from C_1 and C_2

$$= (a+b+c)^2 \begin{vmatrix} b+c-a & 0 & a^2 \\ 0 & c+a-b & b^2 \\ -2b & -2a & 2ab \end{vmatrix},$$ replacing R_2 by $R_3 - R_1 - R_2$

$$= (a+b+c)^2 \begin{vmatrix} b+c & a^2/b & a^2 \\ b^2/a & c+a & b^2 \\ 0 & 0 & 2ab \end{vmatrix},$$ replacing C_1 and C_2 by $C_1 + \frac{1}{a}C_2$ and $C_2 + \frac{1}{b}C_2$ respectively. **(Note)**

$$= 2ab\,(a+b+c)^2 \begin{vmatrix} b+c & a^2/b \\ b^2/a & c+a \end{vmatrix},$$ expanding with respect to R_2

$= 2ab\,(a+b+c)^2\,[(b+c)\,(c+a) - (b^2/a)\,(a^2/b)]$

$= 2ab\,(a+b+c)^2\,[bc + ba + c^2 + ca - ab]$

$= 2ab\,(a+b+c)^2\,[b+a+c] = 2abc\,(a+b+c)^2.$ **Ans.**

Example 4(c):

Prove that $\begin{vmatrix} a & b & b & b \\ a & b & a & a \\ a & a & b & a \\ b & b & b & a \end{vmatrix} = -(b-a)^4$

Solution:

We have

$$= \begin{vmatrix} a & b-a & b-a & b-a \\ a & b-a & 0 & 0 \\ a & 0 & b-a & 0 \\ b & 0 & 0 & a-b \end{vmatrix},$$ replacing C_2, C_3 and C_4 by $C_2 - C_1$, $C_3 - C_1$ and $C_4 - C_1$ respectively.

$$= -(b-a)\begin{vmatrix} a & b-a & 0 \\ a & 0 & b-a \\ b & 0 & 0 \end{vmatrix} + (a-b)\begin{vmatrix} a & b-a & b-a \\ a & b-a & 0 \\ a & 0 & b-a \end{vmatrix},$$

expanding with respect to C_1

$$= -(b-a)\,b\begin{vmatrix} b-a & 0 \\ 0 & b-a \end{vmatrix} + (a-b)\begin{vmatrix} 0 & 0 & b-a \\ a & b-a & 0 \\ a & 0 & b-a \end{vmatrix},$$

expanding first determinant with respect to R_2 and in the second determinant replacing R_1 by $R_1 - R_2$.

$$= -b(b-a)^2 + (a-b)(b-a)\begin{vmatrix} a & b-a \\ a & 0 \end{vmatrix},$$

expanding the second det. with respect to R_1

$$= -b(b-a)^3 + (a-b)(b-a)[(0 - a(b-a)]$$

$$= -b(b-a)^3 + a(b-a)^2 = -(b-a)^3(b-a)$$

$$= -(b-a)^4.$$ **Hence proved.**

Example 5:

Evaluate $\begin{vmatrix} 1 & bc+ad & b^2c^2 + a^2d^2 \\ 1 & ca+bd & c^2a^2 + b^2d^2 \\ 1 & ab+cd & a^2b^2 + c^2d^2 \end{vmatrix}$

Solution:

We have

$$= \begin{vmatrix} 1 & bc + ad & b^2c^2 + a^2d^2 \\ 0 & ca + bd - bc - ad & c^2a^2 + b^2d^2 - b^2c^2 - a^2d^2 \\ 0 & ab + cd - bc - ad & a^2b^2 + c^2d^2 - b^2c^2 - c^2d^2 \end{vmatrix},$$

replacing R_2 and R_3 by $R_2 - R_1$ and $R_3 - R_1$ respectively.

$$= \begin{vmatrix} ca - bc + bd - ad & c^2a^2 - b^2c^2 + b^2d^2 - a^2d^2 \\ ab - bc + cd - ad & a^2b^2 - b^2c^2 + c^2d^2 - a^2d^2 \end{vmatrix},$$

expanding with respect to C_1

$$= \begin{vmatrix} (c - d)(a - b) & (c^2 - d^2)(a^2 - b^2) \\ (b - d)(a - c) & (b^2 - d^2)(a^2 - c^2) \end{vmatrix},$$ factorising the elements

$$= (c - d)(a - b)(b - d)(a - c) \begin{vmatrix} 1 & (c + d)(a + b) \\ 1 & (b + d)(a + c) \end{vmatrix},$$

taking out the common factors

$$= (c - d)(a - b)(b - d)(a - c) \begin{vmatrix} 1 & ca + bc + da + db \\ 0 & ba + dc - ca - db \end{vmatrix},$$

replacing R_2 by $R_3 - R_1$

$$= (c - d)(a - b)(b - d)(a - c)(ba + dc - ca - db)$$

$$= (c - d)(a - b)(b - d)(a - c)(a - d)(b - c)$$ **Ans.**

Example 6(a):

Show that $\begin{vmatrix} 1 & 1 & 1 \\ bc(b + c) & ca(c + a) & ab(a + b) \\ c^2c^2 & c^2a^2 & a^2b^2 \end{vmatrix}$

$= abc\,(a - b)(b - c)(c - a)(a + b + c).$

Solution:

We have, $= \begin{vmatrix} 1 & 1 & 1 \\ b^2c + bc^2 & c^2a + ca^2 & a^2b + ab^2 \\ b^2c^2 & c^2a^2 & a^2b^2 \end{vmatrix}$

$$= \begin{vmatrix} 1 & 0 & 0 \\ b^2c + bc^2 & c(a - b)(a + b + c) & b(a - c)(a + b + c) \\ c^2c^2 & c^2(a - b)(a + b) & b^2(a - c)(a + c) \end{vmatrix},$$

replacing C_2, C_3 by $C_2 - C_1$, $C_3 - C_1$

$$= \begin{vmatrix} c(a - b)(a + b + c) & b(a - c)(a + b + c) \\ c^2(a - b)(a + b) & b^2(a - c)(a + c) \end{vmatrix},$$

expanding with respect to R_1

$$= c(a - b)\, b(a - c) \begin{vmatrix} a+b+c & a+b+c \\ c(a + b) & b(b+c) \end{vmatrix},$$

$$= bc(a - b)(a - c)(a + b + c) \begin{vmatrix} 1 & 1 \\ ca + cb & ba + bc \end{vmatrix},$$

taking out $a + b + c$ common from R_1

$$= bc(a - b)(a - c)(a + b + c) \begin{vmatrix} 1 & 0 \\ ca + cb & ba - bc \end{vmatrix},$$

replacing C_2 by $C_2 - C_1$

$$= bc(a - b)(a - c)(a + b + c)(ab - ca)$$

$$= -abc(a - b)(b - c)(c - a)(a + b + c)$$ **Ans.**

Example 6(b):

Evaluate $\begin{vmatrix} a & b & ax + by \\ b & c & bx + cy \\ ax + by & bx + cy & 0 \end{vmatrix}$

Solution:

We have

$$= \begin{vmatrix} a & b & 0 \\ b & c & 0 \\ ax + by & bx + cy & -x(ax + by) \\ & & - y(bx + cy) \end{vmatrix}, \text{ replacing } C_2 \text{ by } C_3 - x\,C_1 - y\,C_2 \quad \textbf{(Note)}$$

$$= -(ax^2 + 2bxy + cy^2) \begin{vmatrix} a & b \\ b & c \end{vmatrix}, \text{ expanding with respect to } C_2$$

$$= -\,(ax^2 + 2bxy + cy^2)\,(ac - b^2).$$ **Ans.**

Example 6(c):

Prove that $\begin{vmatrix} 0 & -c & b & -l \\ c & 0 & -a & -m \\ -b & a & 0 & -n \\ x & y & z & 0 \end{vmatrix} = (al + bm + cn)\,(ax + by + cz)$

Solution:

We have

$$= \frac{1}{a} \begin{vmatrix} 0 & -ac & ab & -al \\ c & 0 & -a & -m \\ -b & a & 0 & -n \\ x & y & z & 0 \end{vmatrix}, \text{ taking } 1/a \text{ common from } R_1 \quad \textbf{(Note)}$$

$$= \frac{1}{a} \begin{vmatrix} 0 & 0 & 0 & -al - bm - cn \\ c & 0 & -a & -m \\ -b & a & 0 & -n \\ x & y & z & 0 \end{vmatrix}, \text{ replacing } R_1 \text{ by } R_1 + b\,R_2 + c\,R_3$$

$$= \frac{(al + bm + cn)}{a} \begin{vmatrix} c & 0 & -a \\ -b & a & 0 \\ x & y & z \end{vmatrix}, \text{ expanding with respect to } R_1$$

$$= \frac{(al + bm + cn)}{a^2} \begin{vmatrix} ac & 0 & -a \\ -ab & a & 0 \\ ax & y & z \end{vmatrix}, \text{ taking } 1/a \text{ common from } C_1.$$ **(Note)**

$$= \frac{(al + bm + cn)}{a^2} \begin{vmatrix} 0 & 0 & -a \\ 0 & a & 0 \\ ax + by + cz & y & z \end{vmatrix}, \text{ replacing } C_1 \text{ by } C_1 + b\,C_2 + c\,C_2$$

$$= \frac{(al + bm + cn)(ax + by + cz)}{a^2} \begin{vmatrix} 0 & -a \\ a & 0 \end{vmatrix},$$

$= (al + bm + cn)\ (ax + by + cz).$ **Hence proved.**

Example 6(d):

Evaluate $\begin{vmatrix} a^2 & a^2 - (b-c)^2 & bc \\ b^2 & b^2 - (c-a)^2 & ca \\ c^2 & c^2 - (a-b)^2 & ab \end{vmatrix}$

Solution:

We have

$$= \begin{vmatrix} a^2 & -(b-c)^2 & bc \\ b^2 & -(c-a)^2 & ca \\ c^2 & -(a-b)^2 & ab \end{vmatrix}, \text{ replacing } C_2 \text{ by } C_2 - C_1$$

$$= -\begin{vmatrix} a^2 & (b^2 + c^2) - 2bc & bc \\ b^2 & (c^2 - a^2) - 2ca & ca \\ c^2 & (a^2 + b^2) - 2ab & ab \end{vmatrix}$$ **(Note)**

$$= -\begin{vmatrix} a^2 & b^2 + c^2 & bc \\ b^2 & c^2 + a^2 & ca \\ c^2 & a^2 + b^2 & ab \end{vmatrix}$$, replacing C_2 by $C_2 + 2C_3$

$$= -\begin{vmatrix} a^2 & b^2 + c^2 + a^2 & bc \\ b^2 & c^2 + a^2 + b^2 & ca \\ c^2 & a^2 + b^2 + c^2 & ab \end{vmatrix}$$, replacing C_2 by $C_2 + C_1$

$$= -(a^2 + b^2 + c^2)\begin{vmatrix} a^2 & 1 & bc \\ b^2 & 1 & ca \\ c^2 & 1 & ab \end{vmatrix}$$, taking out the common factor from C_2

$$= -(a^2 + b^2 + c^2)\begin{vmatrix} a^2 & 1 & bc \\ b^2 - a^2 & 0 & ca - bc \\ c^2 - a^2 & 0 & ab - bc \end{vmatrix}$$, replacing R_2 and R_3 by $R_2 - R_1$ and $R_3 - R_1$ respectively.

$$= (a^2 + b^2 + c^2)\begin{vmatrix} b^2 - a^2 & c(a - b) \\ c^2 - a^2 & b(a - c) \end{vmatrix}$$, expanding with respect to C_2

$$= (a - b)(a - c)(a^2 + b^2 + c^2)\begin{vmatrix} -(b + a) & c \\ -(c + a) & b \end{vmatrix},$$

taking out the common factors

$$= (a - b)(a - c)(a^2 + b^2 + c^2)[-b(b + a) + (c + a)]$$
$$= (a - b)(a - c)(a^2 + b^2 + c^2)[-b^2 - ab + c^2 + ac]$$
$$= (a - b)(a - c)(a^2 + b^2 + c^2)[(c^2 - b^2) + a(c - b)]$$
$$= (a - b)(a - c)(a^2 + b^2 + c^2)(c - b)(a + b + c)$$
$$= (a - b)(a - c)(a + b + c)(a^2 + b^2 + c^2).$$ **Ans.**

Example 7:

Prove that

$$\begin{vmatrix} a^2+\lambda & ab & ac & ad \\ bd & b^2+\lambda & bc & bd \\ ca & cb & c^2+\lambda & cd \\ da & db & dc & d^2+\lambda \end{vmatrix} = \lambda^2 (a^2 + b^2 + c^2 + d^2 + \lambda)$$

Solution:

We have

$$= abcd \begin{vmatrix} a+\frac{\lambda}{a} & a & a & a \\ b & b+\frac{\lambda}{b} & b & b \\ c & c & c+\frac{\lambda}{c} & c \\ d & d & d & d+\frac{\lambda}{d} \end{vmatrix},$$ taking out a, b, c, d common from C_1, C_2, C_3 and C_4 respectively.

$$= abcd \begin{vmatrix} a+\frac{\lambda}{a} & -\frac{\lambda}{a} & -\frac{\lambda}{a} & -\frac{\lambda}{a} \\ b & \frac{\lambda}{b} & 0 & 0 \\ c & 0 & \frac{\lambda}{c} & 0 \\ d & 0 & 0 & \frac{\lambda}{d} \end{vmatrix},$$ replacing C_2 , C_3 and C_4 by $C_2 - C_1$, $C_3 - C_1$ and $C_4 - C_1$ respectively.

$$= abcd \begin{vmatrix} a & 0 & 0 & -\frac{\lambda}{a} \\ b & \frac{\lambda}{b} & 0 & 0 \\ c & 0 & \frac{\lambda}{c} & 0 \\ d+\frac{\lambda}{d} & -\frac{\lambda}{d} & -\frac{\lambda}{d} & \frac{\lambda}{d} \end{vmatrix},$$ replacing C_1, C_2 and C_3 by $C_1 + C_4$, $C_2 - C_4$ and $C_3 - C_4$ respectively.

$$= a^2bcd \begin{vmatrix} \frac{\lambda}{b} & 0 & 0 \\ 0 & \frac{\lambda}{c} & 0 \\ -\frac{\lambda}{d} & -\frac{\lambda}{d} & \frac{\lambda}{d} \end{vmatrix} + \lambda\, bcd \begin{vmatrix} b & \frac{\lambda}{b} & 0 \\ 0 & 0 & \frac{\lambda}{c} \\ d+\frac{\lambda}{d} & -\frac{\lambda}{d} & -\frac{\lambda}{d} \end{vmatrix},$$

expanding with respect to R_1

$$= a^2bcd.\frac{\lambda}{b} \begin{vmatrix} \frac{\lambda}{c} & 0 \\ -\frac{\lambda}{d} & \frac{\lambda}{d} \end{vmatrix} + \lambda\, bcd.b \begin{vmatrix} 0 & \frac{\lambda}{c} \\ -\frac{\lambda}{d} & -\frac{\lambda}{d} \end{vmatrix}$$

$$- \lambda bcd.\frac{\lambda}{b} \begin{vmatrix} c & \frac{\lambda}{c} \\ d+\frac{\lambda}{d} & -\frac{\lambda}{d} \end{vmatrix},$$

expanding each determinant with respect to R_1

$$= \lambda\, a^2\, cd \left(\frac{\lambda^2}{cd}\right) + \lambda b^2\, cd \left(\frac{\lambda^2}{cd}\right) - \lambda^2\, cd \left(-\frac{\lambda c}{d} - \frac{\lambda d}{c} - \frac{\lambda^2}{cd}\right)$$

$$= \lambda^2\, a^2 + \lambda^2\, b^2 + \lambda^2\, cd \left(\frac{\lambda c^2 + \lambda d^2 + \lambda^2}{cd}\right)$$

$$= \lambda^2\, a^2 + \lambda^2\, b^2 + \lambda^2\, (c^2 + d^2 + \lambda) = \lambda^2\, (a^2 + b^2 + c^2 + d^2 + \lambda)$$

Hence proved.

Example 8:

Prove that

$$\begin{vmatrix} 1 & 0 & x & 0 & x \\ 0 & 1 & 0 & x & 0 \\ x & 0 & x+1 & 0 & x \\ 0 & x & 0 & 1 & 0 \\ x & 0 & x & 0 & 1 \end{vmatrix} = (x-1)^2\, (x+1)\, (1+2x-x^2)$$

Solution:

The given determinant

$$= \begin{vmatrix} 1 & 0 & 0 & 0 & 0 \\ 0 & 1 & 0 & x & 0 \\ x & 0 & x+1-x^2 & 0 & -1 \\ 0 & x & 0 & 1 & 0 \\ x & 0 & x-x^2 & 0 & 1-x \end{vmatrix}$$, replacing C_3 and C_5 by $C_3 - x\,C_1$ and $C_5 - C_3$ respectively.

$$= \begin{vmatrix} 1 & 0 & x & 0 \\ 0 & x+1-x^2 & 0 & -1 \\ x & 0 & 1 & 0 \\ 0 & x-x^2 & 0 & 1-x \end{vmatrix}$$, expanding with respect to R_1

$$= \begin{vmatrix} 1 & 0 & 0 & 0 \\ b & x+1-x^2 & 0 & -1 \\ x & 0 & 1-x^2 & 0 \\ 0 & x-x^2 & 0 & 1-x \end{vmatrix}$$, replacing C_3 by $C_2 - x\,C_1$

$$= \begin{vmatrix} x+1-x^2 & 0 & -1 \\ 0 & 1-x^2 & 0 \\ x(1-x) & 0 & 1-x \end{vmatrix}$$, expanding with respect to R_1

$$= -(1-x) \begin{vmatrix} 0 & 1-x^2 & 0 \\ x+1-x^2 & 0 & -1 \\ x & 0 & 1 \end{vmatrix}$$, interchanging R_1 and R_2 and taking out $(1-x)$ common from R_2

$$= (1-x)(1-x^2) \begin{vmatrix} x+1-x^2 & -1 \\ x & 1 \end{vmatrix}$$, expanding with respect to R_1

$= (1 - x)^2 (1 + x) [(x + 1 - x^2) 1 - (-1).x]$

$= (x - 1)^2 (1 + x) (2x + 1 - x^2)$. **Hence proved.**

Example 9:

Evaluate $\begin{vmatrix} 1^2 & 2^2 & 3^2 & 4^2 \\ 2^2 & 3^2 & 4^2 & 5^2 \\ 3^2 & 4^2 & 5^2 & 6^2 \\ 4^2 & 5^2 & 6^2 & 7^2 \end{vmatrix}$

Solution:

We have

$$= \begin{vmatrix} 1 & 4 & 9 & 16 \\ 4 & 9 & 16 & 25 \\ 9 & 16 & 25 & 36 \\ 16 & 25 & 36 & 49 \end{vmatrix}$$

$$= \begin{vmatrix} 1 & 0 & 0 & 0 \\ 4 & -7 & -20 & -39 \\ 9 & -20 & -56 & -108 \\ 16 & -39 & -108 & -207 \end{vmatrix},$$ replacing C_2, C_3, C_4 by $C_2 - 4C_1, C_2 - 9C_1$ and $C_4 - 16C_1$ respctively.

$$= (-1)^2 \begin{vmatrix} 7 & 20 & 39 \\ 20 & 56 & 108 \\ 39 & 108 & 207 \end{vmatrix},$$

$$= \begin{vmatrix} 7 & -1 & -1 \\ 20 & -4 & -4 \\ 39 & -9 & -9 \end{vmatrix},$$ replacing C_2 and C_3 by $C_2 - 3C_1$ and $C_3 - 2C_3$ respctively.

$= 0$, as two columns are identical. **Ans.**

Example 10:

Evaluate $\begin{vmatrix} 1 & 2 & 3 & 4 \\ 2 & 3 & 4 & 1 \\ 3 & 4 & 1 & 2 \\ 4 & 1 & 2 & 3 \end{vmatrix}$

Solution:

The given determinant

$$= \begin{vmatrix} 10 & 2 & 3 & 4 \\ 10 & 3 & 4 & 1 \\ 10 & 4 & 1 & 2 \\ 10 & 1 & 2 & 3 \end{vmatrix}, \text{ replacing } C_1 \text{ by } C_1 + C_2 + C_3 + C_4$$

$$= 10 \begin{vmatrix} 1 & 2 & 3 & 4 \\ 1 & 3 & 4 & 1 \\ 1 & 4 & 1 & 2 \\ 1 & 1 & 2 & 3 \end{vmatrix}, \text{ taking out 10 common from } C_1$$

$$= 10 \begin{vmatrix} 1 & 2 & 3 & 4 \\ 0 & 1 & 1 & -3 \\ 0 & 2 & -2 & -2 \\ 0 & -1 & -1 & -1 \end{vmatrix}, \text{ replacing } R_2, R_3, R_4 \text{ by } R_2 - R_1, R_3 - R_1 \text{ and } R_4 - R_1 \text{ respectively}$$

$$= 10 \begin{vmatrix} 1 & 1 & -3 \\ 2 & -2 & -2 \\ -1 & -1 & -1 \end{vmatrix}, \text{ expanding with respect to } C_1$$

$$= -20 \begin{vmatrix} 1 & 1 & -3 \\ 1 & -1 & -1 \\ 1 & 1 & 1 \end{vmatrix}, \text{ taking out 2 common from } R_2 \text{ and } -1 \text{ common from } R_3$$

$$= -20 \begin{vmatrix} 1 & 1 & -3 \\ 0 & -2 & 2 \\ 0 & 0 & 4 \end{vmatrix}, \text{ replacing } R_2 \text{ and } R_3 \text{ by } R_2 - R_1 \text{ and } R_3 - R_1 \text{ respectively}$$

$$= -20 \begin{vmatrix} -2 & 2 \\ 0 & 4 \end{vmatrix}, \text{ expanding with respect to } C_1$$

$= -20\ [(-2).4 - 0.2] = -20\ [-8] = 160.$ **Ans.**

Example 11:

Evaluate $\begin{vmatrix} 1 & 1 & 1 \\ 1 & 1+x & 1 \\ 1 & 1 & 1+y \end{vmatrix}$

Solution:

We have $= \begin{vmatrix} 1 & 1 & 1 \\ 0 & x & 0 \\ 0 & 0 & y \end{vmatrix}$, replacing R_3 by $R_2 - R_1$ and R_3 by $R_3 - R_1$

$= 1 \times \begin{vmatrix} x & 0 \\ 0 & y \end{vmatrix}$, expanding with respect to the first column.

$= xy.$ **Ans.**

Example 12:

Evaluate $\begin{vmatrix} a & -a & -a & -a \\ b & -b & -b & -b \\ c & -c & -c & -c \\ d & -d & -d & -d \end{vmatrix}$

Solution:

Since three columns of the given determinant are identical, so the value of the determinant is zero. **Ans.**

Example 13:

Evaluate $\begin{vmatrix} -4 & 1 & 1 & 1 & 1 \\ 1 & -4 & 1 & 1 & 1 \\ 1 & 1 & -4 & 1 & 1 \\ 1 & 1 & 1 & -4 & 1 \\ 1 & 1 & 1 & 1 & -4 \end{vmatrix}$

Solution:

We have

$= \begin{vmatrix} 0 & 1 & 1 & 1 & 1 \\ 0 & -4 & 1 & 1 & 1 \\ 0 & 1 & -4 & 1 & 1 \\ 0 & 1 & 1 & -4 & 1 \\ 0 & 1 & 1 & 1 & -4 \end{vmatrix}$, replacing C_1 by $C_1 + C_2 + C_3 + C_4 + C_5$

= 0, expanding with respect to elements of first column. **Ans.**

Example 14:

Evaluate $\begin{vmatrix} 3 & 2 & 1 & 4 \\ 15 & 29 & 2 & 14 \\ 16 & 19 & 3 & 17 \\ 33 & 39 & 8 & 38 \end{vmatrix}$

Solution:

The given determinant

$$= \begin{vmatrix} 0 & 0 & 1 & 0 \\ 9 & 25 & 2 & 6 \\ 7 & 13 & 3 & 5 \\ 9 & 23 & 8 & 6 \end{vmatrix}, \text{ replacing } C_1, C_2 \text{ and } C_4 \text{ by } C_1 - 3C_3, C_2 - 2C_3, C_4 - 4C_3 \text{ respectively.}$$

$$= \begin{vmatrix} 9 & 25 & 6 \\ 7 & 13 & 5 \\ 9 & 23 & 6 \end{vmatrix}, \text{ expanding with respect to } R_1$$

$$= \begin{vmatrix} 0 & 2 & 0 \\ 7 & 13 & 5 \\ 9 & 23 & 6 \end{vmatrix}, \text{ replacing } R_1 \text{ by } R_1 - R_3$$

$$= -2 \begin{vmatrix} 7 & 5 \\ 9 & 6 \end{vmatrix}, \text{ expanding with respect to } R_1$$

$= -2\ [42 - 45] = (-2) \times (-3) = 6.$ **Ans.**

Example 15:

Prove that

$$\begin{vmatrix} b+c & c+a & a+b \\ q+r & r+p & p+q \\ y+z & z+x & x+y \end{vmatrix} = 2 \begin{vmatrix} a & b & c \\ p & q & r \\ x & y & z \end{vmatrix}$$

Solution:

We have

$$\begin{vmatrix} b+c & c+a & a+b \\ q+r & r+p & p+q \\ y+z & z+x & x+y \end{vmatrix}$$

$$= \begin{vmatrix} b & c+a & a+b \\ q & r+p & p+q \\ y & z+x & x+y \end{vmatrix} + \begin{vmatrix} c & c+a & a+b \\ r & r+p & p+q \\ z & z+x & x+y \end{vmatrix} \quad \textbf{(Note)}$$

$$= \begin{vmatrix} b & c+a & a \\ q & r+p & p \\ y & z+x & x \end{vmatrix} + \begin{vmatrix} b & c+a & b \\ q & r+p & q \\ y & z+x & y \end{vmatrix} + \begin{vmatrix} c & c & a+b \\ r & r & p+q \\ z & z & x+y \end{vmatrix}$$

$$+ \begin{vmatrix} c & a & a+b \\ r & p & p+q \\ z & x & x+y \end{vmatrix}$$

$$= \begin{vmatrix} b & c & a \\ q & r & p \\ y & z & x \end{vmatrix} + \begin{vmatrix} b & a & a \\ q & p & p \\ y & x & x \end{vmatrix} + \begin{vmatrix} c & a & a \\ r & p & p \\ z & x & x \end{vmatrix}$$

$$+ \begin{vmatrix} c & a & b \\ r & p & q \\ z & x & y \end{vmatrix},$$ second and third deter minants vanish as two coumns in each are identical.

$$= \begin{vmatrix} b & c & a \\ q & r & p \\ y & z & x \end{vmatrix} + \begin{vmatrix} c & a & b \\ r & p & q \\ z & x & y \end{vmatrix},$$ second and third deter minants vanish as two columns in each are identical.

$$= - \begin{vmatrix} b & a & c \\ q & p & r \\ y & x & z \end{vmatrix} - \begin{vmatrix} a & c & b \\ p & r & q \\ x & z & y \end{vmatrix},$$ interchanging C_2 and C_3 in first determin ant and C_1 and C_2 in second.

$$= \begin{vmatrix} a & b & c \\ p & q & r \\ x & y & z \end{vmatrix} + \begin{vmatrix} a & b & c \\ p & q & r \\ x & y & z \end{vmatrix},$$ interchanging C_1 and C_2 in first and C_2 and C_3 in second determinat.

$$= 2 \begin{vmatrix} a & b & c \\ p & q & r \\ x & y & z \end{vmatrix}$$

Hence proved.

Example 16:

Evaluate $$\begin{vmatrix} 5 & 7 & 10 & 14 \\ 2 & 3 & 7 & 6 \\ 3 & 3 & 6 & 9 \\ 5 & 6 & 11 & 20 \end{vmatrix}$$

Solution:

The given determinant

$$= \begin{vmatrix} 0 & 1 & -1 & -6 \\ 2 & 3 & 7 & 6 \\ 1 & 0 & -1 & 3 \\ 5 & 6 & 11 & 20 \end{vmatrix},$$ replacing R_1 and R_3 by $R_1 - R_4$ and $R_3 - R_2$ respectively.

$$= \begin{vmatrix} 0 & 0 & -1 & 0 \\ 2 & 10 & 7 & 24 \\ 1 & -1 & -1 & 5 \\ 5 & 17 & 11 & 56 \end{vmatrix},$$ replacing C_2 and C_4 by $C_2 + C_3$ and $C_4 + 6C_2$ respectively.

Now expand with respect to 1st row and proceed. **Ans.**

Example 17:

Show that

$$\begin{vmatrix} a+b+2c & a & b \\ c & b+c+2a & b \\ c & a & c+a+2b \end{vmatrix} = 2(a+b+c)^2$$

Solution:

The given determinant

$$= \begin{vmatrix} 2a+2b+2c & a & b \\ 2a+2b+2c & b+c+2a & b \\ 2a+2b+2c & a & c+a+2b \end{vmatrix}, \text{ replacing } C_1 \text{ by } C_1 + C_2 + C_3$$

$$= (2a+2b+2c)\begin{vmatrix} 1 & a & b \\ 1 & b+c+2a & b \\ 1 & a & c+a+2b \end{vmatrix}, \text{ taking out } 2a+2b+2c \text{ common}$$

$$= 2(a+b+c)\begin{vmatrix} 1 & a & b \\ 0 & b+c+a & b \\ 0 & a & c+a+b \end{vmatrix}, \text{ replacing } R_2 \text{ and } R_3 \text{ by } R_2 - R_1 \text{ and } R_3 - R_1 \text{ respectively}$$

$$= 2(a+b+c)\begin{vmatrix} b+c+a & 0 \\ 0 & c+a+b \end{vmatrix}, \text{ expanding with respect to 1st column.}$$

$$= 2(a+b+c)\,[(b+c+a)(c+a+b)]$$

$$= 2(a+b+c)^3.$$

Example 18(a):

Evaluate $\begin{vmatrix} 1 & 1 & 1 & 1 & 1 \\ 1 & 2 & 3 & 4 & 5 \\ 1 & 3 & 6 & 10 & 15 \\ 1 & 4 & 10 & 20 & 35 \\ 1 & 5 & 15 & 35 & 69 \end{vmatrix}$

Solution:

We have

$$= \begin{vmatrix} 1 & 0 & 0 & 0 & 0 \\ 1 & 1 & 2 & 3 & 4 \\ 1 & 2 & 5 & 9 & 14 \\ 1 & 3 & 9 & 19 & 34 \\ 1 & 4 & 14 & 34 & 68 \end{vmatrix}, \text{ replacing } C_2, C_3, C_4 \text{ and } C_5 \text{ by } C_2 - C_1, C_3 - C_1, C_4 - C_1 \text{ and } C_5 - C_1 \text{ respectively}$$

$$= \begin{vmatrix} 1 & 2 & 3 & 4 \\ 2 & 5 & 9 & 14 \\ 3 & 9 & 19 & 34 \\ 4 & 14 & 34 & 68 \end{vmatrix}, \text{ expanding with respect to first row.}$$

$$= \begin{vmatrix} 1 & 0 & 0 & 0 \\ 2 & 1 & 3 & 6 \\ 3 & 3 & 10 & 22 \\ 4 & 6 & 22 & 52 \end{vmatrix}, \text{ replacing } C_2, C_3 \text{ and } C_4 \text{ by } C_2 - 2C_1, C_3 - 3C_1 \text{ and } C_4 - 4C_1 \text{ respectively.}$$

$$= \begin{vmatrix} 1 & 3 & 6 \\ 3 & 10 & 22 \\ 6 & 22 & 52 \end{vmatrix}, \text{ expanding with respect to first row}$$

$$= \begin{vmatrix} 1 & 0 & 0 \\ 3 & 1 & 4 \\ 6 & 4 & 16 \end{vmatrix}, \text{ replacing } C_2 \text{ and } C_3 \text{ by } C_3 - 3C_1 \text{ and } C_3 - 6C_1 \text{ respectively.}$$

$$= \begin{vmatrix} 1 & 4 \\ 4 & 16 \end{vmatrix}, \text{ expanding with respect to first row.}$$

$= 1\ (16) - 4\ 4 = 0.$ **Ans.**

Example 18(b):

Show that $= \begin{vmatrix} 1 & 1 & 1 & 1 \\ \alpha & \beta & \gamma & \delta \\ \beta + \gamma & \gamma + \delta & \delta + \alpha & \alpha + \beta \\ \delta & \alpha & \beta & \gamma \end{vmatrix} = 0$

Solution:

We have

$$= \begin{vmatrix} 1 & 1 & 1 & 1 \\ \alpha & \beta & \gamma & \delta \\ \beta+\gamma & \gamma+\delta & \delta+\alpha & \alpha+\beta \\ \alpha+\beta+\gamma+\delta & \alpha+\beta+\gamma+\delta & \alpha+\beta+\gamma+\delta & \alpha+\beta+\gamma+\delta \end{vmatrix},$$

replacing R_4 by $R_2 + R_3 + R_4$

$$= (\alpha+\beta+\gamma+\delta) \begin{vmatrix} 1 & 1 & 1 & 1 \\ \alpha & \beta & \gamma & \delta \\ \beta+\gamma & \gamma+\delta & \delta+\alpha & \alpha+\beta \\ 1 & 1 & 1 & 1 \end{vmatrix},$$

taking out $(\alpha + \beta + \gamma + \delta)$ common from R_4

= 0, since two row are identical. **Hence proved.**

Example 18(c):

Evaluate $\begin{vmatrix} a-b-c & 2a & 2a \\ 2b & b-c-a & 2b \\ 2c & 2c & c-a-b \end{vmatrix}$

Solution:

The given determinant

$$= \begin{vmatrix} a+b+c & b+c+a & a+b+c \\ 2b & b-c-a & 2b \\ 2c & 2c & c-a-b \end{vmatrix},$$ replacing R_1 by $R_1 + R_2 + R_3$

$$= (a+b+c) \begin{vmatrix} 1 & 1 & 1 \\ 2b & b-c-a & 2b \\ 2c & 2c & c-a-b \end{vmatrix},$$ taking out $(a + b + c)$ common.

$$= (a + b + c)\begin{vmatrix} 1 & 0 & 0 \\ 2b & -a-b-c & 0 \\ 2c & 0 & -a-b-c \end{vmatrix},$$ replacing C_2 and C_3 by $C_2 = C_1$ and $C_3 - C_1$ respectively.

$$= (a + b + c)\begin{vmatrix} -a-b-c & 0 \\ 0 & -a-b-c \end{vmatrix},$$ expanding with respect to first row.

$$= (a + b + c)\,(-a-b-c)\,(-a-b-c) = (a + b + c)^2.$$ **Ans.**

Example 18(d):

Evaluate $\begin{vmatrix} y+z & x & y \\ z+x & z & x \\ x+y & y & z \end{vmatrix}$

Solution:

We have

$$= \begin{vmatrix} 2x+2y+2z & x+y+z & x+y+z \\ z+x & z & x \\ x+y & y & z \end{vmatrix},$$ replacing R_1 by $R_1 + R_2 + R_3$

$$= (x + y + z)\begin{vmatrix} 2 & 1 & 1 \\ z+x & z & x \\ x+y & y & z \end{vmatrix},$$

$$= (x + y + z)\begin{vmatrix} 0 & 1 & 1 \\ 0 & z & x \\ x-y & y & z \end{vmatrix},$$ replacing C_1 by $C_1 - C_2 - C_3$

$$= (x + y + z)\,(x - z)\begin{vmatrix} 1 & 1 \\ z & x \end{vmatrix},$$ expanding with respect to first column.

$$= (x + y + z)\,(x - z) = (x + y + z)\,(x - z)^2$$ **Ans.**

Example 18(e):

Evaluate $\begin{vmatrix} b+c & a+b & a \\ c+a & b+c & b \\ a+b & c+a & c \end{vmatrix}$

Solution:

We have

$$= \begin{vmatrix} 2a+2b+2c & 2a+2b+2c & a+b+c \\ c+a & b+c & b \\ a+b & c+a & c \end{vmatrix}, \text{ replacing } R_1 \text{ by } R_1 + R_2 + R_3$$

$$= (a+b+c) \begin{vmatrix} 2 & 2 & 1 \\ c+a & b+c & b \\ a+b & c+a & c \end{vmatrix}, \text{ taking out } (a+b+c) \text{ common.}$$

$$= (a+b+c) \begin{vmatrix} 0 & 2 & 1 \\ a-b & b+c & b \\ b-c & c+a & c \end{vmatrix}, \text{ replacing } C_1 \text{ by } C_1 - C_2$$

$$= (a+b+c) \begin{vmatrix} 0 & 0 & 1 \\ a-b & c-b & b \\ b-c & a-c & c \end{vmatrix}, \text{ replacing } C_2 \text{ by } C_2 - 2C_2$$

$= (a + b + c) [(a - b) (a - c) - (b - c) (c - b)]$

$= (a + b + c) [a^2 - ac - ba + bc + b^2 + c^2 - 2bc]$

$= (a + b + c) (a^2 + b^2 + c^2 - ab - bc - ca)$

$= a^2 + b^2 + c^2 - 3abc.$ **Ans.**

Example 19:

Evaluate $\begin{vmatrix} a & 1 & 1 & 1 \\ 1 & a & 1 & 1 \\ 1 & 1 & a & 1 \\ 1 & 1 & 1 & a \end{vmatrix}$

Solution:

We have $= \begin{vmatrix} a+3 & 1 & 1 & 1 \\ a+3 & a & 1 & 1 \\ a+3 & 1 & a & 1 \\ a+3 & 1 & 1 & a \end{vmatrix}$, replacing C_1 by $C_1 + C_2 + C_3 + C_4$

$$= (a+3)\begin{vmatrix} 1 & 1 & 1 & 1 \\ 1 & a & 1 & 1 \\ 1 & 1 & a & 1 \\ 1 & 1 & 1 & a \end{vmatrix}, \text{ taking out } (a+3) \text{ common from } C_1$$

$$= (a+3)\begin{vmatrix} 1 & 1 & 1 & 1 \\ 0 & a-1 & 0 & 0 \\ 0 & 0 & a-1 & 0 \\ 0 & 0 & 0 & a-1 \end{vmatrix}, \text{ replacing } R_2, R_3, R_4 \text{ by } R_2 - R_1, R_3 - R_1 \text{ and } R_4 - R_1 \text{ respectively.}$$

$$= (a+3)\begin{vmatrix} a-1 & 0 & 0 \\ 0 & a-1 & 0 \\ 0 & 0 & a-1 \end{vmatrix}, \text{ expanding with respect to } C_1$$

$$= (a+3)(a-1)\begin{vmatrix} a-1 & 0 \\ 0 & a-1 \end{vmatrix}, \text{ expanding with respect to } R_1$$

$= (a + 3)(a - 1)[(a - 1)(a - 1) = 0.0]$

$= (a + 3)(a - 1)^3$. **Ans.**

Example 20:

Prove that

$$\begin{vmatrix} 1+a_1 & a_2 & a_3 & a_4 \\ a_1 & 1+a_2 & a_3 & a_4 \\ a_1 & a_2 & 1+a_3 & a_4 \\ a_1 & a_2 & a_3 & 1+a_4 \end{vmatrix} = 1 + a_1 + a_2 + a_3 + a_4$$

Solution:

We have

$$= \begin{vmatrix} 1+a_1+a_2+a_3+a_4 & a_2 & a_3 & a_4 \\ 1+a_1+a_2+a_3+a_4 & 1+a_2 & a_3 & a_4 \\ 1+a_1+a_2+a_3+a_4 & a_2 & 1+a_3 & a_4 \\ 1+a_1+a_2+a_3+a_2 & a_2 & a_3 & 1+a_4 \end{vmatrix}, \text{ replacing } C_1 \text{ by } C_1 + C_2 + C_3 + C_4$$

$$= (1 + a_1 + a_2 + a_3 + a_4) \begin{vmatrix} 1 & a_2 & a_3 & a_4 \\ 1 & 1 + a_2 & a_3 & a_4 \\ 1 & a_2 & 1 + a_3 & a_4 \\ 1 & a_2 & a_3 & 1 + a_4 \end{vmatrix}$$, taking out $(1 + a_1 + a_2 + a_3 + a_4)$ common from C_1

$$= (1 + a_1 + a_2 + a_3 + a_4) \begin{vmatrix} 1 & a_2 & a_3 & a_4 \\ 0 & 1 & 0 & 0 \\ 0 & 0 & 1 & 0 \\ 0 & 0 & 0 & 1 \end{vmatrix}$$, replacing R_2, R_3 and R_4 by $R_2 - R_1$, $R_3 - R_1$ and $R_4 - R_1$ respectively.

$$= (1 + a_1 + a_2 + a_3 + a_4) \begin{vmatrix} 1 & 0 & 0 \\ 0 & 1 & 0 \\ 0 & 0 & 1 \end{vmatrix}$$, expanding with respect to C_1

$$= (1 + a_1 + a_2 + a_3 + a_4) \begin{vmatrix} 1 & 0 \\ 0 & 1 \end{vmatrix}$$, expanding with respect to R_1

$= (1 + a_1 + a_2 + a_3 + a_4)$. **Hence proved.**

Example 21:

Evaluate $\begin{vmatrix} 1 + a & 1 & 1 & 1 \\ 1 & 1 + a & 1 & 1 \\ 1 & 1 & 1 + a & 1 \\ 1 & 1 & 1 & 1 + a \end{vmatrix}$

Solution:

We have

$$= \begin{vmatrix} 4 + a & 1 & 1 & 1 \\ 4 + a & 1 + a & 1 & 1 \\ 4 + a & 1 & 1 + a & 1 \\ 4 + a & 1 & 1 & 1 + a \end{vmatrix}$$, replacing C_1 by $C_1 + C_2 + C_3 + C_4$

$$= (4 + a)\begin{vmatrix} 1 & 1 & 1 & 1 \\ 1 & 1+a & 1 & 1 \\ 1 & 1 & 1+a & 1 \\ 1 & 1 & 1 & 1+a \end{vmatrix}, \text{ taking out } (4 + a) \text{ common from first column}$$

$$= (4 + a)\begin{vmatrix} 1 & 0 & 0 & 0 \\ 1 & a & 0 & 0 \\ 1 & 0 & a & 0 \\ 1 & 0 & 0 & a \end{vmatrix}, \text{ replacing } C_1, C_2 \text{ and } C_4 \text{ by } C_2 - C_1, C_3 - C_1 \text{ and } C_4 - C_1 \text{ respectively.}$$

$$= (4 + a)\begin{vmatrix} a & 0 & 0 \\ 0 & a & 0 \\ 0 & 0 & a \end{vmatrix}, \text{ expanding with respect to } R_1.$$

$$= (4 + a)\begin{vmatrix} a & 0 \\ 0 & a \end{vmatrix}, \text{ expanding with respect to } R_1.$$

$= (4 + a)\ a\ (a^2) = (4 + a)\ a^2.$ **Ans.**

Example 22:

Evaluate $\begin{vmatrix} x & a & a & a \\ a & x & a & a \\ a & a & x & a \\ a & a & a & x \end{vmatrix}$

Solution:

We have

$$= \begin{vmatrix} x + 3a & a & a & a \\ x + 3a & x & a & a \\ x + 3a & a & x & a \\ x + 3a & a & a & x \end{vmatrix}, \text{ replacing } C_1 \text{ by } C_1 + C_2 + C_3 + C_4$$

$$= (x + 3a) \begin{vmatrix} 1 & a & a & a \\ 1 & x & a & a \\ 1 & a & x & a \\ 1 & a & a & x \end{vmatrix}, \text{ taking out } (x + 3a) \text{ common}$$

$$= (x + 3a) \begin{vmatrix} 1 & a & a & a \\ 0 & x-a & 0 & 0 \\ 0 & 0 & x-a & 0 \\ 0 & 0 & 0 & x-a \end{vmatrix},$$ replacing R_2, R_3 and R_4 by $R_2 - R_1$, $R_3 - R_1$ and $R_4 - R_1$ respectively.

$$= (x + 3a) \begin{vmatrix} x-a & 0 & 0 \\ 0 & x-a & 0 \\ 0 & 0 & x-a \end{vmatrix}, \text{ expanding with respect to } C_1$$

$$= (x + 3a)(x - a) \begin{vmatrix} x-a & 0 \\ 0 & x-a \end{vmatrix}, \text{ expanding with respect to } C_1$$

$= (x + 3a)(x - a)(x - a)(x - a) = (x + 3a)(x - a)^3$. **Ans.**

Example 23(a):

Evaluate $\begin{vmatrix} 1 & x & 1 & y \\ x & 1 & y & 1 \\ 1 & y & 1 & x \\ y & 1 & x & 1 \end{vmatrix}$

Solution:

We have

$$= \begin{vmatrix} x+y+2 & x+y+2 & x+y+2 & x+y+2 \\ x & 1 & y & 1 \\ 1 & y & 1 & x \\ y & 1 & x & 1 \end{vmatrix},$$ replacing R_1 by $R_1 + R_2 + R_3 + R_4$

$$= (x+y+2)\begin{vmatrix} 1 & 1 & 1 & 1 \\ x & 1 & y & 1 \\ y & 1 & x & 1 \\ y & 1 & x & 1 \end{vmatrix},$$ taking out $(x + y + 2)$ common from R_1

$$= (x+y+2)\begin{vmatrix} 1 & 0 & 0 & 0 \\ x & 1-x & y-x & 1-x \\ 1 & y-1 & 0 & x-1 \\ y & 1-y & x-y & 1-y \end{vmatrix},$$ replacing C_2, C_3 and C_4 by $C_2 - C_1$, $C_3 - C_1$ and $C_4 - C_1$ respectively.

$$= (x+y+2)\begin{vmatrix} 1-x & y-x & 1-x \\ y-1 & 0 & x-1 \\ 1-y & x-y & 1-y \end{vmatrix},$$ expanding with respect to R_1

$$=(x+y+2)\begin{vmatrix} 0 & y-x & 1-x \\ y-x & 0 & x-1 \\ 0 & x-y & 1-y \end{vmatrix},$$ replacing C_1 by $C_1 - C_2$

$$= -(x+y+2)(y-x)\begin{vmatrix} y-x & 1-x \\ x-y & 1-y \end{vmatrix},$$ expanding with respect to C_1

$$= -(x+y+2)(y-x)(y-x)\begin{vmatrix} -1 & 1-x \\ -1 & 1-y \end{vmatrix},$$ taking out $(y - x)$ common

$$= -(x + y + 2)(y - x)^2 [(1 - y) - (-1)(1 - x)]$$
$$= -(x + y + 2)(x - y)^2 (2 - x - y)$$
$$= (x + y + 2)(x - y)^2 (x + y - 2).$$

Example 23(b):

Prove that $\begin{vmatrix} x+a & b & c & d \\ a & x+b & c & d \\ a & b & x+c & d \\ a & b & c & x+d \end{vmatrix}$

$= x^2 (x + a + b + c + d).$

Solution:

We have

$$= \begin{vmatrix} x+a+b+c+d & b & c & d \\ x+a+b+c+d & x+b & c & d \\ x+a+b+c+d & b & x+c & d \\ x+a+b+c+d & b & c & x+d \end{vmatrix}, \text{ replacing } C_1 \text{ by } C_1+C_2+C_3+C_4$$

$$=(x+a+b+c+d)\begin{vmatrix} 1 & b & c & d \\ 1 & x+b & c & d \\ 1 & b & x+c & d \\ 1 & b & c & x+d \end{vmatrix}, \text{ taking out } (x+a+b+c+d) \text{ common from } C_1$$

$$= (x+a+b+c+d)\begin{vmatrix} 1 & b & c & d \\ 0 & x & 0 & 0 \\ 0 & 0 & x & 0 \\ 0 & 0 & 0 & x \end{vmatrix}, \text{ replacing } R_2, R_3, R_4 \text{ by } R_2-R_1, R_3-R_1 \text{ and } R_4-R_1 \text{ respectively.}$$

$$= (x+a+b+c+d)\begin{vmatrix} x & 0 & 0 \\ 0 & x & 0 \\ 0 & 0 & x \end{vmatrix}, \text{ expanding with respect to } C_1$$

$$= (x+a+b+c+d)\begin{vmatrix} x & 0 \\ 0 & x \end{vmatrix}, \text{ expanding with respect to } C_1$$

$= x\ (x + a + b + c + d)\ x^2$

$= x^3\ (x + a + b + c + d).$ **Hence proved.**

Example 23(c):

Evaluate
$$\begin{vmatrix} x & a & b & c & 1 \\ d & x & f & h & 1 \\ d & e & x & k & 1 \\ d & e & g & x & 1 \\ d & e & g & m & 1 \end{vmatrix}$$

Solution:

The given determinant

$$= \begin{vmatrix} x-d & a-x & b-f & c-h & 0 \\ 0 & x-e & f-x & h-k & 0 \\ 0 & 0 & x-g & k-x & 0 \\ 0 & 0 & 0 & x-m & 0 \\ d & e & g & m & 1 \end{vmatrix},$$

replacing R_1, R_2, R_3 and R_4 by $R_1 - R_2$, $R_2 - R_3$, $R_3 - R_4$ and $R_{4-R}5$ respectively.

$$= (x-m)\begin{vmatrix} x-d & a-x & b-f & 0 \\ 0 & x-e & f-x & 0 \\ 0 & 0 & x-g & 0 \\ d & e & g & 1 \end{vmatrix},$$

expanding with respect to R_4 **(Note)**

$$= (x-m)\begin{vmatrix} x-d & a-x & b-f \\ 0 & x-e & f-x \\ 0 & 0 & x-g \end{vmatrix},$$ expanding with respect to C_4

$$= (x-m)(x-d)\begin{vmatrix} x-c & f-x \\ 0 & x-g \end{vmatrix},$$ expanding with respect to C_1

$$= (x-m)(x-d)(x-c)(x-g).$$ **Ans.**

Example 24:

Evaluate $\begin{vmatrix} b+c & a-c & a-b \\ b-c & c+a & b-a \\ c-b & c-a & a+b \end{vmatrix}$

Solution:

$$= \begin{vmatrix} 2b & 2a & 0 \\ b-c & c+a & b-a \\ 0 & 2c & 2b \end{vmatrix},$$ replacing R_1 and R_2 by $R_1 + R_2$ and $R_3 + R_2$ respectively.

$$=4\begin{vmatrix} b & a & 0 \\ b-c & c+a & b-a \\ 0 & c & b \end{vmatrix},$$ taking out 2 common from R_1 and R_2

$$=4\begin{vmatrix} b & a & 0 \\ -c & c & b-a \\ 0 & c & b \end{vmatrix}, \text{ replacing } R_2 \text{ by } R_2 - R_1$$

$$= 4\left[b\begin{vmatrix} c & b-a \\ c & b \end{vmatrix} + c\begin{vmatrix} a & 0 \\ c & b \end{vmatrix}\right], \text{ expanding with respect to } C_1$$

$$= 4\left[b\begin{vmatrix} c & b-a \\ 0 & a \end{vmatrix} + c\begin{vmatrix} a & 0 \\ c & b \end{vmatrix}\right], \text{ replacing } R_2 \text{ by } R_2 - R_1 \text{ in first determinant.}$$

$= 4\ [b\ (ca) + c\ (ab)] = 4\ [2abc] = 8abc.$ **Ans.**

Example 25:

Evaluate $\begin{vmatrix} b+c & a & a \\ b & c+a & b \\ c & c & a+b \end{vmatrix}$

Solution:

We have

$$= \begin{vmatrix} b+c & 0 & 0 \\ b & c+a-b & b \\ c & c-a-b & a+b \end{vmatrix}, \text{ replacing } C_2 \text{ by } C_2 - C_3$$

$$= (b+c)\begin{vmatrix} c+a-b & b \\ c-a-b & a+b \end{vmatrix} + a\begin{vmatrix} b & c+a-b \\ c & c-a-b \end{vmatrix},$$

expanding with respect to R_1

$$= (b+c)\begin{vmatrix} c+a-b & b \\ -2a & a \end{vmatrix} + a\begin{vmatrix} b & c+a-b \\ c-b & -2a \end{vmatrix},$$

Replacing R_2 by $R_2 - R_1$ in each determinant.

$= (b + c)\ [a\ (c + a - b) - b\ (-2a)] + a\ [b - (-2a) - (c - b)\ (c + a - b)]$

$= (b + c)\ (ac + a^2 + ab) + a\ (-2ab - c^2 - ca + bc + bc + ab - b^2)$

$= abc + a^2b + ab^2 + ac^2 + a^2c + abc - 2a^2b - ac^2 - ca^2$
$+ 2abc + a^2b - ab^2$

$= 4\ 4abc.$ **Ans.**

Example 26:

Show that $\begin{vmatrix} 1 & a & bc \\ 1 & b & ca \\ 1 & c & ab \end{vmatrix} = (b - c)(c - a)(a - b)$

Solution:

The given determinant

$= \begin{vmatrix} 1 & a & bc \\ 0 & b - a & ca - bc \\ 0 & c - a & ab - bc \end{vmatrix}$, replacing R_2 and R_3 by $R_2 - R_1$ and $R_3 - R_1$ respective.

$= \begin{vmatrix} b - a & -c(b - a) \\ c - a & -b(c - a) \end{vmatrix}$, expanding with respect to C_1

$= (b - a)(c - a)\begin{vmatrix} 1 & -c \\ 1 & -c \end{vmatrix}$, taking common factors out

$= (b - a)(c - a)(-b + c)$

$= (a - b)(b - c)(c - a)$ **Hence proved.**

Example 27:

Evaluate $\begin{vmatrix} a & x & y & a \\ x & 0 & 0 & y \\ y & 0 & 0 & x \\ a & y & z & a \end{vmatrix}$

Solution:

The given determinant

$= \begin{vmatrix} 0 & x - y & y & a \\ x - y & 0 & 0 & y \\ y - x & 0 & 0 & x \\ a & y + x & x & a \end{vmatrix}$, replacing C_1 and C_2 by $C_1 - C_4$ and $C_2 + C_3$ respectively.

$$= (x - y)(x + y) \begin{vmatrix} 0 & 1 & y & a \\ 1 & 0 & 0 & y \\ -1 & 0 & 0 & x \\ 0 & 1 & x & a \end{vmatrix},$$ taking out $x - y$ and $x + y$ common from C_1 and C_2 respectively.

$$= (x^2 - y^2) \begin{vmatrix} 0 & 1 & y & a \\ 1 & 0 & 0 & y \\ 0 & 0 & 0 & x+y \\ 0 & 0 & x-y & a \end{vmatrix},$$ replacing R_3 and R_4 by $R_2 + R_3$ and $R_4 - R_1$ respectively.

$$= -(x^2 - y^2) \begin{vmatrix} 1 & y & a \\ 0 & 0 & x+y \\ 0 & x-y & 0 \end{vmatrix},$$ expanding with respect to C_1

$$= -(x^2 - y^2) \begin{vmatrix} 0 & x+y \\ x-y & 0 \end{vmatrix}$$

$$= -(x^2 - y^2)\,[-(x + y)(x - y)] = (x^2 - y^2)^2.$$

Example 28:

Evaluate $$\begin{vmatrix} 1 & bc & a(b+c) \\ 1 & ca & b(c+a) \\ 1 & ab & c(a+b) \end{vmatrix}$$

Solution:

The given determinant

$$= \begin{vmatrix} 1 & bc & a(b+c) \\ 0 & c(a-b) & c(b-a) \\ 0 & b(a-c) & b(c-a) \end{vmatrix},$$ replacing R_2 and R_3 by $R_2 - R_1$ and $R_3 - R_1$ respectively.

$$= \begin{vmatrix} c(a-b) & c(b-a) \\ b(a-c) & b(c-a) \end{vmatrix},$$ expanding with respect to C_1

$$= (a - b)(a - c) \begin{vmatrix} c & -c \\ b & -b \end{vmatrix},$$ taking out $(a - b)$ and $(a - c)$ common from R_1 and R_2 respectively

$= -(a-b)(a-c)\begin{vmatrix} c & c \\ b & b \end{vmatrix} = 0$, two columns belong identical. **Ans.**

Example 29:

Evaluate $\begin{vmatrix} 1 & 1 & 1 \\ a & b & c \\ a^2 & b^2 & c^2 \end{vmatrix}$

Solution:

The given determinant

$= \begin{vmatrix} 1 & 0 & 0 \\ a & b-a & c-a \\ a^2 & b^2-a^2 & c^2-a^2 \end{vmatrix}$, replacing R_2 and R_3 by $R_3 - R_1$ and $R_3 - R_1$ respectively.

$= \begin{vmatrix} b-a & c-a \\ b^2-a^2 & c^2-a^2 \end{vmatrix}$, expanding with respect to R_1

$= (b-a)(c-a)\begin{vmatrix} 1 & 1 \\ b+a & c+a \end{vmatrix}$, taking common factors out

$= (b-a)(c-a)[(c+a)-(b+a)]$

$= (a-b)(c-a)(b-c)$. **Ans.**

Example 30:

Prove that, $\begin{vmatrix} 1 & a^2+bc & a^3 \\ 1 & b^2+ca & b^3 \\ 1 & c^2+ab & c^3 \end{vmatrix} = -(b-c)(c-a)(a-b)(a^2+b^2+c^2)$

Solution:

The given determinant

$= \begin{vmatrix} 1 & a^2 & a^3 \\ 1 & b^2 & b^3 \\ 1 & c^2 & c^3 \end{vmatrix} + \begin{vmatrix} 1 & bc & a^3 \\ 1 & ca & b^3 \\ 1 & ab & c^3 \end{vmatrix}$, breaking into two determinante

(Note)

$$= \begin{vmatrix} 1 & 1 & 1 \\ a^2 & b^2 & c^2 \\ a^3 & b^3 & c^3 \end{vmatrix} + \begin{vmatrix} 1 & 1 & 1 \\ bc & ca & ab \\ a^3 & b^3 & c^3 \end{vmatrix},$$ interchanging rows and columns **(Note)** ...(i)

The first determinant

$$= (a - b)(b - c)(c - a)(ab + bc + ca) \qquad ...(ii)$$

The second determinant

$$= \begin{vmatrix} 1 & 0 & 0 \\ ba & ca - bc & ab - bc \\ a^2 & b^3 - a^3 & c^3 - a^3 \end{vmatrix},$$ replacing C_2 and C_3 by $C_2 - C_1$ and $C_3 - C_1$

$$= \begin{vmatrix} c(a - b) & b(a - c) \\ (b - a)(b^2 + ab + a^2) & (c - a)(c^2 + ac + a^2) \end{vmatrix},$$

expanding with respect to R_1

$$= (a - b)(a - c) \begin{vmatrix} c & b \\ -(b^2 + ab + a^2) & -(c^2 + ac + a^2) \end{vmatrix},$$

taking out common factors from C_1 and C_2

$$= (a - b)(a - c) \begin{vmatrix} c - b & b \\ -(b^2 - c^2 + ab - ac) & -(c^2 + ac + a^2) \end{vmatrix},$$

replacing C_1 by $C_1 - C_2$

$$= (a - b)(a - c) \begin{vmatrix} (c - b) & b \\ (c - b)(c + b + a) & -(c^2 + ac + a^2) \end{vmatrix}$$

$$= (a - b)(a - c)(c - b) \begin{vmatrix} 1 & b \\ a + b + c & -(c^2 + ca + a^2) \end{vmatrix},$$

taking $(c - b)$ common from C_1.

$$= (a - b)(b - c)(c - a)[-(c^2 + ca + a^2) - b(a + b + c)]$$

$$= -(a - b)(b - c)(c - a)(a^2 + b^2 + c^2 + ab + bc + ca) \qquad ...(iii)$$

Substituting values from (ii) and (iii) in (i), we find the given determinant

$= (a - b)(b - c)(c - a)(ab + bc + ca)$

$- (a - b)(b - c)(c - a)(a^2 + b^2 + c^2 + ab + bc + ca)]$

$= (a - b)(b - c)(c - a)(a^2 + b^2 + c^2)$. **Ans.**

Example 31:

Evaluate $\begin{vmatrix} 1 & 1 & 1 \\ a^2 & b^2 & c^2 \\ a^3 & b^3 & c^3 \end{vmatrix}$

Solution:

We have

$$= \begin{vmatrix} 1 & 0 & 0 \\ a^2 & b^2 - a^2 & c^2 - a^2 \\ a^3 & b^3 - a^3 & c^3 - a^3 \end{vmatrix}$$, replacing C_2 and C_3 by $C_2 - C_1$ and $C_3 - C_1$ respectively.

$$= \begin{vmatrix} b^2 - a^2 & c^2 - a^2 \\ b^3 - a^3 & c^3 - a^3 \end{vmatrix}$$, expanding with respect to R_1

$$= (b - a)(c - a) \begin{vmatrix} b + a & c + a \\ b^2 + ab + a^2 & c^2 + ac + a^2 \end{vmatrix},$$

taking out common factors of C_1, C_2

$$= (b - a)(c - a) \begin{vmatrix} b - c & c + a \\ b^2 + ab - c^2 - ca & c^2 + ca + a^2 \end{vmatrix},$$

replacing C_1 by $C_1 - C_2$

$$= (b - a)(c - a) \begin{vmatrix} b - c & c + a \\ (b^2 - c^2) + a(b - c) & c^2 + ca + a^2 \end{vmatrix}$$

$$= (b - a)(c - a) \begin{vmatrix} 1 & c + a \\ b + c + a & c^2 + ca + a^2 \end{vmatrix}$$

$= -(a-b)(b-c)(c-a)[1(c^2+ca+a^2)-(c+a)(b+c+a)$

$= -(a-b)(b-c)(c-a)$

$[c^2+ca+a^2-cb-c^2-ac-ab-ac-a^2]$

$= -(a-b)(b-c)(c-a)[-ab-bc-ca]$

$= (a-b)(b-c)(c-a)(ab+bc+ca).$ **Ans.**

Example 32:

Show that $\begin{vmatrix} a & b & c \\ a^2 & b^2 & c^2 \\ a^3 & b^3 & c^3 \end{vmatrix} = abc(a-b)(b-c)(c-a)$

Solution:

The given determinant

$= abc \begin{vmatrix} 1 & 1 & 1 \\ a & b & c \\ a^2 & b^2 & c^2 \end{vmatrix}$, taking out a, b and c common from C_1, C_2 and C_3 respectively.

Ans. $abc(a-b)(b-c)(c-a)$

Example 33:

Evaluate $\begin{vmatrix} 1 & 1 & 1 \\ a & b & c \\ a^3 & b^3 & c^3 \end{vmatrix}$

Solution:

The given determinant

$= \begin{vmatrix} 1 & 0 & 0 \\ a & b-a & c-a \\ a^3 & b^3-a^3 & c^3-a^3 \end{vmatrix}$, replacing C_2, C_3 by $C_3 - C_1$ and $C_3 - C_1$ respectively.

$= \begin{vmatrix} b-a & c-a \\ b^3-a^3 & c^3-a^3 \end{vmatrix}$, expanding with respect to R_1

$$= (b - a)(c - a)\begin{vmatrix} 1 & 1 \\ b^2 + ab + a^2 & c^2 + ac + a^2 \end{vmatrix},$$

taking common factors out from C_1 and C_2

$= (b - a)(c - a)[(c^2 + ac + a^2) - (b^2 + ab + a^2)]$

$= (b - a)(c - a)[(c^2 - b^2) + a(c - b)]$

$= (b - a)(c - a)(c - b)[(c + b) + a]$

$= (a - b)(b - c)(c - a)(a + b + c).$ **Ans.**

Example 34:

If a, b, c are all different and

$$\begin{vmatrix} a & a^2 & 1 + a^3 \\ b & b^2 & 1 + b^3 \\ c & c^2 & 1 + c^3 \end{vmatrix} = 0, \text{ prove that } 1 + abc = 0.$$

Solution:

The given determinant

$$= \begin{vmatrix} a & a^2 & 1 \\ b & b^2 & 1 \\ c & c^2 & 1 \end{vmatrix} + \begin{vmatrix} a & a^2 & a^3 \\ b & b^2 & b^3 \\ c & c^2 & c^3 \end{vmatrix},$$ breaking into two determinants. **(Note)**

$$= \begin{vmatrix} a & b & c \\ a^2 & b^2 & c^2 \\ 1 & 1 & 1 \end{vmatrix} + \begin{vmatrix} a & b & c \\ a^2 & b^2 & c^2 \\ a^3 & b^3 & c^3 \end{vmatrix},$$ interchanging rows and columns in each determinant. **(Note)**

The value of given determinant $= (1 + abc)[(a - b)(b - c)(c - a)]$

Example 35(a):

Prove that $\begin{vmatrix} a^3 & 3a^2 & 3a & 1 \\ a^2 & a^2 + 2a & 2a + 1 & 1 \\ a & 2a + 1 & a + 2 & 1 \\ 1 & 3 & 3 & 1 \end{vmatrix} = (a - 1)^6$

Solution:

The given determinant

$$= \begin{vmatrix} a^3 - 3a^2 + 3a - 1 & 3a^2 & 3a & 1 \\ 0 & a^2 + 2a & 2a + 1 & 1 \\ 0 & 2a + 1 & a + 2 & 1 \\ 0 & 3 & 3 & 1 \end{vmatrix},$$

replacing C_1 by $C_1 - C_2 + C_3 - C_4$ **(Note)**

$$= (a^3 - 3a^2 + 3a - 1) \begin{vmatrix} a^2 + 2a & 2a + 1 & 1 \\ 2a + 1 & a + 2 & 1 \\ 3 & 3 & 1 \end{vmatrix},$$ expanding with respect to C_1

$$= (a - 1)^3 \begin{vmatrix} (a^2 + 2a) - (2a + 1) + 1 & 2a + 1 & 1 \\ (2a + 1) - (a + 2) + 1 & a + 2 & 1 \\ 3 - 3 + 1 & 3 & 1 \end{vmatrix},$$

replacing C_1 by $C_1 - C_2 + C_3$

$$= (a - 1)^3 \begin{vmatrix} a^2 & 2a + 1 & 1 \\ a & a + 2 & 1 \\ 1 & 3 & 1 \end{vmatrix},$$ on simplifying

$$= (a - 1)^3 \begin{vmatrix} a^2 - 1 & 2a - 2 & 0 \\ a - 1 & a - 1 & 0 \\ 1 & 3 & 1 \end{vmatrix},$$ replacing R_1 and R_2 by $R_1 - R_2$ and $R_2 - R_3$ respectively.

$$= (a - 1)^3 \begin{vmatrix} a^2 - 1 & 2(a - 1) \\ a - 1 & a - 1 \end{vmatrix},$$ expanding with respect to C_3

$$= (a-1)^3(a-1)(a-1)\begin{vmatrix} a+1 & 2 \\ 1 & 1 \end{vmatrix},$$ taking out common factors.

$$= (a-1)^3\,[(a+1)-2] = (a-1)^6.$$ **Hence proved.**

Example 35(b):

Show that

$$\begin{vmatrix} x & y & z \\ x^2 & y^2 & z^2 \\ yz & zx & xy \end{vmatrix} = \begin{vmatrix} 1 & 1 & 1 \\ x^2 & y^2 & z^2 \\ x^3 & y^3 & z^3 \end{vmatrix}$$

$= (y-x)(z-x)(x-y)(yz+zx+xy).$

Solution:

$$\begin{vmatrix} x & y & z \\ x^2 & y^2 & z^2 \\ yz & zx & xy \end{vmatrix} = \frac{1}{xyz}\begin{vmatrix} x^2 & y^2 & z^2 \\ x^3 & y^3 & z^3 \\ xyz & xyz & xyz \end{vmatrix},$$ multiplying C_1, C_2 C_3 by x, y, z respectively **(Note)**

$$= \frac{1}{xyz}.xyz\begin{vmatrix} x^2 & y^2 & z^2 \\ x^3 & y^3 & z^3 \\ 1 & 1 & 1 \end{vmatrix},$$ taking out xyz common from R_2

$$= \begin{vmatrix} 1 & 1 & 1 \\ x^2 & y^2 & z^2 \\ x^3 & y^3 & z^3 \end{vmatrix},$$ interchanging R_2 and R_3 and then R_1 and R_2 **(Note)**

For the 2nd part do your self.

Example 35(c):

Evaluate $$\begin{vmatrix} 1+a & 1 & 1 & 1 \\ 1 & 1+b & 1 & 1 \\ 1 & 1 & 1+c & 1 \\ 1 & 1 & 1 & 1+d \end{vmatrix}$$

Solution:

We have

$$= abcd \begin{vmatrix} \frac{1}{a}+1 & \frac{1}{a} & \frac{1}{a} & \frac{1}{a} \\ \frac{1}{b} & \frac{1}{b}+1 & \frac{1}{b} & \frac{1}{b} \\ \frac{1}{c} & \frac{1}{c} & \frac{1}{c}+1 & \frac{1}{c} \\ \frac{1}{d} & \frac{1}{d} & \frac{1}{d} & \frac{1}{d}+1 \end{vmatrix},$$ taking a, b, c, d common from R_1, R_2, R_3 and R_4 respectively.

$$= abcd \left(\frac{1}{a}+\frac{1}{b}+\frac{1}{c}+\frac{1}{d}+1\right) \begin{vmatrix} 1 & 1 & 1 & 1 \\ \frac{1}{b} & \frac{1}{b}+1 & \frac{1}{b} & \frac{1}{b} \\ \frac{1}{c} & \frac{1}{c} & \frac{1}{c}+1 & \frac{1}{c} \\ \frac{1}{d} & \frac{1}{d} & \frac{1}{d} & \frac{1}{d}+1 \end{vmatrix},$$

replacing R_1 by $R_1 + R_2 + R_3 + R_4$ and taking $\left(\frac{1}{a}+\frac{1}{b}+\frac{1}{c}+\frac{1}{d}+1\right)$ common from R_1

$$= abcd \left(\frac{1}{a}+\frac{1}{b}+\frac{1}{c}+\frac{1}{d}+1\right) \begin{vmatrix} 1 & 0 & 0 & 0 \\ \frac{1}{b} & 1 & 0 & 0 \\ \frac{1}{c} & 0 & 1 & 0 \\ \frac{1}{d} & 0 & 0 & 1 \end{vmatrix},$$

replacing C_2, C_3 and C_4 by $C_2 - C_1$, $C_3 - C_1$ and $C_4 - C_1$ respectively.

$$= abcd \left(\frac{1}{a}+\frac{1}{b}+\frac{1}{c}+\frac{1}{d}+1\right) \begin{vmatrix} 1 & 0 & 0 \\ 0 & 1 & 0 \\ 0 & 0 & 0 \end{vmatrix},$$ expanding with respect to R_1

$$= abcd \left(\frac{1}{a}+\frac{1}{b}+\frac{1}{c}+\frac{1}{d}+1\right) \begin{vmatrix} 1 & 0 \\ 0 & 1 \end{vmatrix},$$ expanding with respect to R_1

$$= abcd\left(\frac{1}{a}+\frac{1}{b}+\frac{1}{c}+\frac{1}{d}+1\right)$$ **Ans.**

EXERCISES

1. *Evalute* $\begin{vmatrix} x+a & x+2a & x+3a \\ x+2a & x+3a & x+4a \\ x+4a & x+5a & x+6a \end{vmatrix}$

[**Hint:** Replace C_2 and C_3 by $C_2 - C_1$ and $C_3 - C_2$]. (**Ans.** 0)

2. *Calculate the value of the determinant* $\begin{vmatrix} 7 & 13 & 10 & 6 \\ 5 & 9 & 7 & 4 \\ 8 & 12 & 11 & 7 \\ 4 & 10 & 6 & 3 \end{vmatrix}$ (**Ans.** 0)

3. *Show that* $\begin{vmatrix} 21 & 17 & 7 & 10 \\ 24 & 22 & 6 & 10 \\ 6 & 8 & 2 & 3 \\ 5 & 7 & 1 & 2 \end{vmatrix} = 0$

4. *Evaluate* $\begin{vmatrix} 1 & a & b+c \\ 1 & b & c+a \\ 1 & c & a+c \end{vmatrix}$ (**Ans.** 0)

4. *Show that* $\begin{vmatrix} 29 & 26 & 22 \\ 25 & 31 & 27 \\ 63 & 54 & 46 \end{vmatrix} = 132$

5

Forms of the Differential Equations

INTRODUCTION

$$f'(y) \frac{dy}{dx} + f(x)\, P(x) = Q(x) \qquad ...(A)$$

To solve it, we substitute v = f(y) so that $\frac{dv}{dx} = f'(y) \frac{dy}{dx}$.

The given equation (A) becomes

$$\frac{dv}{dx} + P(x)\, v + Q(x).$$

which being a linear equation can be solved.

Remark:

The differential equation of

$$\frac{dy}{dx} + Py = Qy^n$$

$$\Rightarrow \qquad \frac{1}{y^n}\frac{dy}{dx} + \frac{1}{y^{n-1}}P = Q$$

is a particular case of equation (A). As seen earlier, we take $z = 1/y^{n-1} = y^{-n+1}$ *to reduce it to linear form.*

INTEGRATING FACTORS

We have seen that a differential equation of the form:

$$M\, dx + n\, dy = 0$$

(M and N are functions of x and y) is *exact* if $\partial M/\partial y = \partial N/\partial x$.

If a differential equation is not exact, then any factor (a function of x and y) which when multiplied by the given equation converts it an exact differential equation is called an Integrating Factor.

Illustration:

y dx – x dy = 0 is not an exact differential equation.

When multiplied by $\frac{1}{xy}$, the above equation becomes

$$\frac{dx}{x} - \frac{dy}{y} = 0,$$ which is exact ($\because$ $\partial M/\partial y = 0 = \partial N/\partial x$) and

log x – log y = c is its solution.

RULES FOR FINDING INTEGRATING FACTORS

Rule I. *If* $\frac{1}{N}\left[\frac{\partial M}{\partial y} - \frac{\partial N}{\partial x}\right]$ *is a function of x alone, say f(x), then* $e^{\int f(x)dx}$ *is an integrating factor.*

Rule II. *If* $\frac{1}{M}\left[\frac{\partial M}{\partial y} - \frac{\partial N}{\partial x}\right]$ *is a function of y alone, say g(y), then* $e^{-\int g(y)dy}$ *is an integrating factor.*

LINEAR EQUATIONS OF THE FORM $\frac{dx}{dy}$ + RX = S, R AND S BEING FUNCTIONS OF Y ALONE

Sometimes a differential equation $f_1(x, y)\,dx + f_2(x, y)\,dy = 0$ may be expressible as a linear equation of the form:

$$\frac{dx}{dy} + Rx = S. \qquad ...(A)$$

where R and S are functions of y alone.

I.F. = $e^{\int R\,dy}$ and the solution of (A) is given by

$$x\,(\text{I.F.}) = \int S\,(\text{I.F.})dy + c.$$

NON-HOMOGENEOUS EQUATIONS

Non-homogeneous equations of the first degree in x and y are of the form:

$$\frac{dy}{dx} = \frac{ax + by + c}{Ax + By + C}. \qquad ...(1)$$

The above equation can be reduced to the homogeneous form as follows:

Put x = X + h, y = Y + k; where h and k are arbitrary constants.

$$\therefore \quad \frac{dy}{dx} = 1, \frac{dy}{dY} = 1; \frac{dy}{dx} = \frac{dY}{dY}.\frac{dX}{dX}.\frac{dX}{dx} = 1.\frac{dY}{dX}.1 = \frac{dY}{dX}.$$

The equation (1) now becomes

$$\frac{dY}{dX} = \frac{aX + bY + (ah + bk + c)}{AX + BY + (Ah + Bk + C)}. \qquad ...(2)$$

We choose the constants h and k is such a way that

$$ah + bk + c = 0 \ Ah + Bk + C = 0. \qquad ...(3)$$

From (2) and (3), we obtain

$$\frac{dY}{dX} = \frac{aX + bY}{AX + BY}.\left(\frac{a}{A} \neq \frac{b}{B}\right) \qquad ...(4)$$

The above equation (4) is a homogeneous equation is X and Y, which can be solved by the substitution Y = VX, as described in section 3.3.

Exceptional Case

The above method fails if $\frac{a}{A} = \frac{b}{B}$, for then the values of h and k (given by (3)) become infinite or indeterminate.

Suppose $\frac{a}{A} = \frac{b}{B} = \frac{1}{m}$ so that A = ma and B = nb.

Now, equation (1) reduces to

$$\frac{dy}{dx} = \frac{ax + by + c}{m(ax + by) + C}. \qquad ...(5)$$

To solve equation (5), we put ax + by = z so that

$$a + b\frac{dy}{dx} = \frac{dz}{dx}. \qquad ...(6)$$

From (5) and (6), we obtain

$$\frac{1}{b}\left(\frac{dz}{dx} - a\right) = \frac{z + c}{mz + C}.$$

This equation can be solved by the method of separation of variables in terms of x and z.

LINEAR EQUATIONS

Definition: *A differential equation of the form*

$$\frac{dy}{dx} + Py = Q, \quad \text{...(A)}$$

where P and Q are functions of x, is called a Linear Differential Equation of the First Order

OR

A differential equation in which the dependent variable y and its derivative occur in first degree and are not multiplied together (Their co-efficient may be any function of nos. is called a *Linear Differential Equation.*

We shall prove that the solution of (A) is given by

$$ye^{\int P\,dx} = \int Qe^{\int P\,dx}dx + c.$$

Multiply both sides of (A) by $e^{\int P\,dx}$, we obtain

$$\frac{dy}{dx}e^{\int P\,dx} + Pye^{\int P\,dx} = Qe^{\int P\,dx}$$

$$\Rightarrow \quad \frac{d}{dx}\left[ye^{\int P\,dx}\right] = Qe^{\int P\,dx}.$$

Integrating, we get

$$ye^{\int P\,dx} = \int Qe^{\int P\,dx}\,dx + c.$$

The expression $e^{\int P\,dx}$ is called the **Integrating Factor (I.F.).**

Rule: y (I.F) = $\int$Q (I.F) dx + c **is the solution of (A).**

Note: The fact $e^{\log t} = t$ will be used in many problems. Some time the equation become linear if we take x as the dependent variable and y as the independent variable. It is of the form dx/dy + P'(x) = Q' where P' and Q' are function of y alone the I.F.M. This case will be $e^{\int P'dy}$

HOMOGENEOUS EQUATIONS

Definition:

A differential equation of the form

$$\frac{dy}{dx} = \frac{f(x, y)}{g(x, y)}, \quad \text{...(1)}$$

where f(x, y) and (x, y) are homogeneous functions of x and y of the same degree, is called a *Homogeneous Differential Equation.*

To solve such an equation, we substitute

$$y = Vx, \text{ where V is a function of x.}$$

$$\therefore \quad \frac{dy}{dx} = V + x\frac{dV}{dx}.$$

Substituting the values of y and dy/dx in (1), we can verify that (1) reduces to an equation with separated variables x and V.

DIFFERENTIAL EQUATIONS OF FIRST ORDER AND FIRST DEGREE

A differential equation of the form

$$F\,(x,\ y,\ dy/dx) = 0, \quad \text{...(1)}$$

is said be of First Order and First Degree.

The equation y = f (x, c), where c is an arbitrary constant is called the *General Solution* of equation (1) if the given equation y f(x, c) and the derivative obtained from it satisfy (1).

Now, we give various methods of solving equation (1).

EQUATIONS REDUCIBLE TO THE LINEAR FORM

Consider a differential equation of the form:

$$\frac{dy}{dx} + Py = Q\,y^n, \quad \text{...(1)}$$

where P and Q are functions of x. We can reduced (1) to the linear form as follows:

Dividing both sides of (1) by y^n, we get

$$y^{-n}\frac{dy}{dx} + Py^{-n+1} = Q. \quad \text{...(2)}$$

Put $\qquad y^{-n+1} = z$

$$\Rightarrow \quad (-n + 1)\,y^{-n}\,\frac{dy}{dx} = \frac{dz}{dx}.$$

Substituting in (2), we obtain

$$\frac{dz}{dx} + (1 - n)Pz = (1 - n)Q,$$

which is now a linear equation with z as dependent variable.

RULES FOR FINDING INTEGRATING FACTORS

Rule III. *If $M\,dx + N\,dy = 0$ is homogeneous and $Mx + ny \neq 0$, then* $\dfrac{1}{Mx + Ny}$ *is an integrating factor of $Mdx + Ndy = 0$.*

Rule IV. *If $M - Ny \neq 0$ and the equation is of the form*

$f_1(xy)\,ydx + f_2(xy)\,x\,dy = 0$, then $\dfrac{1}{Mx - Ny}$ *is an integrating factor.*

Rule V. *Sometimes $Mdx + Ndy = 0$ may have an integrating factor $x^\alpha y^\beta$, where α are β are so chosen that after multiplying by $x^\alpha y^\beta$, the given equation becomes exact.*

EQUATIONS WITH SEPARATED VARIABLES

Definition:

A differential equation of the form

$$f(x)\,dx + g(y)\,dy = 0$$

is said to have separated variables.

Its solution is given as

$\int f(x)\,dx + \int g(y)\,dy = c$, c being an arbitrary constant.

OR

An equation of type $M + \dfrac{Ndy}{dx} = 0$ where M and N are function of x and y is called a differential equation of first ordered first degree.

OR

An equation which can be so written that the coefficient of dy is a function y alone and the coefficient of dx is a function of x alone is said to belong to this category.

SOLVED EXAMPLES

Example 1:

Solve $(x^2 - yx^2)\,dy + (y^2 + xy^2)\,y\,dx = 0$.

Solution:

We have $(x^2 - yx^2)\,dy + (y^2 + xy^2)\,y\,dx = 0$.

We get $x^2(1 - y)\,dy + y^3(1 + x)dx = 0$

Separate the variables, we obtain

$$\Rightarrow \quad \frac{1 + x}{x^2}dx + \frac{1 - y}{y^3}dy = 0.$$

Integrating, we get

$$\int\left(\frac{1}{x^2}+\frac{1}{x}\right)dx+\int\left(\frac{1}{y^3}-\frac{1}{y^2}\right)dy=c$$

$$\Rightarrow \qquad -\frac{1}{x}+\log x-\frac{1}{2y^2}+\frac{1}{y}=c$$

Hence, $2x\,y^2 \log x - 2y^2 + 2xy - 2\,c\,xy^2 - x = 0$ is the solution.

Ans.

Example 2:

Solve $\dfrac{dy}{dx} = e^{x+y} + x^2 e^y.$

Solution:

We have $\qquad \dfrac{dy}{dx} = c^{x+y} + x^2 e^y.$

We get $\qquad \dfrac{dy}{dx} = e^x \,.\, e^y + x^2 e^y = (e^x + x^2)\, e^y$

Separate the variables, we obtain

$$\Rightarrow \qquad (e^x + x^2)dx = e^{-y}\, dy.$$

Integrating, we obtain

$$e^x + \frac{1}{3}x^3 = -e^{-y} + c,$$ c being an arbitrary constant.

Hence, $e^x + e^{-y} + \dfrac{1}{3}x^3 = c$ is the required solution. **Ans.**

Example 3:

Solve $y - x\dfrac{dy}{dx} = 3\left(1 + x^2\dfrac{dy}{dx}\right).$

Solution:

We have $y - x\dfrac{dy}{dx} = 3\left(1 + x^2\dfrac{dy}{dx}\right).$

$$\Rightarrow \qquad y - 3 = x\,(3x + 1)\,\frac{dy}{dx}$$

Separate the variables, we have

$$\Rightarrow \qquad \int\frac{dy}{y-3} = \int\frac{dx}{x(3x+1)} + \log c,$$ log c being a constant.

$$\Rightarrow \qquad \log(y-3) = \int\left(\frac{1}{x} - \frac{3}{3x+1}\right) dx + \log c$$

$$\Rightarrow \qquad \log(y-3) = \log x - \log(3x+1) + \log c$$

$$\Rightarrow \qquad \log(y-3) + \log(3x+1) = \log c + \log x$$

$$\Rightarrow \qquad \log(1+3x)(y-3) = \log c\, x.$$

Hence $cx = (1+3x)(y-3)$ is the required solution. **Ans.**

Example 4:

Solve $(e^x + 1)\, y\, dy = (y+1)\, e^x\, dx$

Solution:

We have $(e^x +)\, y\, dy = (y+1)\, e^x\, dx$.

Separate the variables, we have

$$\therefore \qquad \int \frac{e^x\, dx}{e^x + 1} = \int \frac{y\, dy}{y+1} + \log c, \text{ log c being a constant.}$$

$$\Rightarrow \qquad \log(e^x + 1) = \int\left(1 - \frac{1}{y+1}\right) dy + \log c$$

$$\Rightarrow \qquad \log(e^x + 1) = y - \log(y+1) + \log c$$

$$\Rightarrow \qquad \log(e^x + 1) + \log(y+1) + \log c = y$$

$$\Rightarrow \qquad \log \frac{(e^x+1)(y+1)}{c} = y$$

$$\Rightarrow \qquad \frac{(e^x+1)(y+1)}{c} = e^y.$$

Hence, $(1+y)(1+e^x) = c\, e^x$ is the required solution. **Ans.**

Example 5:

Solve $(x+y)^2 \dfrac{dy}{dx} = a^2.$

Solution:

We have $(x+y)^2 \dfrac{dy}{dx} = a^2.$

Put $x + y = z \Rightarrow 1 + \dfrac{dy}{dx} = \dfrac{dz}{dx}.$

From (2) and (3), we obtain

$$z^2\left(\frac{dz}{dx} - 1\right) = a^2 \Rightarrow \frac{dz}{dx} = \frac{a^2}{z^2} + 1 = \frac{a^2 + z^2}{z^2}.$$

Separate the variables, we have

$$\therefore \qquad \int \frac{z^2\,dz}{a^2 + z^2} = \int dx + c$$

$$\Rightarrow \qquad x + c = \int\left(1 - \frac{a^2}{a^2 + z^2}\right)dz$$

$$= z - a^2\left(\frac{1}{a}\tan^{-1}\frac{z}{a}\right) = z - a\tan^{-1}\frac{z}{a}$$

$$\therefore \qquad x + c = x + y\ a\tan^{-1}\frac{x + y}{a}.$$

Hence, $y = c + a\tan^{-1}\dfrac{x + y}{a}$ is the solution. **Ans.**

Example 6:

Solve $(x - y)^2 \dfrac{dy}{dx} = a^2$.

Solution:

We have $(x - y)^2 \dfrac{dy}{dx} = a^2$.

Put $x - y = z \Rightarrow 1 \dfrac{dy}{dx} = \dfrac{dz}{dx} \Rightarrow \dfrac{dy}{dx} = 1 - \dfrac{dz}{dx}$

Using in (1), $z^2\left(1 - \dfrac{dz}{dx}\right) = a^2$

$$\Rightarrow \qquad 1 - \frac{dz}{dx} = \frac{a^2}{z^2} \Rightarrow \frac{dz}{dx} = \frac{z^2 - a^2}{z^2}$$

Separate the variables, we have

$$\Rightarrow \qquad \int \frac{z^2\,dz}{z^2 - a^2} = \int dx + c, \ c \text{ being a constant.}$$

$$\Rightarrow \qquad x + c = \int \frac{z^2 - a^2 + a^2}{z^2 - a^2}dz$$

$$= \int \left(1 + \frac{a^2}{z^2 - a^2}\right) dz$$

$$= z + a^2 . \frac{1}{2a} \log \frac{z - a}{z + a}, \quad z = x - y.$$

Hence, $\frac{a}{2} \log \frac{x - y - a}{x - y + a} - y = c$ is the solution. **Ans.**

Example 7:

Solve $3e^x \tan y \, dx + (1 - e^x) \sec^2 y \, dy = 0$.

Solution:

We have $3e^x \tan y \, dx + (1 - e^x)\sec^2 y \, dy = 0$.

Separate the variables, we have

$$\Rightarrow \quad \frac{3e^x}{1 - e^x} dx + \frac{\sec^2}{\tan y} dy = 0.$$

Integrating, we obtain

$$-3\log(1 - e^x) + \log \tan y = \log c.$$

Hence, $\frac{\tan y}{(1 - e^x)^3} = c$ or $\tan y = c(1 - e^x)^3$ is the solution.

Example 8:

Solve $y - x\frac{dy}{dx} = a\left(y^2 + \frac{dy}{dx}\right)$.

Solution:

We have $y - x\frac{dy}{dx} = a\left(y^2 + \frac{dy}{dx}\right)$.

We get $\quad y - ay^2 = \frac{dy}{dx}(a + x)$

Separate the variables, we obtain

$$\Rightarrow \quad \int \frac{dx}{a + x} = \int \frac{dy}{y - ay^2} + k, \qquad \text{(k being a constant)}$$

Integrating both side, we have

$$\therefore \quad \log(a + x) = \int \frac{dy}{y(1 - ay)} + k$$

$$= \int\left(\frac{1}{y} + \frac{a}{1 - ay}\right)dy + k$$

$$= \log y - \log(1 - ay) + \log c, \qquad (k = \log c)$$

$$\therefore \qquad \log(a + x) = \log \frac{cy}{1 - ay}$$

Hence, $(1 - ay)(a + x) = cy$ is the required solution. **Ans.**

Example 9:

Solve $\frac{dy}{dx} = (x + y)^2$.

Solution:

We have $\frac{dy}{dx} = (x + y)^2$. ...(4)

From (3) and (4), we obtain

$$\frac{dz}{dx} - 1 = z^2 \qquad \Rightarrow \qquad \frac{dz}{dx} = 1 + z^2.$$

Separate the variables, we have

$$\therefore \qquad \int\frac{dz}{1 + z^2} = \int dx + c \quad \Rightarrow \quad \tan^{-1} z = x + c.$$

Integrating, both side we have

Hence, $x + c \tan^{-1}(x + y)$ is the solution. **Ans.**

Example 10:

Solve $\frac{dy}{dx} = (4x + y + 1)^2$.

Solution:

We have $\frac{dy}{dx} = (4x + y + 1)^2$ ...(5)

Let $4x + y + 1 = z$ so that $4 + \frac{dy}{dx} = \frac{dz}{dx}$.

From (5) and (6), we obtain

$$\frac{dz}{dx} - 4 = z^2 \Rightarrow \frac{dz}{dx} = 4 + z^2.$$

Separate the variables, we have

$$\therefore \qquad \int\frac{dz}{4+z^2} = \int dx + c \Rightarrow \frac{1}{2}\tan^{-1}\frac{z}{2} = x + c.$$

Integrating, both side we have

Hence, $\tan^{-1}\left(\frac{4x+y+1}{2}\right) = 2x + a$ is the solution. (a = 2c). **Ans.**

Example 11:

Solve $\frac{dy}{dx} = \frac{y}{x} + \frac{y^2}{x^2}$.

Solution:

We have $\frac{dy}{dx} = \frac{y}{x} + \frac{y^2}{x^2}$.

It is a homogeneous diff equation

$$\Rightarrow \qquad \frac{dy}{dx} = \frac{xy + y^2}{2x^2}$$

$$\Rightarrow \qquad V + x\frac{dV}{dx} = \frac{Vx^2 + V^2x^2}{2x^2} \qquad \left(y = Vx \Rightarrow \frac{dy}{dx} = V + x\frac{dV}{dx}\right)$$

Separate the variables, we have

$$\Rightarrow \qquad x\frac{dV}{dx} = \frac{V+V^2}{2} - V = \frac{V^2 - V}{2} = \frac{V(V-1)}{2}$$

$\Rightarrow \qquad \frac{dV}{V(V-1)} = \frac{dV}{2x}$. Integrating, both side we have

$$\Rightarrow \qquad \int\left(\frac{1}{V-1} - \frac{1}{V}\right)dV = \frac{1}{2}\int\frac{dx}{x} + \log c$$

$$\Rightarrow \qquad \log(V-1) - \log V = \frac{1}{2}\log x + \log c$$

$$\Rightarrow \qquad \log(V-1) = \log V + \log\sqrt{x} + \log c = \log c\,V\sqrt{x}$$

$$\Rightarrow \qquad V - 1 = cV\sqrt{x}$$

$$\Rightarrow \qquad \frac{y}{x} - 1 = c\,\frac{y}{x}.\,\sqrt{x}\,.$$

Hence, $y - x = cy\sqrt{x}$ is the required solution. **Ans.**

Example 12:

Solve $(x^2 + xy)dy = (x^2 + y^2)dx$.

Solution:

We have $(x^2 + xy)dy = (x^2 + y^2)dx$.

Which is a homogeneous equation

$$\Rightarrow \quad \frac{dy}{dx} = \frac{x^2 + y^2}{x^2 + xy}$$

$$\Rightarrow \quad V + x\frac{dV}{dx} = \frac{1 + V^2}{1 + V} \qquad \left(y = Vx, \frac{dy}{dx} = V + x\frac{dV}{dx}\right)$$

$$\Rightarrow \quad x\frac{dV}{dx} = \frac{1 + V^2}{1 + V} - V = \frac{1 - V}{1 + V}$$

Separate the variables, we have

$$\Rightarrow \quad \frac{1 + V}{1 - V}dV = \frac{dx}{x}\cdot \text{ Integrating, we get}$$

$$\int\frac{V + 1}{V - 1}dV + \int\frac{dx}{x} + \log c = 0$$

$$\Rightarrow \quad \int\left(1 + \frac{2}{V - 1}\right)dV + \log x + \log c = 0$$

$$\Rightarrow \quad V + 2\log(V - 1) + \log cx = 0$$

$$\Rightarrow \quad V + \log[cx(V - 1)^2 = 0, \; V = y/x.$$

Hence, $y + x\log\left[\frac{c}{x}(y - x)^2\right] = 0$ is the required solution.

Example 13:

Solve $\frac{1}{2x}\frac{dy}{dx} + \frac{x + y}{x^2 + y^2} = 0.$

Solution:

We have $\quad \frac{1}{2x}\frac{dy}{dx} + \frac{x + y}{x^2 + y^2} = 0.$

$$\Rightarrow \qquad \frac{dy}{dx} + \frac{2x^2 + 2xy}{x^2 + y^2} = 0$$

It is a homogeneous equation

$$\Rightarrow \qquad V + x\frac{dV}{dx} + \frac{2 + 2V}{1 + V^2} = 0 \quad \left(y = Vx, \frac{dy}{dx} = V + x\frac{dV}{dx}\right)$$

Separate the variables, we have

$$\Rightarrow \qquad x\frac{dV}{dx} + \frac{V + V^3 + 2 + 2V}{1 + V^2} = 0$$

Integrating, we have

$$\Rightarrow \qquad \int\frac{(1 + v^2)\, dV}{V^3 + 3V + 2} + \int\frac{dx}{x} = \log c$$

$$\Rightarrow \qquad \frac{1}{3}\int\frac{dt}{t} + \log x = \log c$$

$[\because\ V^3 + 3V + 2 = t \ \Rightarrow\ 3(V^2 + 1)dV = dt]$

$$\Rightarrow \qquad \frac{1}{3}\log t + \log x = c \quad \Rightarrow \quad tx^3 = k, \text{ where } k = e^{3c}.$$

$$\Rightarrow \qquad (V^3 + 3V + 2)\, x^3 = k.$$

Hence, $y^3 + 3yx^2 + 2x^2 = k$ is the required solution.

Example 14:

Solve $\dfrac{dy}{dx} - y = \sqrt{x^2 + y^2}$.

Solution:

We have $\dfrac{dy}{dx} - y = \sqrt{x^2 + y^2}$.

Then given equation may be written as

$$\frac{dy}{dx} = \frac{y + \sqrt{x^2 + y^2}}{x}.$$

Put $y = V\,x$ so that $\dfrac{dy}{dx} = V + x\dfrac{dV}{dx}$. Now (1) becomes

$$V + x\frac{dV}{dx} = \frac{Vx + \sqrt{x^2 + V^2x^2}}{x} = V + \sqrt{1 + V^2}$$

Separate the variables, we have

$$\Rightarrow \qquad x\frac{dV}{dx} = \sqrt{1+V^2} \Rightarrow \frac{dV}{\sqrt{1+V^2}} = \frac{dx}{x}.$$

Integrating, we obtain

$$\log[V + \left[V + \sqrt{1+V^2}\right] = \log x + \log c$$

$$\Rightarrow \qquad V + \sqrt{1+V^2} = cx \quad \Rightarrow \quad \frac{y}{x} + \sqrt{1+\frac{y^2}{x^2}} = cx.$$

Hence, $y + \sqrt{x^2+y^2} = c\,x^2$ is the required solution. **Ans.**

Example 15:

Solve x sin y/x dy = (y sin y/x – x)dx.

Solution:

We have x sin y/x dy = (y sin y/x – x)dx.

$$\Rightarrow \qquad \frac{dy}{dx} = \frac{y\sin y/x - x}{x\sin y/x}$$

$$\Rightarrow \qquad V + x\frac{dV}{dx} = \frac{V\sin V - 1}{\sin V}$$

$$\left(\because y = Vx, \frac{dy}{dx} = V + x\frac{dV}{dx}\right)$$

$$\Rightarrow \qquad x\frac{dV}{dx} = \frac{V\sin V - 1}{\sin V} - V = -\frac{1}{\sin V}$$

Separate the variables, we have

$$\Rightarrow \qquad \sin V dV + \frac{dx}{x} = 0.\ \text{Integrating, we obtain}$$

$$-\cos V + \log x = c.$$

Hence, log x = cos y/x + c is the required solution. **Ans.**

Example 16:

Solve $\frac{dy}{dx} + \frac{x-2y}{2x-y} = 0.$

Solution

We have $\dfrac{dy}{dx} + \dfrac{x - 2y}{2x - y} = 0.$

It is a homogeous diff. equation

Putting y = Vx and $\dfrac{dy}{dx} = V + x\dfrac{dV}{dx}$, the given equation becomes

$$V + x\frac{dV}{dx} + \frac{1 - 2V}{2 - V} = 0$$

Separate the variables, we have

$$\Rightarrow \qquad x\frac{dV}{dx} + \frac{1 - 2V}{2 - V} = 0 \Rightarrow \frac{2 - V}{1 - V^2}dV + \frac{dx}{x} = 0.$$

Integrating, we obtain

$$-2\int\frac{dV}{V^2 - 1} + \int\frac{V\,dV}{V^2 - 1} + \int\frac{dx}{x} = \log c$$

$$\Rightarrow \qquad -2.\frac{1}{2}\log\frac{V - 1}{V + 1} + \frac{1}{2}\log(V^2 - 1) + \log x = \log c$$

$$\Rightarrow \qquad \sqrt{V^2 - 1}.\left(\frac{V + 1}{V - 1}\right).\, x = c$$

$$\Rightarrow \qquad \frac{\sqrt{V^2 - 1}\,(V + 2)}{\sqrt{V - 1}}.\, x = c.$$

Squaring both sides, we obtain

$$(V + 1)^3 .x^2 = a(V - 1),\ a = c^2$$ **Ans.**

Hence $(x + y)^3 = a\,(y - x)$ is the required solution. $(\because V = y/x)$

Example 17:

Solve $(x^3 + y^3)dx = (x^2y + xy^2)dy$.

Solution:

We have $(x^3 + y^3)dx = (x^2y + xy^2)dy$.

$$\Rightarrow \qquad \frac{dy}{dx} = \frac{x^3 + y^3}{x^2y + xy^2}$$

It is a homogeneous diff. equation

$$\Rightarrow \quad V + x\frac{dV}{dx} = \frac{1+V^3}{V+V^2} \qquad \left(y = Vx, \frac{dy}{dx} = V + x\frac{dV}{dx}\right)$$

$$\Rightarrow \quad x\frac{dV}{dx} = \frac{1+V^3}{V+V^2} - V = \frac{1-V^2}{V+V^2} = \frac{(1-V)(1+V)}{V(1+V)}$$

Separate the variables, we have

$$\Rightarrow \quad \frac{V\,dV}{(1-V)} = \frac{dx}{x}\cdot \text{ Integrating, we obtain}$$

$$\int \frac{V\,dV}{(V-1)} + \int \frac{dx}{x} + \log c = 0$$

$$\Rightarrow \quad \int \left(1 + \frac{1}{V-1}\right) dV + \log x + \log c = 0$$

$$\Rightarrow \quad V + \log(V-1) + \log x + \log c = 0$$

$$\Rightarrow \quad \frac{y}{x} + \log\left[\left(\frac{y}{x} - 1\right).x.c\right] = 0$$

Hence, y + x log[c(y = x)] = 0 is the required solution. **Ans.**

Example 18:

Solve $x^2 y\,dx - (x^3 + y^3)\,dy = 0$.

Solution:

We have

The given equ. can be written as

Putting y = Vx and dy/dx = V + x dV/dx

$$\frac{dy}{dx} = \frac{x^2 y}{x^3 + y^2} \Rightarrow V + x\frac{dV}{dx} = \frac{Vx^3}{x^3 + V^3 x^3} \quad (y = yx)$$

$$\Rightarrow \quad x\frac{dV}{dx} = \frac{V}{1+V^3} - V = \frac{V^4}{1+V^3}.$$

Separate the variables, we have

$$\therefore \quad \int \frac{1+V^3}{V^4} dV + \int \frac{dx}{x} = \log c$$

Integrating, we have

$$\Rightarrow \quad \int \left(\frac{1}{V^4} + \frac{1}{V}\right) dV + \log x = \log c$$

$$\Rightarrow \qquad -\frac{1}{3V^3} + \log V + \log x = \log c$$

$$\Rightarrow \qquad -\frac{1}{3}\frac{x^3}{y^3} + \log x = \log c \qquad \left(\because V = \frac{y}{x}, Vx = y\right)$$

$$\Rightarrow \qquad \log\frac{y}{c} = \frac{x^3}{3y^3} \quad \Rightarrow \quad \frac{y}{x} = e^{x^3/3y^3}.$$

Hence, $y = c\, e^{x^3/3y^3}$ is the required solution. **Ans.**

Example 19:

Solve y – xp = x + yp (p = dy/dx).

Solution:

The given equation can be written as

$$\frac{dy}{dx} = \frac{y-x}{y+x} \Rightarrow V + x\frac{dV}{dx} = \frac{V-1}{V+1}, \; y = Vx$$

Putting $y = Vx$ and $dy/dx = x + dV/dx$ we have

$$\Rightarrow \qquad x\frac{dV}{dx} = \frac{V-1}{V+1} - V = \frac{-(1+V^2)}{V+1}$$

Separate the variables, we have

$$\therefore \qquad \int\left(\frac{V+1}{1+V^2}\right)dV + \int\frac{dx}{x} = \log c$$

Integrating, we have

$$\frac{1}{2}\log(1 + V^2) + \tan^{-1} V + \log x = \log c$$

$$\Rightarrow \qquad \log\frac{x\sqrt{1+V^2}}{c} = -\tan^{-1} V, \; V = \frac{y}{x}.$$

Hence, $\sqrt{x^2 + y^2} = c\, e^{-\tan^{-1} y/x}$ is the required solution.

Example 20:

Solve $\frac{dy}{dx} - 2y \tan x = y^2 \tan^2 x$.

Solution:

We have $\frac{dy}{dx} - 2y \tan x = y^2 \tan^2 x$.

Dividing by y^2, we obtain

$$\frac{1}{y^2}\frac{dy}{dx} - \frac{2}{y}\tan x = \tan^2 x. \text{ Put } -\frac{1}{y} = z.$$

The given equation becomes

$$\frac{dz}{dx} + 2 \tan x \,.\, z = \tan^2 x.$$

$$\text{I.F.} = e^{\int \tan x\, dy} = e^{2\log \sec x} = e^{\log \sec^2 x} = \sec^2 x.$$

the solution of (6) is given by

$$z \sec^2 x = \int \tan^2 x \sec^2 x\, dx + c = \frac{\tan^3 x}{3} + c, \; z = -\frac{1}{y}$$

$$\Rightarrow \quad -\frac{1}{y}\sec^2 x = \frac{\tan^3 x}{3} + c. \text{ Hence } y \tan^3 x + 3 \sec^2 x + ay = 0.$$

Example 21:

Solve $\sin y \frac{dy}{dx} = \cos x(2\cos y - \sin^2 x)$.

Solution:

We have $\sin y \frac{dy}{dx} = \cos x(2\cos y - \sin^2 x)$.

Put $v = \cos y$ so that $-\sin y \frac{dy}{dx} = \frac{dv}{dx}$.

The given equation now becomes

$$\frac{dv}{dx} + 2 \cos x.\, v = \cos x \sin^2 x, \text{ which is linear.}$$

$\text{I.F} = e^{\int 2\cos x\, dx} = e^{2\sin x}$. The solution is given by

$$v e^{2\sin x} = \int \cos x \sin^2 x\, e^{2\sin x} dx + c$$

$$= \int t^2 e^{2t}\, dt + c, \text{ where } t = \sin x$$

$$= \frac{1}{2}t^2 e^2 t - \frac{1}{2}\int 2t\, e^{2t}\, dt + c$$

$$= \frac{1}{2}t^2e^{2t} - \left[\frac{1}{2}te^{2t} - \frac{1}{2}\int e^{2t}.1dt\right] + c$$

$$= \left(\frac{1}{2}t^2 - \frac{1}{2}t + \frac{1}{4}\right)e^{2t} + c.$$

Hence, $\cos y = \frac{1}{2}\sin^2 x - \frac{1}{2}\sin x + \frac{1}{4} + ce^{-2\sin x}$ is the solution. **Ans.**

Example 22:

Solve $x^2 \cos y \frac{dy}{dx} = 2x \sin y - 1.$

Solution:

We have $x^2 \cos y \frac{dy}{dx} = 2x \sin y - 1.$

The given equation can be written as

$$\cos y \frac{dy}{dx} + \sin y\left(-\frac{2}{x}\right) = -\frac{1}{x^2}.$$

Put $v = \sin y$ so that $\frac{dv}{dx} = \cos y \frac{dy}{dx}.$

The given equation becomes

$$\frac{dv}{dx} - \frac{2}{x}v = -\frac{1}{x^2}, \text{ which is linear.}$$

I.F. $= e^{-2\int 1/x\,dx} = e^{-2\log x} = x^{-2}$. The solution is given by

$$vx^{-2} = -\int \frac{1}{x^2}.\frac{1}{x^2}dx + c = \frac{1}{3x^3} + c.$$

Hence, is $y = \frac{1}{3x} + cx^2$ is the required solution. **Ans.**

Example 23:

Solve $(x^2 + y^2 + 2x)dx + 2y\,dy = 0.$

Solution:

We have $(x^2 + y^2 + 2x)dx + 2y\,dy = 0.$

$M = x^2 + y^2 + 2x$, $N = 2y$; $\partial M/\partial y = 2y$, $\partial N/\partial x = 0.$

Now, $\frac{1}{N}\left(\frac{\partial M}{\partial y} - \frac{\partial N}{\partial x}\right) = \frac{2y}{2y} = 1.$

$\therefore \qquad \text{I.F} = e^{\int 1\, dy} = e^x.$

Multiplying by e^x, the given equation becomes

$$(x^2 + y^2 + 2x)\, e^x\, dx + 2y\, e^x\, dy = 0, \text{ which is exact}$$

$$\Rightarrow \quad (x^2 + 2x)\, e^x\, dx + (y^2 e^x\, dx + e^x\, 2y\, dy) = 0$$

$$\Rightarrow \quad d(x^2 e^x) + d(y^2 e^x) = 0.$$

Integrating, $x^2 e^x + y^2 e^x = c$.

Hence $(x^2 + y^2)\, e^x = c$ is the required solution. **Ans.**

Example 24:

Solve $(y^4 + 2y)\, dx + (xy^3 + 2y^4 - 4x)\, dy = 0.$

Solution:

We have $(y^4 + 2y)\, dx + (xy^3 + 2y^4 - 4x)\, dy = 0.$

$$M = y^4 + 2y,\; N = xy^3 + 2y^4 - 4x,$$

$$\partial M/\partial y = 4y^3 + 2,\; \partial N/\partial x = y^3 - 4. \text{ Not exact.}$$

Now, $\frac{1}{M}\left(\frac{\partial M}{\partial y} - \frac{\partial N}{\partial x}\right) = \frac{3y^3 + 6}{y^4 + 2x} = \frac{3}{y}$, a function of y.

By **Rule II.** I.f. $= e^{-\int \frac{3}{y} dy} = e^{-3\log x} = y^{-3}.$

Multiplying the given equation by $1/y^3$, we get

$$\left(y + \frac{2}{y^2}\right)dx + \left(x + 2y - \frac{4x}{y^3}\right)dy = 0 \Rightarrow M_1 dy + N_1 dy = 0.$$

The above equation is exact.

Now, $\int M_1 dx = \int (y + 2/y^2)\, dx = (y + 2/y^2)x$, $\int N_1 dy = \int 2y\, dy = y^2.$

Hence the required solution is

$$\left(y + \frac{2}{y^2}\right)x + y^2 = c.$$ **Ans.**

Example 25:

Solve $(x^2 + y^2 + 1)\, dx - 2xy\, dy = 0.$

Solution:

We have $(x^2 + y^2 + 1)\,dx - 2xy\,dy = 0$.

$$M\ x^2 + y^2 + 1,\ N = -2xy. \text{ Then}$$

$\partial M/\partial y = 2y$, $\partial N/\partial x = -2y$. The given equation is not exact.

Now, $\dfrac{1}{N}\left(\dfrac{\partial M}{\partial y} - \dfrac{\partial N}{\partial x}\right) = \dfrac{4y}{-2xy} = \dfrac{2}{x}$.

which is a function of x alone.

$\therefore\quad$ I.F. $= e^{-\int \frac{2}{x}dx} = e^{-2\log x} = e^{\log x^{-2}} = x^{-2}$.

Multiplying the given equation by $1/x^2$, we obtain

$$\left(1 + \frac{y^2}{x^2} + \frac{1}{x^2}\right)dx - \frac{2y}{x}dy = 0 \Rightarrow M_1dx + N_1dy = 0. \qquad ...(1)$$

The above equation is exact, since

$$\frac{\partial M_1}{\partial y} = \frac{2y}{x^2}, \quad \frac{\partial N_1}{\partial x} = \frac{2y}{x^2}.$$

Now $\int M_1\,dx = \int\left(1 + \dfrac{y^2}{x^2} + \dfrac{1}{x^2}\right)dx = x - \dfrac{y^2}{x} - \dfrac{1}{x}$.

As $N_1 = -2y/x$ is not independent of x, $\int N_1\,dy = 0$.

Hence, the solution is $x - y^2/x - 1/x = c$

$\Rightarrow\qquad x^2 - y^2 = cx + 1.$ **Ans.**

Example 26:

Solve $(x^2 + y^2 + x)\,dx + xy\,dy = 0$.

Solution:

We have $(x^2 + y^2 + x)\,dx + xy\,dy = 0$.

$M = x^2 + y^2 + x$, $N = xy$; $\partial M/\partial y = 2y$, $\partial N/\partial x = y$.

$$\therefore\quad \frac{1}{N}\left(\frac{\partial M}{\partial y} - \frac{\partial N}{\partial x}\right) = \frac{y}{xy} = \frac{1}{x},$$

$\therefore\quad$ I.F. $= e^{\int \frac{1}{x}dx} = e^{\log x} = x$.

Multiplying by x, the given equation becomes

$(x^3 + xy^2 + x^2)\,dx + x^2y\,dy = 0$, which is exact.

Now, $\int (x^3 + xy^2 + x^2)dx = \frac{1}{4}x^4 + \frac{1}{2}x^2y^2 + \frac{1}{3}x^3.$

Since $N_1 = x^2y$ is not independent of x, $\int N_1 dy = 0.$

Hence, the solution is $\frac{1}{4}x^4 + \frac{1}{2}x^2y^2 + \frac{1}{3}x^3 = c.$ **Ans.**

Example 27:

Solve $(3x^2y^4 + 2xy)dx + (2x^2y^{3| - x}2)dy = 0.$

Solution:

We have $(3x^2y^4 + 2xy)dx + (2x^2y^{3| - x}2)dy = 0.$

$M = 3x^2y^4 + 2xy,\ N = 2x^3y^3 - x^2,$

$\partial M/\partial y = 12x^2y^3 + 2x,\ \partial N/\partial x = 6x^2\,y^3 - 2x.$

Now, $\frac{1}{M}\left(\frac{\partial M}{\partial y} - \frac{\partial N}{\partial x}\right) = \frac{6x^2y^3 + 4x}{y(3x^2y^3 + 2x)} = \frac{2}{y}.$

By **Rule II.** I.F. $= e^{-2\int 1/y\,dy} = e^{-2\log y} = y^{-2}.$

Multiplying the given equation by y^{-2}, we obtain

$$\left(3x^2y^2 = \frac{2x}{y}\right)dx + \left(2x^3y - \frac{x^2}{y^2}\right)dy = 0 \text{ or } M_1\,dx + N_1\,dy = 0.$$

The above equation is exact.

$$\int M_1\,dx = x^3y^2 + x^2/y,$$

$\int N_1\,dy = 0$, as N_1 is not independent of x.

Hence, the solution is $x^3y^2 + x^2/y = c.$ **Ans.**

Example 28:

Solve $(2xy^2\,e^y + 2xy^3 + y)dx + (x^2y^4e^y - x^2y^2 - 3x)\,dy = 0.$

Solution:

We have $(2xy^2\,e^y + 2xy^3 + y)dx + (x^2y^4e^y - x^2y^2 - 3x)\,dy = 0.$

$\partial M/\partial y = 8xy^3e^y + 2xy^4e^y + 6xy^2 + 1,$

$\partial N/\partial x = 2xy^4e^y - 2xy^2 - 3.$

$$\therefore \quad \frac{1}{M}\left(\frac{\partial M}{\partial y} - \frac{\partial N}{\partial x}\right) = \frac{4}{y} \text{ and I.F.} = e^{-4\int 1/y\, dy} = y^{-4}.$$

Multiplying the given equation by y^{-4}, we obtain

$$\left(2x\, e^x + 2\frac{x}{y} + \frac{1}{y^3}\right)dx + \left(x^2 e^y - \frac{x^2}{y^2} - \frac{3x}{y^4}\right)dy = 0.$$

$\Rightarrow$ $M_1\, dx + N_1\, dy = 0$, which is exact

The required solution is given by

$$\int(2xe^y + 2x/y + 1/y^3)\, dx + 0 = c$$

$$\Rightarrow \quad x^2e^y + x^2/y + x/y^3 = c.$$ **Ans.**

Example 29:

Solve $x^2y\, dx - (x^3 + y^3)dy = 0$.

Solution:

We have $x^2y\, dx - (x^3 + y^3)dy = 0$.

The given equation is homogeneous and

$$Mx + Ny + x^3y - y^4 = -y^4 \neq 0.$$

$$\therefore \quad \text{I.F.} = \frac{1}{Mx - Ny} = -\frac{1}{y^4}.$$

Multiplying the given equation by $-1/y^4$, we obtain

$$-\frac{x^2}{y^3}dx + \left(\frac{x^3}{y^4} + \frac{1}{y}\right)dy = 0, \text{ which is exact.}$$

The required solution is given by

$$\int -\frac{x^2}{y^3}dx + \int \frac{1}{y}dy = \log c$$

$$\Rightarrow \quad -\frac{x^3}{3y^3} + \log y = \log c$$

$$\Rightarrow \quad \log y = \log c + -\frac{x^3}{3y^3}.$$

Hence, $y = e\, e^{x^3/3y^3}$ is the required solution. **Ans.**

Example 30(a):

Solve $y(x^2y^2 + 2)dx + x(2 - 2x^2y^2)dy = 0.$

Solution:

We have $y(x^2y^2 + 2)dx + x(2 - 2x^2y^2)dy = 0.$

The given equation is of the form

$f_1(xy)\, y\, dx + f_2(xy)\, x\, dy = 0.$ **By Rule IV.**

$$\text{I.F.} = \frac{1}{Mx - Ny} = -\frac{1}{3x^3y^3},$$

Multiplying the given equation by $\frac{1}{3x^3y^3}$, we get

$$\left(\frac{1}{3x} + \frac{2}{3x^3y^2}\right)dx + \left(\frac{2}{3x^2y^3} - \frac{2}{3y}\right)dy = 0,$$

which is exact. The required solution is given by

$$\int\left(\frac{1}{3x} + \frac{2}{3x^3y^2}\right)dx + \int\frac{2}{3y}dy = \log c$$

$$\Rightarrow \quad \frac{1}{3}\log x - \frac{1}{3x^2y^2} - \frac{2}{3}\log y = \log c$$

$$\log\frac{x}{y^2} = \log c^3 + \frac{1}{x^2y^2} \Rightarrow x = ky^2\, e^{1/x^2y^2}.$$

Example 30(b):

Solve $(x^2y - 2xy^2)dx - (x^3 - 3x^2y)dy = 0.$

Solution:

We have $(x^2y - 2xy^2)dx - (x^3 - 3x^2y)dy = 0.$

The given equation is homogeneous and

$$Mx + Ny = x^3y - 2x^2y^2 - x^3y + 3x^3y^2 = x^2y^2 \neq 0.$$

$$\therefore \quad \text{I.F.} = \frac{1}{Mx - Ny} = \frac{1}{x^2y^2}. \qquad \textbf{(Rule III)}$$

Multiplying the given by $1/x^2y^2$, we get

$\left(\frac{1}{y} - \frac{2}{x}\right)dx - \left(\frac{x}{x^2} - \frac{3}{y}\right)dy = 0$, which is exact.

The required solution is given by

$$\int\left(\frac{1}{y}-\frac{2}{x}\right)dx+\int\frac{3}{y}dy=c.$$

$\Rightarrow$ $$x/y-2\log x+3\log y=c.$$

Hence, $\dfrac{x}{y}+\log\dfrac{y^3}{x^2}=c$ is the required solution. **Ans.**

Example 30(c):

Solve $(xy\sin xy+\cos xy)\,y\,dx+(xy\sin xy-\cos xy)x\,dy=0$.

Solution:

We have $(xy\sin xy+\cos xy)\,y\,dx+(xy\sin xy-\cos xy)x\,dy=0$.

The given equation is of the form

Now, $Mx-Ny=xy\,(xy\sin xy+\cos xy)-sy(xy\sin xy-\cos xy)$

$$=2xy\cos xy.$$

$$\text{I.F.}=\frac{1}{Mx-Ny}=\frac{1}{2xy\cos xy}.$$

Multiplying by the above I.F., the given equation becomes

$$\frac{1}{2}\left(y\tan xy+\frac{1}{x}\right)dx+\frac{1}{2}\left(x\tan xy-\frac{1}{y}\right)dy=0\text{, which is exact.}$$

The required solution is given by

$$\int\left(y\tan xy+\frac{1}{x}\right)dx+\int\left(-\frac{1}{y}\right)dy=\log c$$

$$-\log\cos xy+\log x-\log y=\log c$$

$\dfrac{x}{y\cos xy}=c$. Hence $x=cy\cos(xy)$ is the required solution. **Ans.**

Example 30(d):

Solve $y(xy-2x^2y^2)dx+x(xy-x^2y^2)dy=0$.

Solution:

We have $y(xy-2x^2y^2)dx+x(xy-x^2y^2)dy=0$.

The given equation is of the form

$$f_1(x\ y)y\,dx+f_2(x\ y)x\,dy=0.$$

By Rule IV

$$\text{I.F.} = \frac{1}{Mx - Ny} = \frac{1}{xy(xy + 2x^2y^2) - xy(xy - x^2y^2)} = \frac{1}{3x^3y^3}.$$

Multiplying the given equation by $\frac{1}{3x^3y^3}$, we get

$$\left(\frac{1}{3x^2} = \frac{2}{3x}\right)dx + \left(\frac{1}{3xy^2} - \frac{1}{3y}\right)dy = 0, \text{ which is exact.}$$

The required solution is given by

$$\int\left(\frac{1}{3x^2} = \frac{2}{3x}\right)dx - \int\frac{1}{3y}dy = c$$

$$\Rightarrow \qquad -\frac{1}{3xy} + \frac{2}{3}\log x - \frac{1}{3}\log y = \log c$$

$$\Rightarrow \qquad \log\frac{x^2}{y} = \log c^3 + \frac{1}{xy} \Rightarrow \frac{x^2}{y} = ke^{1/xy} \Rightarrow x^2 = kye^{1/xy}.$$

Ans.

Example 31:

Solve $y(2xy + 1)dx + x(1 + 2xy - x^3y^3)dy = 0$.

Solution:

We have $y(2xy + 1)dx + x(1 + 2xy - x^3y^3)dy = 0$.

$$\text{I.F.} = \frac{1}{Mx - Ny} = \frac{1}{x^4y^4}.$$

Multiplying the given equation by $1/x^4y^4$, we get

$$\left(\frac{2}{x^3y^2} + \frac{1}{x^4y^4}\right)dx + \left(\frac{1}{x^3y^4} + \frac{2}{x^2y^3} - \frac{1}{y}\right)dy = 0,$$

which is exact. The required solution is

$$\int\left(\frac{2}{x^3y^2} + \frac{1}{x^4y^3}\right)dx - \int\frac{1}{y}dy = a$$

Hence, $\frac{1}{x^2y^2} + \frac{1}{3x^3y^3} + \log y = c$ is the required solution.

Example 32:

Solve $y\,dx - x\,dy + 3x^2y^2\,e^{x^3}\,dx = 0.$

Solution:

We have $y\,dx - x\,dy + 3x^2y^2\,e^{x^3}\,dx = 0.$

The given equation can be written as

$$\frac{y\,dx - x\,dy}{y^2} + 3x^2\,e^{x^3} = 0 \Rightarrow d\left(\frac{x}{y}\right) + e^{x^3}\,d(x^3) = 0.$$

On integrating, the required solution is $\frac{x}{y} + e^{x^3} = c.$ **Ans.**

Example 33:

Solve $y\,dx - x\,dy + \log x\,dx = 0.$

Solution:

We have $y\,dx - x\,dy + \log x\,dx = 0.$

The given equation can be written as

$$x\frac{dy}{dx} - y = \log x \Rightarrow \frac{dy}{dx} - \frac{1}{x}y = \frac{\log x}{x},\ \text{which is linear}$$

$\text{I.F.} = e^{\int 1/x\,dx} = e^{-\log x} = x^{-1}$. The solution is given by

$$yx^{-1} = \int \frac{1}{x^2}\log x\,dx - c = -\frac{1}{x}(1 + \log x) - c.$$

Hence, $y + 1 + \log x + cx = 0$ is the required solution. **Ans.**

Example 34:

Solve $x\,dy - y\,dx - x(x^2 - y^2)^{1/2}\,dx = 0.$

Solution:

We have $x\,dy - y\,dx - x(x^2 - y^2)^{1/2}\,dx = 0.$

The given equation can be written as

$$\frac{x\,dy - y\,dx}{\sqrt{x^2 - y^2}} - x\,dx = 0$$

$$\Rightarrow \quad \frac{(x\,dy - y\,dx)}{x^2\sqrt{1-(y/x)^2}} - dx = 0 \Rightarrow \frac{d(y/x)}{\sqrt{1-(y/x)^2}} - dx = 0$$

Integrating, we get

$\sin^{-1}(y/x) = x + c$ as the required solution. **Ans.**

Example 35:

Solve $x^2 \dfrac{dy}{dx} + \sqrt{1 - x^2y^2} = 0.$

Solution:

We have $x^2 \dfrac{dy}{dx} + \sqrt{1 - x^2y^2} = 0.$

The given equation can be written as

$$x(x\,dy + y\,dx) + \sqrt{1 - x^2y^2}\;dx = 0.$$

Dividing throughout by $x\sqrt{1 - x^2y^2}$, we obtain

$$\frac{x\,dy + y\,dx}{\sqrt{1 - x^2y^2}} + \frac{dx}{x} = 0 \Rightarrow \frac{d(xy)}{\sqrt{1 - x^2y^2}} + \frac{dx}{x} = 0.$$

Integrating, we get

$\sin^{-1} xy + \log x = c$ as the required solution. **Ans.**

Example 36:

Solve $3yx^2\,dx - (x^3 + 2y^4)dy = 0.$

Solution:

We have $3yx^2\,dx - (x^3 + 2y^4)dy = 0.$

The given equation can be written as

$$x^2(3y\,dx - x\,dy) - 2y^4\,dy = 0. \qquad ...(1)$$

It will help to remember that $x^{a-1}y^{b-1}$ *is an I.F. for* $ay\,dx + bx\,dy = 0$, *for all real values of a and b.*

In the expression $3y\,dx - x\,dy$; we have $a = 3$ and $b = -1$.

Thus, I.F. $= a^{a-1}\,y^{b-1} = x^2y^{-2}$.

Multiplying the equation (1) by y^{-2} (x^2 is already there), we get

$$x^2\left(\frac{3y\,dx - x\,dy}{y^2}\right) - 2\,y^2\,dy = 0$$

$$\Rightarrow \qquad d\left(\frac{d^3}{y}\right) - 2y^2\,dy = 0 \Rightarrow \frac{x^3}{y} - \frac{2}{3}y^3 = c.$$

Hence $3x^3 - 2y^4 = ky$ is the required solution. **Ans.**

Example 37:

Solve $(x + y^2)dy + (y - x^2)dx = 0.$

Solution:

We have $(x + y^2)dy + (y - x^2)dx = 0.$

The given equation can be written as

$$x\,dy + y\,dx + y^2\,dy - x^2\,dx = 0$$

$$\Rightarrow \qquad d(xy) + y^2\,dy - x^2\,dx = 0.$$

Integrating, we have $xy + \frac{1}{3}y^3 - \frac{1}{3}x^3 = 0$

$\Rightarrow \qquad y^3 - x^3 + 3xy = k.$ **Ans.**

Example 38(a):

Solve $(1 + xy)y\,dx + (1 - xy)x\,dy = 0$

Solution:

We have $(1 + xy)y\,dx + (1 - xy)x\,dy = 0$

The given equation can be written as

$$x\,dy + y\,dx + xy(y\,dx - x\,dy) = 0$$

$$\Rightarrow \qquad d(xy) + xy(y\,dx - x\,dy) = 0.$$

Dividing by x^2y^2, we obtain

$$\frac{d(xy)}{x^2y^2} + \frac{y\,dx - x\,dy}{xy} = 0$$

$$\Rightarrow \qquad \frac{d(xy)}{(xy)^2} + \left(\frac{dx}{x} - \frac{dy}{y}\right) = 0.$$

Integrating, we obtain

$$\frac{1}{xy} + \log x - \log y = a \Rightarrow x = cy\,e^{1/xy}.$$ **Ans.**

Example 38(b):

Solve $\frac{1}{\sqrt{y}}\frac{dy}{dx} + \frac{x}{1-x^2} = xy^{-1/2}$.

Solution:

We have $\frac{1}{y}\frac{dy}{dx} + \frac{x}{1-x^2} = xy^{-1/2}$.

$$\Rightarrow \quad \frac{1}{\sqrt{y}}\frac{dy}{dx} + \frac{x}{1-x^2}.\sqrt{y} = x$$

$$\Rightarrow \quad 2\frac{dy}{dx} + \frac{x}{1-x^2}.z = x \quad \left(\text{Put } \sqrt{2} = z \Rightarrow \frac{1}{2\sqrt{y}}\frac{dy}{dx} = \frac{dz}{dx}\right)$$

$$\Rightarrow \quad \frac{dz}{dx} + \frac{1}{2}.\frac{x}{(1-x^2)}.z = \frac{x}{2}. \qquad ...(7)$$

Now, $\int \frac{x}{2(1-x^2)}dx = -\frac{1}{4}\log(1-x^2) = \log(1-x^2)^{-1/4}$.

$\therefore$ I.F. $= e^{\log(1-x^2)^{-1/4}} = (1-x^2)^{-1/4}$.

The solution of equation (7) is given by

$$z(1-x^2)^{-1/4} = \int \frac{x}{2}(1-x^2)^{-1/4}\,dx + c$$

$$= \frac{1}{4}\int t^{-1/4}dt + c$$

(Put $1 - x^2 = t$ so that $-2x\,dx = dt$)

$$= -\frac{1}{4}.\frac{4}{3}t^{3/4} + c = -\frac{1}{3}(1-x^2)^{3/4} + c$$

$$z = -\frac{1}{3}(1-x^2) + c(1-x^2)^{1/4}, \text{ where } z = \sqrt{y}.$$

Hence, $3\sqrt{y} + 1 - x^2 = c\,(1-x^2)^{1/4}$ is the required solution.

Example 38(c):

Solve $y^2dx + (x^2 - xy - y^2)dy = 0$.

Solution:

We have $y^2dx + (x^2 - xy - y^2)dy = 0$.

The given equation is homogeneous.

$$\text{I.F.} = \frac{1}{Mx - Ny} = \frac{1}{y(x^2 - y^2)}.$$

Multiplying the given equation by the above I.F. we obtain

$$\frac{y}{x^2 - y^2}dx + \left(\frac{1}{y} - \frac{x}{x^2 - y^2}\right)dy = 0, \text{ which is exact.}$$

Its solution is $\int \frac{y\,dx}{x^2 - y^2} + \int \frac{1}{y}dy = 0$

$$\Rightarrow \qquad y.\frac{1}{2y}\log\frac{x - y}{x + y} + \log y = \log c$$

$$\Rightarrow \qquad y\sqrt{\frac{x - y}{x + y}} = c.$$

Hence, $(x - y)y^2 = c(x + y)$ is the required solution. **Ans.**

Example 38(d):

Solve $x(4y\,dx + 3x\,dy) + y^3(3y\,dx + 5x\,dy) = 0.$

Solution:

We have $x(4y\,dx + 3x\,dy) + y^3(3y\,dx + 5x\,dy) = 0$.

We shall apply Rule V to solve this equation.

The given equation is

$$(4xy + 3y^4)dx + (2x^2 + 5xy^3)dy = 0.$$

Multiply the above equation by $x^\alpha y^\beta$ to obtain

$$(4x^{\alpha+1}\,y^{\beta+1} + 3x^\alpha\,y^{\beta+4})dx + (2x^{\alpha+2}\,y^\beta + 5x^{\alpha+1}\,y^{\beta+3})dy = 0. \qquad ...(1)$$

This is exact for values of α and β for which $\partial M/\partial y = \partial N/\partial x$

$$\Rightarrow \qquad 4(\beta + 1)x^{\alpha+1}\,y^\beta + 3(\beta + 4)x^\alpha\,x^{\alpha+3}$$

$$= 2(a + 2)\,x^{\alpha+1}\,x^\beta + 5(\alpha + 1)x^\alpha\,x^{\alpha+3}$$

$$\Rightarrow \qquad 4(\beta + 1) = 2(\alpha + 2) \text{ and } 3(\beta + 4) = 5(\alpha + 1).$$

Solving these equations for α and β; we get $\alpha = 2$, $\beta = 1$.

Thus (1) becomes

$$(4x^3y^2 + 3x^2y^5)dx + (2x^4y + 5x^3y^4)dy = 0.$$

The solution of this exact equation is

$$\int(4x^2y^2 + 3x^2y^5)dx + 0 = c$$

$$\Rightarrow \quad x^4y^2 + x^3y^5 = c.$$

Example 38(e):

Solve $(x^3y^3 + x^2y^2 + xy + 1)y + (x^3y^3 - x^2y^2 - xy + 1)x \dfrac{dy}{dx} = 0.$

Solution:

We have $(x^3y^3 + x^2y^2 + xy + 1)y + (x^3y^3 - x^2y^2 - xy + 1)x \dfrac{dy}{dx} = 0.$

The given equation is of the form

$$f_1(xy)y\,dx + f_2(xy)x\,dy = 0.$$

$$Mx - Ny = (x^3y^3 + x^2y^2 + xy + 1)xy + (x^3y^3 - x^2y^2 - xy + 1)xy$$

$$= 2x^3y^3 + 2x^2y^2 = 2x^2y^2(xy + 1).$$

$$\text{I.F.} = \frac{1}{Mx - Ny} = \frac{1}{2x^2y^2(xy + 1)}.$$

Multiplying the given equation by the above I.F., we obtain

$$\frac{1}{2}\left(1 + \frac{1}{x^2y^2}\right)y\,dx + \frac{1}{2x^2y^2(xy + 1)}\,[(x^3y^3 - x^2y^2 - xy + 1)]x\,dy = 0$$

$$\Rightarrow \quad \left(y + \frac{1}{x^2y}\right)dx + \frac{1}{x^2y^2(xy + 1)}\,[(x^3y^3 + y^2y^2) + (xy + 1)$$

$$- 2(x^2y^2 + xy]\,xdy = 0.$$

$$\Rightarrow \quad \left(y + \frac{1}{x^2y}\right)dx + \left(\frac{1}{x^2y^2} - \frac{2xy}{x^2y^2}\right)xdy = 0$$

$$\Rightarrow \quad \left(y + \frac{1}{x^2y}\right)dx + \left(x + \frac{1}{xy^2} - \frac{2}{y}\right)dy = 0, \text{ which is exact.}$$

The required solution is

$$\int\left(y + \frac{1}{x^2y}\right)dx + \int\left(-\frac{2}{y}\right)dy = \log a$$

$$\Rightarrow \quad xy - \frac{1}{xy} - 2\log y - \log a = 0.$$

Taking log a = 2log c and simplifying, we see that $x^2y^2 - 2xy \log cy = 1$ is the required solution. **Ans.**

Example 39:

Solve $x\frac{dy}{dx} + \frac{y^2}{x} = y.$

Solution:

The given equation can be written as

$$\frac{dy}{dx} = \frac{xy - y^2}{x^2}$$

Putting y = Vx, and dy/dx = V + x dy/dx we have

$$\Rightarrow \quad V + x\frac{dV}{dx} = \frac{Vx^2 - V^2x^2}{x^2}, \ y = Vx$$

Separate the variables, we have

$$\Rightarrow \quad x\frac{dV}{dx} = y^2 \Rightarrow \int\frac{dV}{V^2} + \int\frac{dx}{x} + \log c = 0$$

$$\Rightarrow \quad -\frac{1}{V} + \log x + \log + \log c = 0$$

$$\Rightarrow \quad \text{or } \log cx = \frac{1}{V} = \frac{x}{y}.$$

Hence $cx = e^{x/y}$ is the required solution. **Ans.**

Example 40:

Solve $y^2\,dx + (xy + x^2)\,dy = 0.$

Solution:

We have

$$\frac{dy}{dx} = \frac{-y^2}{xy + x^2}$$

If is a homogeneous diff. equation

putting y = Vx and dy/dx = V + x dV/dx we have

$$\Rightarrow \quad V + x\frac{dV}{dx} = \frac{-V^2x^2}{Vx^2 + x^2}, \text{ where } y = Vx$$

Separate the variables, we have

$$\Rightarrow \quad x\frac{dV}{dx} = \frac{-V^2}{V+1} - V = \frac{-V(1+2V)}{(V+1)}$$

$$\therefore \quad \int\frac{(V+1)dV}{V(1+2V)} + \int\frac{dx}{x} = \log c. \qquad ...(1)$$

Integrating, we have

Let $\frac{V+1}{V(1+2V)} = \frac{A}{V} + \frac{B}{1+2V}$.

$$\therefore \quad V + 1 = A(1 + 2V) + BV.$$

Putting $V = 0 \Rightarrow\ = 1$ and $V = -\frac{1}{2} \Rightarrow B = -1$.

Thus (1) reduces to $\int\frac{dV}{V} - \int\frac{dV}{1+2V} + \log x = \log c$

$$\Rightarrow \quad \log V - \frac{1}{2}\log(1+2V) + \log x = \log c$$

$$\Rightarrow \quad \log Vx = \log c\ \sqrt{1+2V}$$

$$\Rightarrow \quad Vx = c\ \sqrt{1+2V}$$

$$\Rightarrow \quad y = c\ \sqrt{1+2\frac{y}{x}}.$$

Hence, $xy^2 = c^2(x + 2)$ is the required solution. **Ans.**

Example 41:

Solve $(4y + 3x)dy + (y - 2x)dx = 0$.

Solution:

The equ. be written as

$$\frac{dy}{dx} = \frac{2x - y}{3x + 4y}$$

If is a homogeneous diff. equation

putting $y = Vx$ and $dy/dx = V + x\ dV/dx$ we have

$$\Rightarrow \quad V + x\frac{dV}{dx} = \frac{2x - Vx}{3x + 4y}, \text{ where } y = Vx$$

$$\Rightarrow \quad x\frac{dV}{dx} = \frac{2 - V}{3 + 4V} - V = \frac{2 - 4V - 4V^2}{3 + 4V}$$

Separate the variables, we have

$$\Rightarrow \qquad \int\frac{(3+4\,V)dV}{4V^2+4V-2}+\int\frac{dx}{x}=0. \qquad ...(1)$$

Intigrating we have

Put $2V^2 + 2V - 1 = t \Rightarrow (4V + 2)dV = dt$. Thus (1) becomes

$$\Rightarrow \qquad \frac{1}{2}\int\frac{dt}{t}+\int\frac{dV}{4V^2+4V-2}+\log x+\log c=0$$

$$\Rightarrow \qquad \frac{1}{2}\log(2V^2+2V-1)+\int\frac{dV}{(2V^2+1)^2-3}+\log x+\log c=0$$

$$\Rightarrow \qquad \frac{1}{2}\log\left[\frac{2y^2}{x^2}+\frac{2y}{x}-1\right]+\frac{1}{4\sqrt{3}}\log\frac{2V+1-\sqrt{3}}{2V+1+\sqrt{3}}+\log x+\log c=0$$

$$\Rightarrow \qquad \log c=2\sqrt{3}\log\left(\frac{2y^2+2xy-x^2}{x^2}\right)+4\sqrt{3}\log x$$

$$=\log\frac{2V+1+\sqrt{3}}{2V+1-\sqrt{3}}$$

$$\therefore \qquad c\left(\frac{2y^2+2xy-x^2}{x^2}\right).(x)^{4\sqrt{3}}=\frac{2V+1+\sqrt{3}}{2V+1-\sqrt{3}}.$$

Hence, $c(2y^2+2xy-x^2)^{2\sqrt{3}}=\dfrac{(\sqrt{3}+1)\,x+2y}{(1-\sqrt{3})\,x+2y}$ $\qquad \left(\because V=\dfrac{y}{x}\right)$

is the required solution. **Ans.**

Example 42:

Solve the following: differential Equation

(i) $(2x + y + 1)\,dx + (4x + 2y - 1)\,dy = 0.$

(ii) $(4x + 6y + 5)\,dx + (2x + 3y + 4)\ 0.$

Solution:

(i) We have $(2x + y + 1)\,dx + (4x + 2y - 1)\,dy = 0$

$$\Rightarrow \qquad \frac{dy}{dx}=\frac{-(2x+y+1)}{(4x+2y\ \ 1)}=\frac{-(2x+y+1)}{[2(2x+y)-1]} \qquad ...(1)$$

Note that $\qquad \dfrac{a}{A}=\dfrac{2}{4}=\dfrac{1}{2}=\dfrac{b}{B}.$

Put 2x + y = z so that 2 + dy/dx = dz/dx.

Substituting in (1), we obtain

$$\frac{dz}{dx} - 2 = \frac{-(z+1)}{(2z-1)}$$

$$\Rightarrow \qquad \frac{dz}{dx} = \frac{3z-3}{2z-1} \qquad \Rightarrow \qquad \frac{2z-1}{3(z-1)}\,dz = dx.$$

Integrating, we get

$$\int dx + c = \frac{1}{3}\int \frac{2z-1}{z-1}\,dz$$

$$\therefore \qquad x + c = \frac{2}{3}\int \left[1 + \frac{1}{2(z-1)}\right] dz$$

$$= \frac{2}{3}\left[z + \frac{1}{2}\log(z-1)\right]$$

$$= \frac{1}{3}\,[2(2x+y) + \log(2x+y-1)]$$

Hence, log(2x + y − 1) + x + 2y = k is the required solution.

(ii) We have (4x + 6y + 5) dx = (2x + 3y + 4) = 0

$$\frac{dy}{dx} = \frac{4x+6y+5}{2x+3y+4} = \frac{2(2x+3y)+5}{2x+3y+4}. \qquad ...(2)$$

Let 2x + 3y = z so that 2 + 3 (dy/dx) = dz/dz.

Substituting is (2), we obtain

$$\frac{1}{3}\left(\frac{dz}{dx} - 2\right) = \frac{2z+5}{z+4}$$

$$\Rightarrow \qquad \frac{dz}{dx} = \frac{6z+15}{z+4} + 2 = \frac{8z+23}{z+4}$$

$$\Rightarrow \qquad \int \frac{z+4}{8z+23}\,dz = \int dx + c$$

$$\Rightarrow \qquad x + c = \frac{1}{8}\int \frac{(8z+23)+9}{8z+23}\,dz$$

$$= \frac{1}{8}\int dz + \frac{9}{8}\int \frac{dz}{8z+23} = \frac{z}{8} + \frac{9}{64}\log(8z+23).$$

Hence, $x + c = \frac{1}{8}(2x + 3y) + \frac{9}{64}\log(16x + 24y + 23)$ is the solution.

Ans.

Example 43:

Solve the following diff. equ.

(i) $\frac{dy}{dx} = \frac{x - 2y + 5}{2x + y - 1}$.

(ii) $(3y - 7x + 7)\,dx + (7y - 3x + 3)\,dy = 0$.

Solution:

We have $\frac{dy}{dx} = \frac{x - 2y + 5}{2x + y - 1}$

On putting $x = X + h$, $y = Y + k$, the given equation becomes

$$\frac{dY}{dX} = \frac{(X + h) - 2(Y + k) + 5}{2(X + h) + (Y + k) - 1}$$

$$\Rightarrow \quad \frac{dY}{dX} = \frac{X - 2Y + (h - 2k + 5)}{2X - Y + (2h + k - 1)}. \quad \text{...(1)}$$

Choose h and k so that

$$h - 2k + 5 = 0 \quad \text{and} \quad 2h + k - 1 = 0 \quad \text{...(2)}$$

From (1) and (2), we obtain

$$\frac{dY}{dX} = \frac{X - 2Y}{2X + Y}. \quad \text{...(3)}$$

which is a homogeneous equation.

We substitute $Y = VX$ so that $\frac{dY}{dX} = V + X\frac{dV}{dX}$.

Putting in (3), we get

$$V + X\frac{dV}{dX} = \frac{1 - 2V}{2 + V}$$

$$\Rightarrow \quad X\frac{dV}{dX} = \frac{1 - 2V}{2 + V} - V = \frac{1 - 4V - V^2}{2 + V}$$

$$\Rightarrow \quad \int \frac{V + 2}{V^2 + 4V - 1}\,dV + \int \frac{dX}{X} = \log c$$

$$\Rightarrow \quad \frac{1}{2}\int \frac{dt}{t} + \log X = \log c \quad \left[t = V^2 + 4V - 1 \right]$$

$\Rightarrow \quad \frac{1}{2}\log t + \log X = \log c$

$\Rightarrow \quad tX^2 = k, k = c^2$

$\Rightarrow \quad (V^2 + 4V - 1) X^2 = k$

$\Rightarrow \quad Y^2 + 4XY - X^2 = k.$ ...(4)

Solving (2) for h and k, we obtain

$h = -3/5$ $k = 11/5$ and so $X = x + (3/5)$, $Y = y - (11/5)$.

Putting in (4) and simplifying, the required solution is

$x^2 - y^2 - 4xy + 10x + 2y = a.$

(2) $\frac{dy}{dx} = \frac{7x - 3y - 7}{7y - 3x + 3}.$ ...(5)

Putting $x + X + h$, $y = Y + k$ (5), we obtain

$$\frac{dY}{dX} = \frac{(7x - 3Y) + (7h - 3k - 7)}{(7Y - 3X) + (7k - 3h + 3)}. \quad ...(6)$$

Choose h and k so that

$7h - 3k - 7 = 0$ and $7k - 3h + 3 = 0.$ ...(7)

From (6) and (7), $\frac{dY}{dX} = \frac{7X - 3Y}{7Y - 3X}$

$$\Rightarrow \quad V + X\frac{dV}{dX} = \frac{7 - 3V}{7V - 3}, \text{where } Y = VX$$

$$\Rightarrow \quad X\frac{dV}{dX} = \frac{7(1 - V^2)}{7V - 3}$$

$$\Rightarrow \quad \int\frac{(7V - 3)dV}{(V^2 - 1)} + 7\int\frac{dX}{X} = \log c$$

$$\Rightarrow \quad \frac{7}{2}\log(V^2 - 1) - \frac{3}{2}\log\frac{V - 1}{V + 1} + 7\log X = \log c$$

$$\Rightarrow \quad \left[(V^2 - 1)^7\left(\frac{V + 1}{V - 1}\right)^3\right]^{1/2} . X^7 = c$$

$$\Rightarrow \quad (V + 1)^5 (V - 1)^2 X^7 = c, V = Y/X$$

$$\Rightarrow \quad (X + Y)^5 (Y - X)^2 = c. \quad ...(8)$$

Solving (7) for h and k, we get h = 1, k = 0.

$\therefore \quad x = X + h = X + 1$

$\Rightarrow \quad X = x - 1, y = Y + k = Y.$

Putting in (8), $(x + y - 1)^5 (y - x + 1)^2 = c.$

This is the required solution. **Ans.**

Example 44:

Solve the differential equation

$$(x^4 - 2xy^2 + y^4)dx = (2x^2y - 4xy^3 + \sin y)dy.$$

Solution:

We have $(x^4 - 2xy^2 + y^4)dx = (2x^2y - 4xy^3 + \sin y)dy.$

We get $M = x^4 - 2xy^2 + y^4$, $N = -(2x^2y - 4xy^3 + \sin y).$

$\therefore \quad \partial M/\partial y = -4xy + 4y^3.\ \partial N/\partial x = -4xy + 4y^3.$

Here $\dfrac{\partial M}{\partial y} = \dfrac{\partial N}{\partial x}$

Thus, the given equation is exact.

Now, we have

$$\int M\,dx = \int (x^4 - 2xy^2 + y^4)dx = \frac{x^5}{5} - x^2 y^2 + y^4 x, \qquad \text{...(i)}$$

and $\int -\sin y\,dy = \cos y.$ (ii)

[$\because$ N ~ x = − sin y]

Integrating both side, we have

Adding (i) and (ii)

Hence, the required solution is

$$\frac{1}{5}x^5 - x^2y^2 + xy^2 + \cos y = c.$$ **Ans.**

Example 45:

Solve the differential equation $(2x - y + 1)dx + (2y - x - 1)dy = 0.$

Solution:

We have $(2x - y + 1)dx + (2y - x - 1)dy = 0$

Here, $M = 2x - y + 1$, $N = 2y - x - 1$

$$\therefore \quad \frac{\partial M}{\partial y} = -1, \frac{\partial N}{\partial x} = -1.$$

Here $\frac{\partial M}{\partial y} = \frac{\partial N}{\partial x}$

Thus, the given equation is exact.

Now, $\int M\, dx = \int (2x - y + 1)\, dx = x^2 - xy + x,$ ...(i)

$\int (N \sim x) dx = \int (2y - 1) dy = y^2 - y.$ ...(ii)

Adding (i) and (ii)

we have $x^2 - xy + x + y^2 - y = c.$

Which is required solution. **Ans.**

Example 46:

Solve the differential equation

$(x^2 - 4xy - 2y^2)dx + (y^2 - 4xy - 2x^2)dy = 0.$

Solution:

We have $(x^2 - 4xy - 2y^2)dx + (y^2 - 4xy - 2x^2)dy = 0.$

$M = x^2 - 4xy - 2y^2,\ N = y^2 - 4xy - 2x^2.$

$\therefore \quad \frac{\partial M}{\partial y} = -4x - 4y, \quad \frac{\partial N}{\partial x} = -4y - 4x.$

we have $\frac{\partial M}{\partial y} = \frac{\partial N}{\partial x}$

Thus, the given equation is exact.

Now, $\int M\, dx = \int (x^2 - 4xy - 2y^2)dx = \frac{1}{3}x^3 - 2x^2y - 2y^2x,$ ...(i)

$\int (N \sim x)\, dy = \int y^2 dy = \frac{1}{3}y^3.$ (ii)

Adding (i) and (ii) we have

$\frac{1}{3}x^3 - 2x^2y - 2y^2x + \frac{1}{3}y^3 = a$ or $x^3 - 6xy^2 + y^3 = c.$

Example 47:

Solve the differential equation

$(ax + hy + g)dx + (hx + by + f)dy = 0.$

Solution:

We have $(ax + hy + g)dx + (hx + by + f)dy = 0.$

Here, we obtain M = ax + hy + g. N = hx + by + f.

$\therefore$ $\partial M/\partial y = h$, $\partial N/\partial x = h$

so that $$\frac{\partial M}{\partial y} = \frac{\partial N}{\partial x}.$$

Thus, the given equation is exact.

Now, $\int M\,dx = \int (ax + hy + g)dx = \frac{ax^2}{2}$ hyx + gx. ...(i)

The terms of N which do not contain x are by + f.

Now, $\int (by + f)dy = \frac{by^2}{2} + fy.$ ...(ii)

Adding (i) and (ii)

Hence, the required solution is

$$\frac{a}{2}a^2 + hxy + gx + \frac{b}{2}y^2 + fy = k.$$

$\Rightarrow$ $$ax^2 + 2hxy + by^2 + 2gx + 2fy + c = 0.$$ **Ans.**

EXERCISES

1. $\frac{xdx + ydy}{xdy - ydx} = \sqrt{\frac{a^2 - x^2 - y^2}{x^2 + y^2}}.$

2. $(2x + y + 1)dx + (4x + 2y - 1)dy = 0.$

   Ans. $e^{x+2y}(2x + y - 1) = c.$

3. $\frac{dy}{dx} = \frac{6x - 4y + 3}{3x - 2y + 1}$ **Ans.** $2x - y = \log(3x - 2y + 3) + \frac{1}{2}c$

4. $(2x + y + 3)dy = (x + 2y + 3)dx.$

   Ans. $(x - y)^2 = c(x + y - 2)$

5. $(2x - 2y + 5)dy = (x - y + 3)dx.$

   Ans. $2y - x = \log(x - y + 2) + c$

6. $(x^2 + y^2 + 1)dx - 2xy.\ dy = 0.$ **Ans.** $x^2 - y^2 = 1 + ax$

7. $x(x^3 + 3y^3)dx = y(y^2 + 3x^2)dy.$

   Ans. $(x^2 - y^2)^4 (x^2 + y^2)^{-2} = c.$

8. $y(x^2 + y^2 + a^2)dy + x(x^2 + y^2 - a^2)dx = 0.$

9. $(2x^3 + 3y)dx + (3x + y = 1)\ dy = 0.$

Ans. $x^2 + 6xy + y^2 - 2y = c$

10. $y - x\dfrac{dy}{dx} = b\left(1 + x^2\dfrac{dy}{dx}\right).$ **Ans.** $c(y - b) = \dfrac{x}{1 + bx}$

11. $(2x + 4y + 3)dy = (2y + x + 1)dx.$

Ans. $\log(4x + 8y + 5) = 4x - 8y + c$

12. $(2x + 3y - 5)dy = (3x + 2y - 5)dx = 0.$

Ans. $3(x^2 + y^2) + 4xy - 10(x + y) = c$

13. $(xy^2 + x)dx + (yx^2 + y)dy = 0.$

Ans. $x^2 + y^2 + x^2y^2 = c$

14. $(2ax + by + g)dx + (2cy + bx + e)dy = 0.$

Ans. $Ax^2 + bxy + cy^2 + gx + ey = k$

15. $(x^2 + y^2 - a^2)x\ dx + (x^2 - y^2 - b^2)y\ dy = 0.$

Ans. $x^4 - y^4 + 2x^2y^2 - 2a^2x^2 - 2b^2y^2 = c$

16. $(x^2 + 2xy + 3y^2)dx + (4y^3 + 6xy - x^2)dy = 0.$

Ans. $\frac{1}{3}x^3 - x^2y + 3xy^2 + y^4 = c$

17. $(1 + 6y^2 - 3x^2y)\dfrac{dy}{dx} = 3xy^2 - x^2.$ **Ans.** $y + 2y^3 + \frac{1}{3}x^3 - \frac{3}{2}x^2y^2 = c$

18. $(e^y + 1)\cos x\ dx + e^y \sin x\ dy = 0.$ **Ans.** $\sin x(e^y + 1) = c$

19. $y(1 + x)dx + x(1 + y)dy = 0$ **Ans.** $x + y + \log xy = c$

20. $(x + y)^2\dfrac{dy}{dx} = a^2.$ **Ans.** $\dfrac{x + y}{a} = \tan\dfrac{y + c}{a}$

21. $xdy - ydx - (1 - x^3)dx = 0.$ **Ans.** $y + x^2 + 1 = cx$

22. $xdx + ydy + 4y^3(x^2 + y^2)dy = 0.$ **Ans.** $(x^2 + y^2)e^{2y^4} = c.$